HSC Year 12
BIOLOGY

TIMOTHY SLOANE | MATHEW SLOANE

- summary notes
- revision questions
- detailed sample answers
- study and exam preparation advice

STUDY NOTES

A+ HSC Biology Study Notes
1st Edition
Timothy Sloane
Mathew Sloane
ISBN 9780170465267

Publisher: Alice Wilson
Series editor: Catherine Greenwood
Copyeditor: Kay Waters
Reviewers: Kathryn Fraser, Fiona McCrossin
Series text design: Nikita Bansal
Series cover design: Nikita Bansal
Series designer: Cengage Creative Studio
Artwork: MPS Limited and Straive
Production controller: Karen Young
Typeset by: Nikki M Group Pty Ltd

ISBN 978 0 17 046526 7

Cengage Learning Australia
Level 7, 80 Dorcas Street
South Melbourne, Victoria Australia 3205

Cengage Learning New Zealand
Unit 4B Rosedale Office Park
331 Rosedale Road, Albany, North Shore 0632, NZ

For learning solutions, visit **cengage.com.au**

Printed in Singapore by C.O.S. Printers Pte Ltd.
1 2 3 4 5 6 7 26 25 24 23 22

CONTENTS

CONTENTS

CHAPTER 2

MODULE 6: GENETIC CHANGE

CHAPTER

MODULE 7: INFECTIOUS DISEASE

CHAPTER 4

MODULE 8: NON-INFECTIOUS DISEASE AND DISORDERS

HOW TO USE THIS BOOK

The *A+ Biology* resources are designed to be used year-round to prepare you for your HSC Biology exam. *A+ HSC Biology Study Notes* includes topic summaries of all key knowledge in the New South Wales HSC Biology syllabus that you will be assessed on during your exam. Each chapter of this book addresses one module. This section gives you a brief overview of each chapter and the features included in this resource.

Module summaries

The module summaries at the beginning of each chapter give you a high-level overview of the essential knowledge and key science skills you will need to demonstrate during your exam.

Concept maps

The concept maps at the beginning of each chapter provide a visual summary of each module outcome.

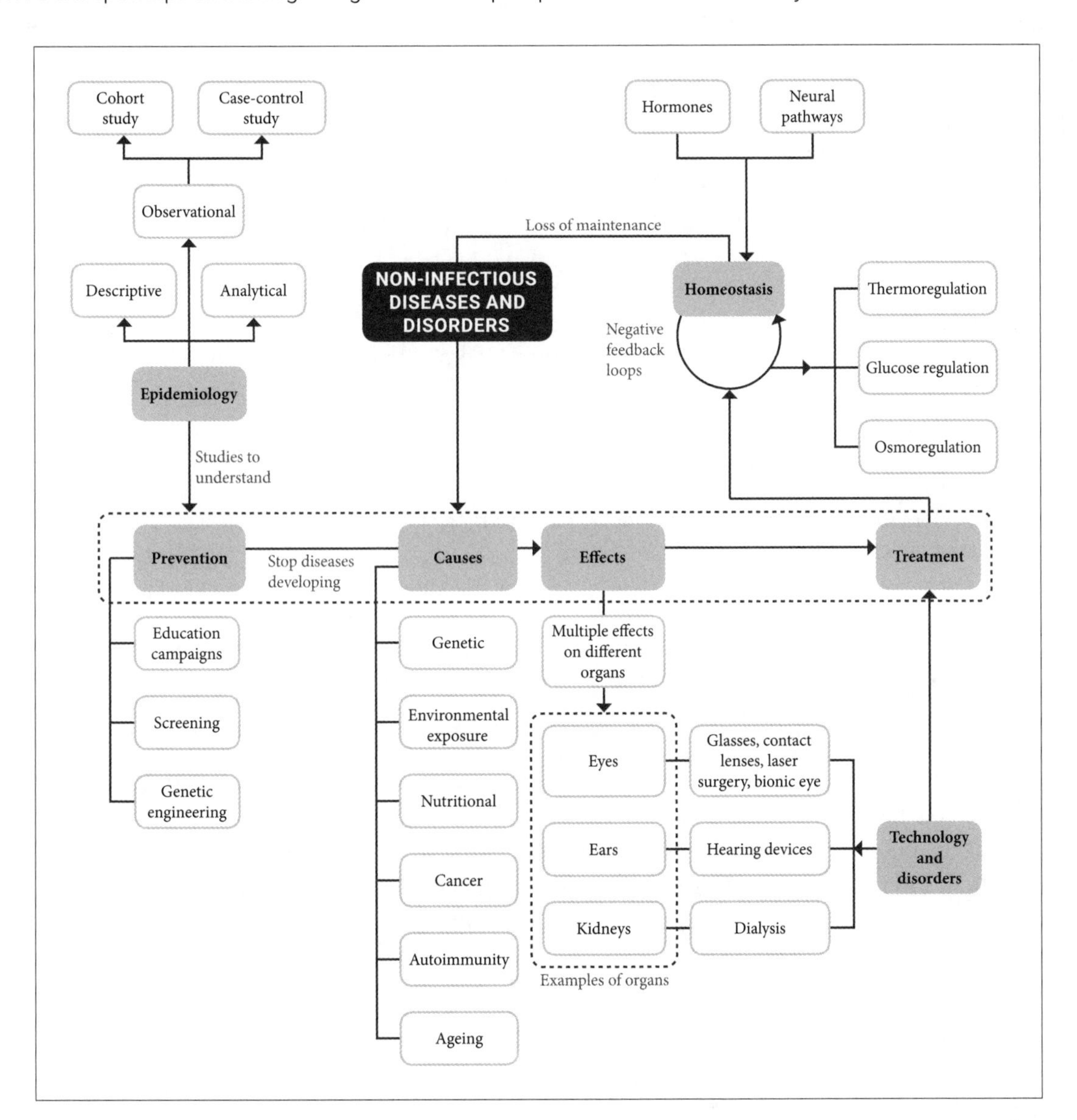

Inquiry question summaries

All the dot points under each inquiry question are summarised sequentially throughout inquiry section summaries.

Exam practice

Exam practice questions are at the end of each chapter to test you on what you have just reviewed in the chapter. These are written in the same style as the questions in the actual HSC Biology exam. There are some official past exam questions in each chapter.

Multiple-choice questions

Each chapter has approximately 20 multiple-choice questions.

Short-answer questions

There are approximately 15 short-answer questions in each chapter, often broken into parts. These questions require you to apply your knowledge across multiple concepts. Mark allocations have been provided for each question.

Solutions

Solutions to practice questions are supplied at the back of the book. They have been written to reflect a high-scoring response and include explanations of what makes an effective answer.

Explanations

The solutions section includes explanations of each multiple-choice option, both correct and incorrect. Explanations of written response items explain what a high-scoring response looks like and signpost potential mistakes.

1 A

Budding refers to an outgrowth of a body region that separates from the original organism, resulting in two individuals.

B is incorrect because binary fission is a type of asexual reproduction where the parent cell divides into two approximately equal daughter cells. **C** is incorrect because fragmentation involves a body part detaching and developing into a new organism, and the original organism regenerates the lost body part. **D** is incorrect because external fertilisation relates to the union of a sperm and egg in sexual reproduction.

53 a The distribution of dengue fever appears to have increased markedly since 1950. Many more parts of the world, such as South America and Africa, are now affected (1 mark). The number of countries with reported cases of malaria decreased significantly between 1900 and 2010, from 140 to 88 (1 mark). However, a growing number of people are at risk, though they represent a smaller percentage of the global population. The actual numbers of at-risk people has increased from 0.9 to 3.4×10^9, which equates to an actual drop from 75% to 50% (1 mark).

It is essential that you respond with the necessary depth associated with the verb in the question stem. In this case 'analyse' requires you to identify different components and relationships shown in *both* the table *and* the image and discuss the implications.

Icons

The icons below occur in the summaries and exam practice sections of each chapter.

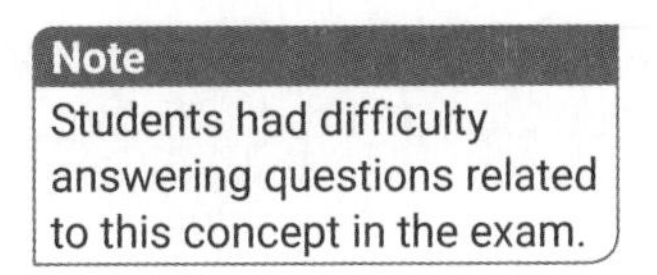

Note boxes appear throughout the summaries to provide additional tips and support.

©NESA 2018 SI Q4

This icon appears with past NESA exam questions.

These icons indicate whether the question is easy, medium or hard.

A+ HSC Biology Practice Exams

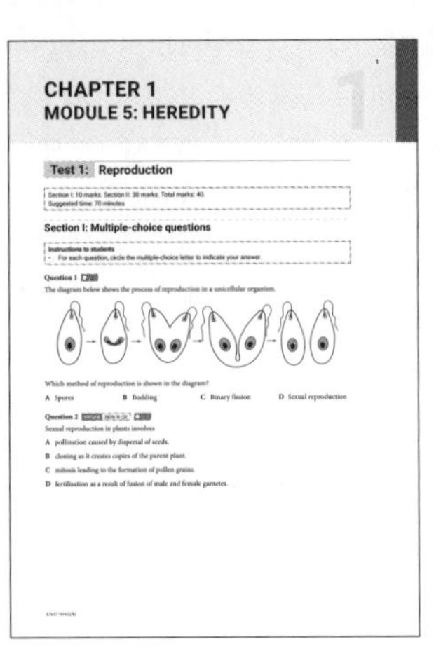

A+ HSC Biology Study Notes can be used independently or alongside the accompanying resource *A+ HSC Biology Practice Exams*. *A+ HSC Biology Practice Exams* features 17 topic tests comprising original HSC-style questions, official HSC questions and two pull-out practice exams. Each topic test includes multiple-choice and short-answer questions, and focuses on one inquiry question of the New South Wales HSC Biology syllabus. There are two complete Module 5–8 practice exams following the tests. Like the *A+ Biology Study Notes*, detailed solutions are included at the end of the book, demonstrating and explaining how to craft high-scoring exam responses.

A+ DIGITAL

Just scan the QR code or type the URL into your browser to access:

- A+ Flashcards: revise key terms and concepts online
- Revision summaries of all concepts from each inquiry question.

Note: You will need to create a free NelsonNet account.

PREPARING FOR THE END-OF-YEAR EXAM

Exam preparation is a year-long process. It is important to keep on top of the theory and consolidate often, rather than leaving work until the last minute. You should aim to have the theory learned and your notes complete so that by the time you reach STUVAC, the revision you do is structured, efficient and meaningful.

Effective preparation involves the following steps.

Study tips

To stay motivated to study, try to make the experience as comfortable as possible. Have a dedicated study space that is well lit and quiet. Create and stick to a study timetable, take regular breaks, reward yourself with social outings or treats, and use your strengths to your advantage. For example, if you are great at art, turn your Biology notes into cartoons, diagrams or flow charts. If you are better with words or lists, create flash cards or film yourself explaining tricky concepts and watch the videos back.

Another strong recommendation is to engage with the performance band descriptors published by NESA. Clear information is provided about what is expected of a student performing at band 6, band 5 and so on, mapped against the Knowledge and Understanding and Working Scientifically course outcomes. Have an honest conversation with yourself as to what level you are currently performing at. This will in turn provide you with guidance on what you need to do to improve. For example, a band 6 student will:

- demonstrate an extensive knowledge and understanding of complex and abstract ideas
- apply knowledge and information to unfamiliar situations to propose comprehensive solutions and explanations
- communicate scientific understanding succinctly, logically and consistently, using correct and precise scientific terms.

Revision techniques

Here are some useful revision methods to help information **'STIC'**.

Spaced repetition	This technique helps to move information from your short-term memory into your long-term memory by spacing out the time between your revision and recall flash card sessions. As the time between retrieving information is slowly extended, the brain processes and stores the information for longer periods.
Testing	Testing is necessary for learning and is a proven method for exam success. If you test yourself continually before you learn all the content, your brain becomes primed to retain the correct answer when you learn it. As part of this process, engage with the marking criteria provided to help decide on areas where improvement is needed.
Interleaving	This is a revision technique that sounds counterintuitive but is very effective for retaining information. Most students tend to revise a single topic in a session, and then move onto another topic in the next session. With interleaving, you choose three topics (1, 2, 3) and spend 20–30 minutes on each topic. You may choose to study 1-2-3 or 2-1-3 or 3-1-2, 'interleaving' the topics and repeating the study pattern over a long period of time. This strategy is most helpful if the topics are from the same subject and are closely related.
Chunking	An important strategy is breaking down large topics into smaller, more manageable 'chunks' or categories. Essentially, you can think of this as a branching diagram or mind map where the key theory or idea has many branches coming off it that get smaller and smaller. By breaking down the topics into these chunks, you will be able to revise the topic systematically.

These strategies take cognitive effort, but that is what makes them much more effective than re-reading notes or trying to cram information into your short-term memory the night before the exam!

Time management

It is important to manage your time carefully throughout the year. Make sure you are getting enough sleep, that you are getting the right nutrition, and that you are exercising and socialising to maintain a healthy balance so that you don't burn out.

To help you stay on target, plan your study timetable. Here is one way to do this.

1. Assess your current study time and social time. How much will you dedicate to each?
2. List all your commitments and deadlines, including sport, work, assignments etc.
3. Prioritise the list and reassess your time, to ensure you can meet all your commitments.
4. Decide on a format, whether weekly or monthly, and schedule in a study routine.
5. Keep your timetable where you can see it.
6. Be consistent.

Studies suggest that 1-hour blocks with a 10-minute break is most effective for studying, and remember you that can interleave three topics during this time. You will also have free periods during the school day that you can use for study, note-taking, assignments, meeting with your teachers and group study sessions. Studying does not have to take hours if it is done effectively. Use your timetable to schedule short study sessions often.

The exam

The examination is held at the end of the year and contributes 50% to your HSC mark. You will have 180 minutes plus 10 minutes of reading time. You are required to attempt multiple-choice questions (Section I) and short-answer questions (Section II), covering all areas of study in Modules 5–8. The following strategies will help you prepare for the exam conditions.

Practise using past papers

To help prepare, download past papers from the NESA website and attempt as many as you can in the lead-up to the exam. These will show you the types of questions to expect and give you practice in writing answers. It is a good idea to make the trial exams as much like the real exam as possible (conditions, time constraints, materials etc.). You can also use *A+ HSC Biology Practice Exams*.

Use trial papers, school-assessed coursework, and comments from your teacher to pinpoint weaknesses, and work to improve these areas. Do not just tick or cross your answers; look at the suggested answers and try to work out why your answer was different. What misunderstandings do your answers show? Are there gaps in your knowledge? Read the examiners' reports to find out the common mistakes students make.

Make sure you understand the material, rather than trying to rote learn information. Most questions are aimed at your understanding of concepts and your ability to apply your knowledge to new situations.

The day of the exam

The night before your exam, try to get a good rest and avoid cramming, as this will only increase stress levels. On the day of the exam, arrive at the venue early and bring everything you will need with you. If you rush to the exam, your stress levels will increase, thereby lowering your ability to do well. Further, if you are late, you will have less time to complete the exam, which means you may not be able to answer all the questions or may rush to finish and make careless mistakes. If you are more than 30 minutes late, you may not be allowed to enter the exam. Do not worry too much about exam jitters. A certain amount of stress is required to help you concentrate and achieve an optimum level of performance. However, if you are feeling very nervous, breathe deeply and slowly. Breathe in for a count of 6 seconds, and out for 6 seconds until you begin to feel calm.

Important information from the syllabus

Sixty per cent of your school-based assessment will have addressed the skills required for Working Scientifically. This will have included a mandatory depth study. You are strongly encouraged to engage with the Working Scientifically syllabus outcomes as part of your study, as they will also constitute a significant component of the HSC exam.

Outcome	Description
BIO12-1	**Questioning and predicting** develops and evaluates questions and hypotheses for scientific investigation
BIO12-2	**Planning investigations** designs and evaluates investigations in order to obtain primary and secondary data and information
BIO12-3	**Conducting investigations** conducts investigations to collect valid and reliable primary and secondary data and information
BIO12-4	**Processing data and information** selects and processes appropriate qualitative and quantitative data and information using a range of appropriate media
BIO12-5	**Analysing data and information** analyses and evaluates primary and secondary data and information
BIO12-6	**Problem solving** solves scientific problems using primary and secondary data, critical thinking skills and scientific processes
BIO12-7	**Communicating** communicates scientific understanding using suitable language and terminology for a specific audience or purpose

Section I of the exam

Section I consists of a question book and an answer sheet. The answers for multiple-choice questions must be recorded on the answer sheet provided. A correct answer scores 1, and an incorrect answer scores 0. There is no deduction for an incorrect answer, so attempt every question. Read each question carefully and underline key words. If you are given a graph or a diagram, make sure you understand it before you read the answer options. You may make notes on the diagrams or graphs.

Section II of the exam

Section II consists of a question book with space to write your answers. The space provided is an indication of the detail required in the answer. Most questions will have several parts, and each part will be testing new information; so read the entire question carefully to ensure you do not repeat yourself. Use correct biological terminology and make an effort to spell it correctly.

Look at the mark allocation. Generally, if two or three marks are allocated to the question, you will be expected to make two or three relevant points. If you make a mistake, cross out any errors but do not write outside the space provided; instead, ask for another booklet and re-write your answer. Mark clearly on your paper which questions you have answered where.

Make sure your handwriting is clear and legible, and attempt all questions. Marks are not deducted for incorrect answers, and you might get some marks if you make an educated guess. You will definitely not get any marks if you leave a question blank!

Do not be put off if you do not recognise an example or context; questions will always be about the concepts that you have covered. In fact, top-performing students are expected to apply learned knowledge to an unfamiliar context (see band 6 on page x).

Reading time

Use your time wisely. *Do not* use the reading time to try and figure out the answers to any of the questions until you have read the whole paper. The exam will not ask you a question testing the same knowledge twice, so look for hints in the stem of the question and avoid repeating yourself. Plan your approach so that when you begin writing, you know which section, and ideally which question, you are going to start with. You do not have to start with Section I.

Strategies for answering Section I

Read the question carefully and underline any important information, to help you break the question down and avoid misreading it. Read all the possible solutions and eliminate any clearly wrong answers. You can annotate or write on any diagrams or infographics and make notes in the margins. Fill in the multiple-choice answer sheet carefully and clearly. Check your answer and move on. Do not leave any answers blank.

Strategies for answering Section II

The examiners' reports always highlight the importance of planning responses before writing. Remember, you have 3 hours to complete 100 marks. This means you have an average of 1.8 minutes per mark. For a 5-mark question, this equals 9 minutes. You should spend a good proportion of this time planning your response.

To do this, **CUBE** the question:

Circle the verb (e.g. identify, describe, explain, evaluate).

Underline the key biological concepts to be covered in your response (e.g. meiosis, polypeptide synthesis, mutation).

Box important information (e.g. plurals, and, with examples, with reference to the stimulus).
Elaborate at the depth required to answer the question.

Many questions require you to apply your knowledge to unfamiliar situations, so it is okay if you have never heard of the context before. You should, however, know which part of the course you are being tested on and what the question is asking you to do. Plan your response in a logical sequence based on the level of detail required by the verb of the question.

Another useful acronym to remember is based on the **ALARM** scaffold (**A** **L**earning **A**nd **R**esponding **M**atrix) developed by Max Woods. We typically accumulate knowledge in a hierarchical nature. For example, before you can evaluate the impact of transgenic species on society, you must first be able to identify and describe what a transgenic species is, and explain how and why they are produced.

Plan your responses following the same logic, and scaffold your responses using **IDEA/E**. If the questions require an assessment or evaluation, first **Identify**, then **Describe**, then **Explain** and finish with the **Assessment/Evaluation**. If the question requires an explanation, stop at IDE.

Rote-learned answers are unlikely to receive full marks, so you must relate the concepts of the syllabus back to the question and ensure that you answer the question that is being asked, *not* the question you think they are asking. Planning your responses to include the relevant information and the key terminology will help you avoid writing too much, contradicting yourself, or 'waffling on' and wasting time. If you have time at the end of the paper, go back and re-read your answers.

ABOUT THE AUTHORS

Timothy Sloane

Tim Sloane is currently Head Teacher Science at Concord High School in Sydney. He has taught HSC Biology for over 20 years in a range of New South Wales state schools. He provides professional development for organisations including the Science Teachers Association of NSW and the Centre for Professional Learning. Tim's focus is on supporting students to write high-quality responses and depth studies mapped to performance band descriptors. He has created high-quality educational resources across print and digital for teachers and students. Prior to his teaching career, Tim was a successfully published research scientist working in the field of cardiovascular disease.

Mathew Sloane

Mat Sloane has been a scientist and educator for 20 years in universities, research institutes and museums. Since receiving his PhD in molecular genetics from Vienna University in 2008, he has applied his research in the fields of cancer, epigenetics, evolutionary biology and wildlife conservation. Mat completed his Bachelor of Teaching (secondary) in 2014. From 2016 to 2021, he was an educator at the Australian Museum in Sydney, where he designed and delivered lessons and digital resources for primary and high school students. Mat is continuing his passion for learning and teaching through the practice of science as the Education Program Manager at the Australian National Maritime Museum.

CHAPTER 1
MODULE 5: HEREDITY

Chapter 1
Module 5: Heredity

Module summary

Module 5 focused on reproduction and heredity, both of which are essential for the continuity of life. 'Reproduction' refers to the mechanisms by which organisms make 'like' copies of themselves, and 'heredity' refers to inheritance and the passing of genetic information from one generation to the next. As scientific understanding of reproduction and heredity has increased, so have technological advancements. Scientists can now rapidly and cheaply sequence the whole genomes of species in evolutionary studies, or sequence specific regions of DNA to determine inheritance patterns associated with various diseases and disorders.

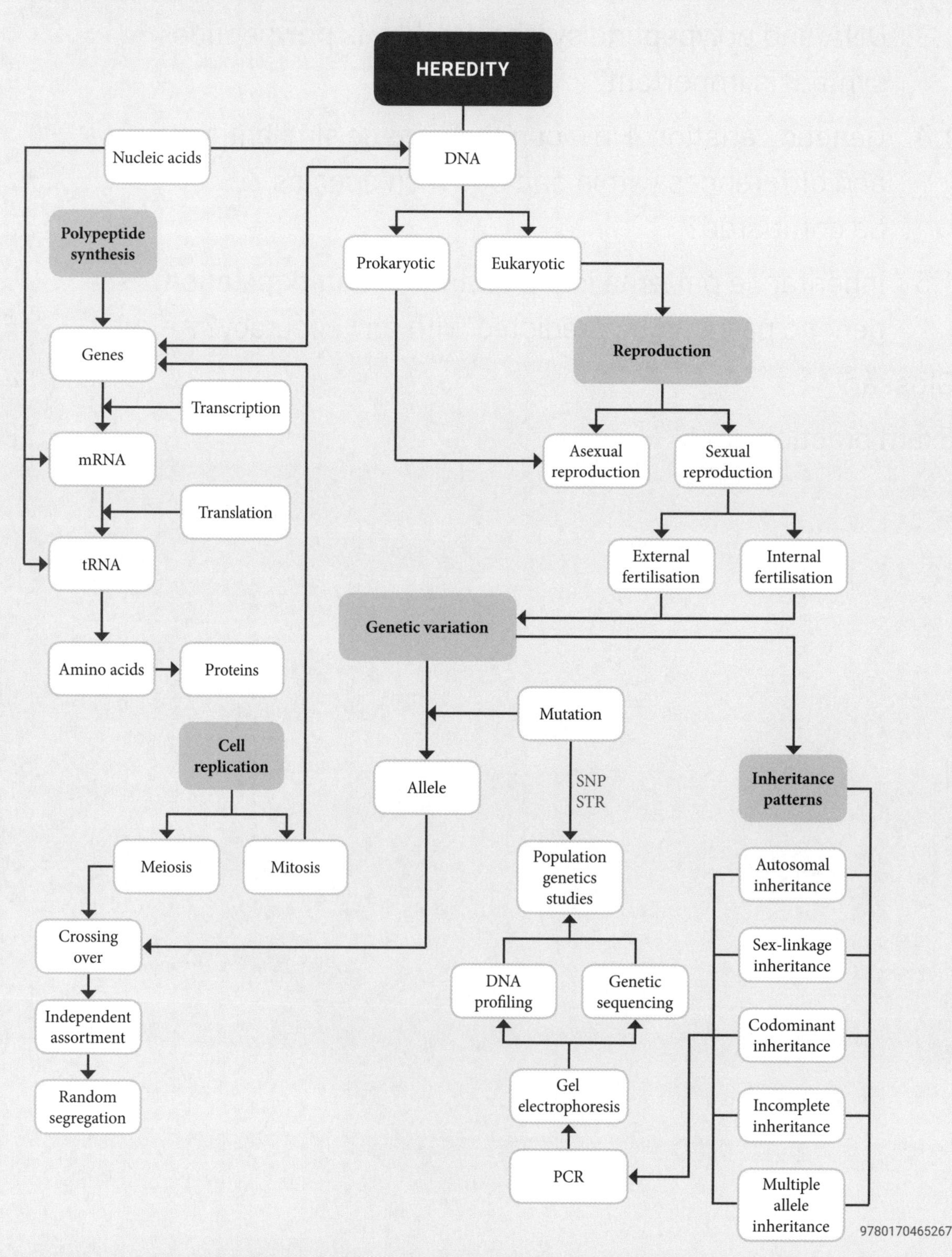

9780170465267

Outcomes

On completing this module, you should be able to:

- explain the structures of DNA
- analyse the mechanisms of inheritance and how processes of reproduction ensure continuity of species.

Key science skills

In this module, you are required to demonstrate the following key science skills:

- select and process appropriate qualitative and quantitative data and information using a range of appropriate media
- analyse and evaluate primary and secondary data and information
- solve scientific problems using primary and secondary data, critical thinking skills and scientific processes.

1.1 How does reproduction ensure the continuity of a species?

Note

If you are asked to compare the two modes of reproduction, this is best done in a table. List similarities and differences in separate columns.

Reproduction is essential for the continuity of life. **Asexual reproduction** involves the generation of genetically identical offspring from a single parent, while **sexual reproduction** requires the union of genetically unique male and female **gametes** at **fertilisation**. Given the respective advantages of asexual and sexual reproduction, many organisms can use both methods to increase their likelihood of survival.

1.1.1 Explaining the mechanisms of asexual and sexual reproduction in different organisms that ensure the continuity of a species

Note

This dot point is an 'including but not limited to' statement. You must be prepared to answer questions on all listed types of asexual and sexual reproduction in animals, plants, fungi, bacteria and protists, and have additional examples to draw upon.

Asexual and sexual reproduction each have advantages and disadvantages. You should aim to recall these and apply them to each kingdom of life covered in this section.

TABLE 1.1 Advantages and disadvantages of sexual and asexual reproduction

	Advantages	Disadvantages
Asexual reproduction	• Offspring are genetically identical to the parent, resulting in individuals that are better suited to stable environments • Population size can increase rapidly in stable environments • Efficient form of reproduction with no energy investment required to search for a mate	• Lack of genetic variation means species are restricted to specific habitats • If environmental conditions change, the species is at risk of extinction due to lack of genetic variation • Rapid increases in population size can result in overcrowding and greater intraspecific competition for resources

››

TABLE 1.1 cont.

››	Advantages	Disadvantages
Sexual reproduction	• Increased genetic variation results from recombination of alleles during meiosis • Individual species can exist in a wider range of environments • If environmental conditions change, the species is better equipped to adapt and has a reduced risk of extinction due to the higher levels of genetic variation • Unfavourable genetic variation is eliminated more efficiently from a population	• Energy must be invested to produce reproductive structures and gametes • Reproductive rate is slower, with fewer offspring produced over a longer period • Recombination of alleles during meiosis can break apart beneficial gene combinations and introduce harmful variations into a population • Often requires energy and time investment to find and court a prospective partner **Specific to internal fertilisation:** • Requires ongoing energy and time investment during **gestation** and rearing of young • Potential for spread of sexually transmitted diseases through a population

Advantages of external and internal fertilisation

The kingdom Animalia is a large and diverse group of multicellular **eukaryotic** organisms that have a **diploid-dominant** life cycle. The only **haploid** cells are the gametes. Most animals reproduce sexually; however, some also use asexual reproduction. Forms of asexual reproduction in animals include **fission**, **budding**, **fragmentation** and **parthenogenesis**. Typically, these animals use asexual reproduction when environmental conditions are favourable and resources are abundant. In contrast, sexual reproduction ensures a mixing of the gene pool when conditions are less desirable and variation is advantageous.

Asexual reproduction in animals

TABLE 1.2 Types of asexual reproduction in animals

Type of asexual reproduction	Description	Examples	Diagram
Budding	Outgrowth of a body region that separates from the original organism, giving rise to two individuals	Common in corals and hydras	
Fission	Organism splits along its longitudinal axis, forming two separate organisms	Sea anemone and planaria (flatworms)	
Fragmentation	A body part detaches and develops into a new organism, and the original organism regenerates the lost body part	Sea sponges and starfish	
Parthenogenesis	An unfertilised egg develops into a complete individual. The resulting offspring can be haploid or **diploid**, depending on the species	Common in invertebrates including ants, aphids and stick insects. Some vertebrate species, including certain reptiles, amphibians and fish	

Sexual reproduction in animals

Fertilisation, which requires the union of genetically unique male and female gametes, can occur internally or externally to the female's body, and is primarily determined by the environment in which a species is found.

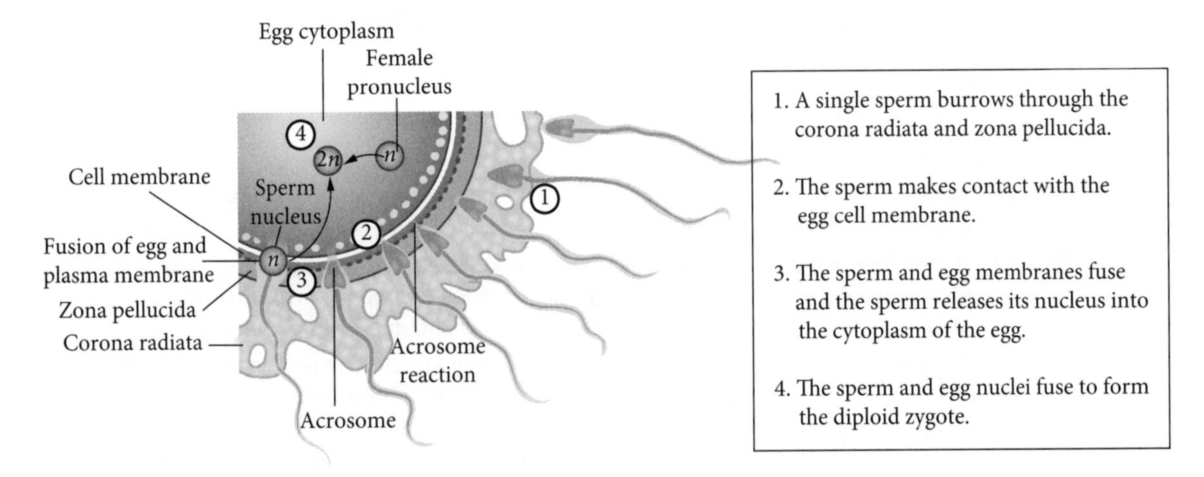

FIGURE 1.1 The process of fertilisation

External fertilisation typically occurs in aquatic environments, where there is little risk of the egg and sperm desiccating (drying out) when they are released. Examples are corals, amphibians and most fish species, which release large numbers of gametes into the water at the same time. These spawning events are determined by environmental cues such as temperature, and maximise the likelihood of fertilisation.

In contrast, terrestrial animals rely upon **internal fertilisation** to increase the chances of a successful union between sperm and egg. The sperm are deposited directly into the female reproductive tract relatively close to the egg(s). The female reproductive tract provides the necessary moist environment to avoid desiccation of the gametes and subsequent **embryo**.

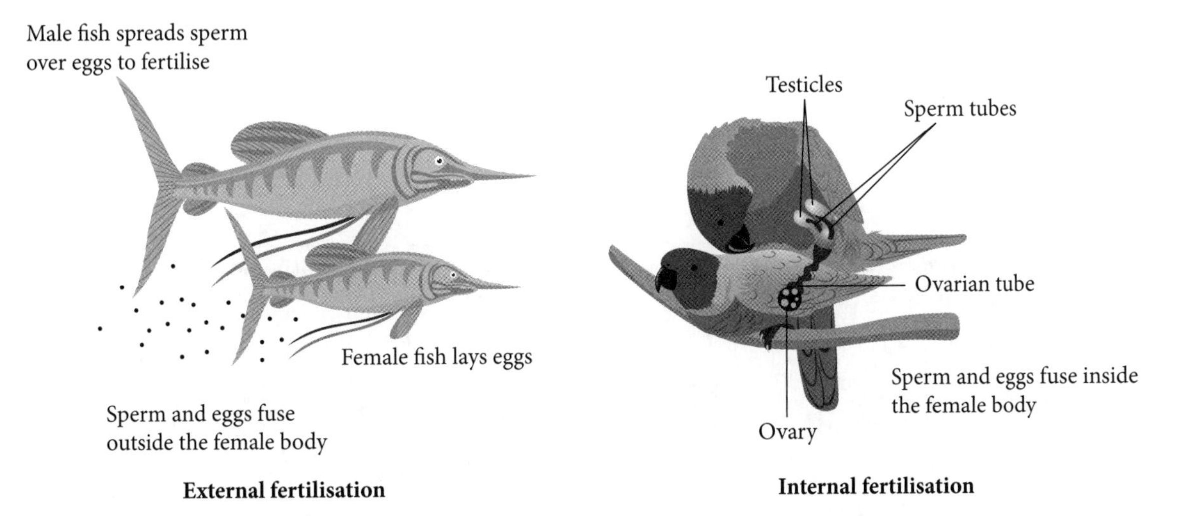

FIGURE 1.2 External versus internal fertilisation

Note

Remember: both internal and external fertilisation relate to sexual reproduction. Students often confuse external fertilisation with asexual reproduction.

TABLE 1.3 Advantages and disadvantages of internal and external fertilisation

	Advantages	Disadvantages
Internal fertilisation	• Successful fertilisation more likely, due to positioning of gametes near each other • Often, fewer gametes are produced due to higher likelihood of successful fertilisation and offspring survival • Can take place in terrestrial environments	• Fewer offspring produced • Large investment of energy and time in mating rituals and rearing of offspring • Parents at greater risk of predation while rearing offspring • Potential of sexually transmitted diseases
External fertilisation	• Large numbers of offspring produced • Offspring are typically dispersed widely, reducing intraspecific competition • Little investment of energy and time in mating rituals and rearing of offspring • Adults can continue to reproduce without pausing to rear each new generation • Parents at reduced risk of predation • No exposure to sexually transmitted diseases	• Successful fertilisation less likely due to the dependence on environmental conditions. Mitigated by synchronised release of gametes • Greater proportion of offspring die due to lack of parental protection and care • More gametes must be produced to increase the likelihood of successful fertilisation and offspring survival • Can only occur in aquatic environments

Plants: asexual and sexual reproduction

The kingdom Plantae consists of four major groups: **angiosperms** (flowering plants), gymnosperms (cone-bearing plants), **pteridophytes** (ferns) and **bryophytes** (mosses and liverworts). Angiosperms, gymnosperms and pteridophytes are vascular plants (they have xylem and phloem), while bryophytes are non-vascular. All plants can reproduce sexually, and many are also able to reproduce asexually. The mechanism by which each group does so differs significantly.

Note

While it is important to be aware of all plant groups, the focus is nearly always on angiosperms. However, with an appropriate stimulus, you could be asked a question on bryophytes, pteridophytes or gymnosperms.

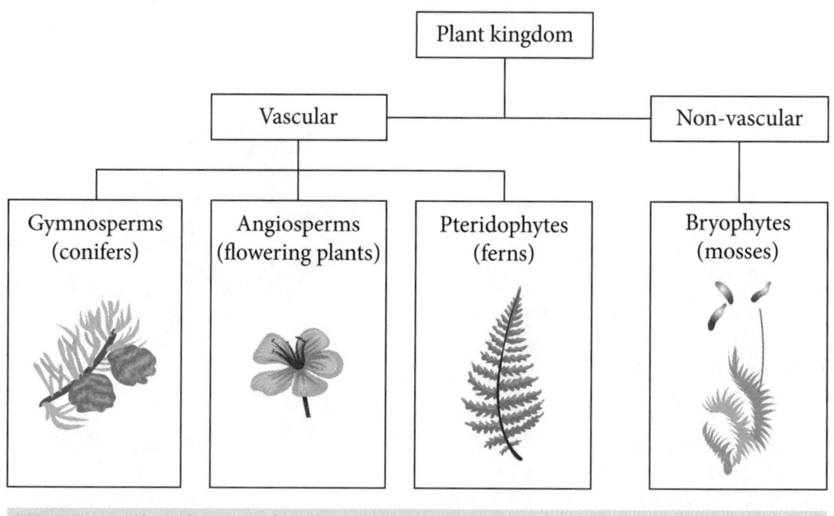

FIGURE 1.3 Classification of the kingdom Plantae

Asexual reproduction in plants

Plants use a wide range of mechanisms to reproduce asexually. All forms are observed in the angiosperm group, while in bryophytes, pteridophytes and gymnosperms, various forms of fragmentation are more common.

TABLE 1.4 Types of asexual reproduction in plants

Type of asexual reproduction	Description	Examples
Bulb	Underground bulbous stem surrounded by fleshy leaves. Used to provide nourishment during dormant period	Onion, daffodil 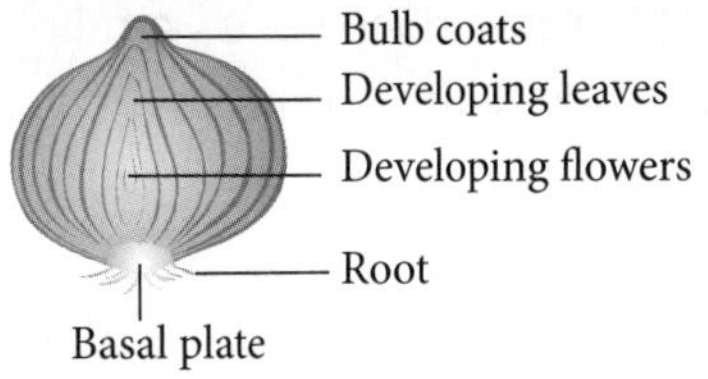
Fragmentation	Broken pieces of parent plant can regenerate and grow into independent plants	Bryophytes, willow tree 
Rhizome	Stem that grows horizontally underground, with suckers giving rise to new shoots and roots	Ginger, bamboo 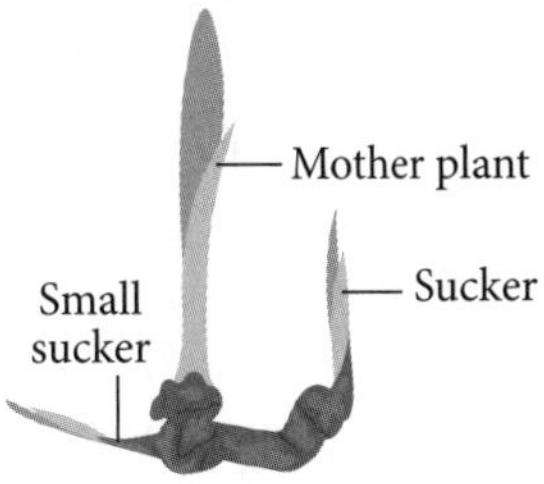
Runner	Stem that runs along the ground surface, producing new shoots and roots from growing nodes	Strawberry, peppermint 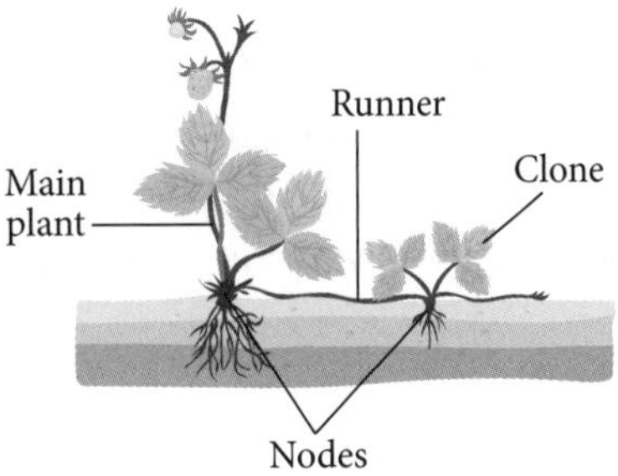
Sucker	Growth that develops directly from the rootstock of a plant	Banana, casuarina 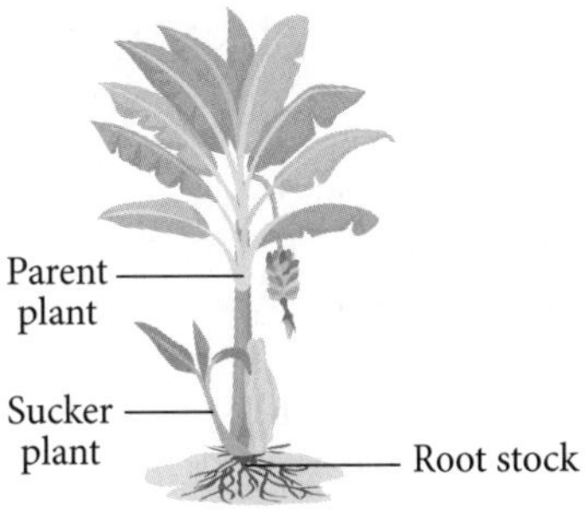
Tuber	Underground modified stem specialised to store starch and provide nourishment to the plant during dormancy	Potato, yam 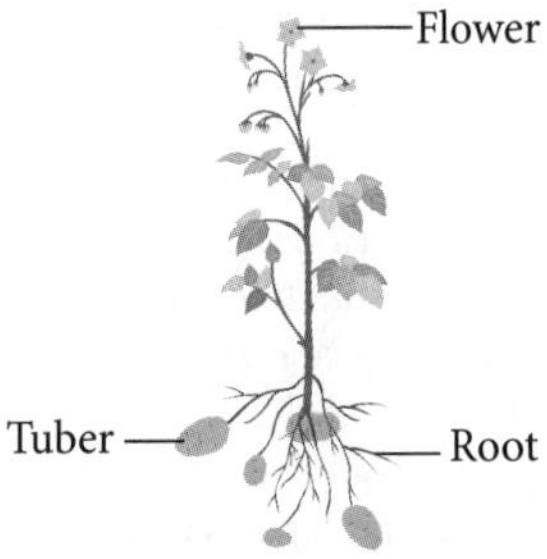

Sexual reproduction in plants

The life cycle of all plants consists of two phases (diploid **sporophyte** and haploid **gametophyte**). This is known as an **alternation of generations**, where both haploid and diploid phases are multicellular and replicate via mitosis. The four major groups of plants differ in the degree to which each of these phases dominates, the nature of the gametes produced, and the mechanisms of reproduction.

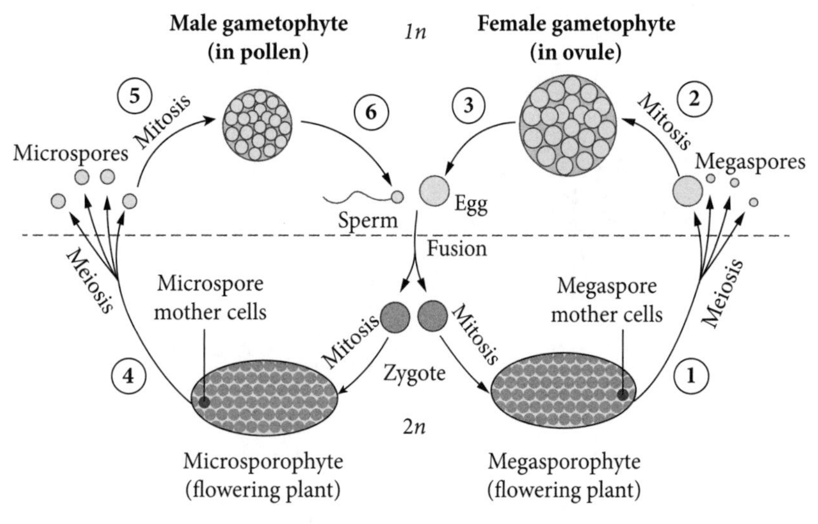

1. One or more egg-producing ovules inside the ovary. Diploid megaspore mother cells are located within ovule and undergo meiosis to produce four haploid megaspores, of which only one survives.
2. The successful megaspore undergoes several mitotic cell divisions to produce a multicellular haploid female gametophyte.
3. One of the cells within the gametophyte becomes the egg, while the others either break down or provide nutrients to the egg.
4. Inside the anther are diploid microspore mother cells. These mother cells undergo meiosis to produce haploid microspores.
5. The microspores develop through mitosis to produce pollen, the male gametophyte which contains the haploid gamete (sperm) and other specialised cells required for successful pollination and fertilisation.
6. The gametophyte contains the haploid gamete (sperm) and other specialised cells required for successful pollination and fertilisation.

FIGURE 1.4 Alternation of generations in angiosperms with heterosporous spores

Angiosperms (flowering plants)

The distinguishing features of angiosperms are that they produce flowers, and bear their seeds in fruit, and most species use other organisms, or mechanisms such as wind, for pollination. Angiosperms (and gymnosperms) are **heterosporous**, with the sporophyte phase of the life cycle dominating, appearing as what we know as the plant or tree.

> **Note**
> The prefix *hetero-* means 'different'. The prefix *homo-* means 'the same'.

The male and female reproductive organs of angiosperms are contained in the flowers. To avoid self-pollination and the associated reduction in genetic variation, the male and female reproductive structures of most angiosperms mature at different times.

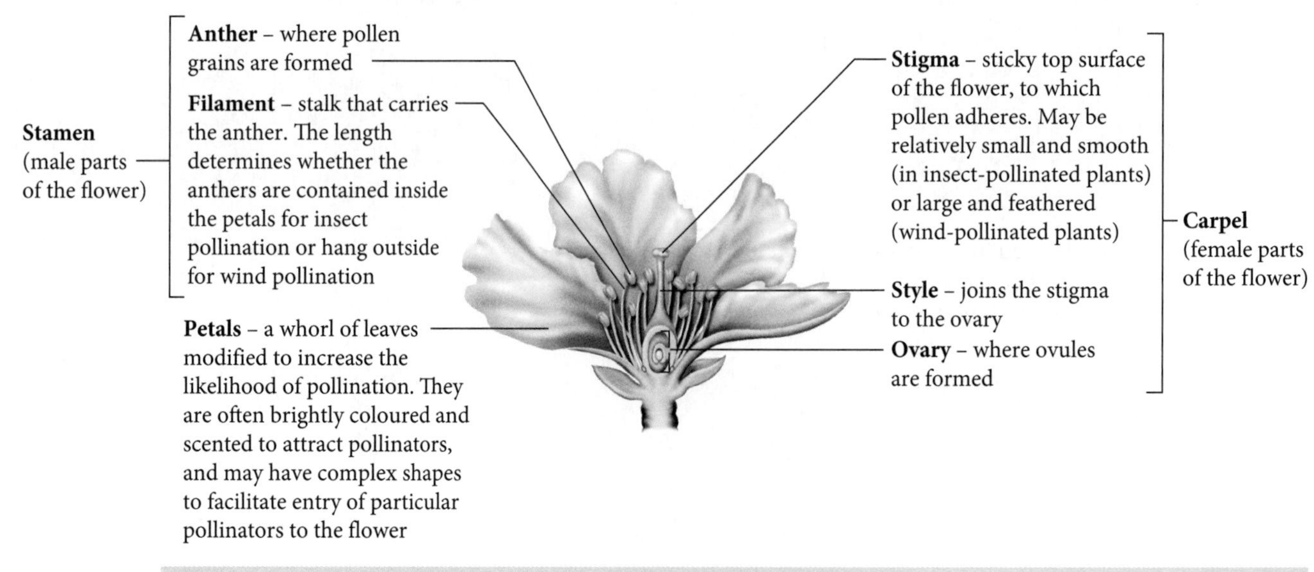

FIGURE 1.5 An angiosperm flower with male and female reproductive organs

Angiosperms have a diverse range of mechanisms to ensure pollination. These include abiotic factors such as wind and water, but what is unique to angiosperms is their use of biotic pollinators: insects, birds and small mammals. Pollination occurs when pollen lands on the **stigma**, no matter what mechanism is involved. Once pollination has occurred, a tube grows from the pollen grain through the **style** to the ovule. A sperm is released from the pollen grain and travels through the pollen tube to the ovule, where it fertilises an egg, forming a **zygote**, the first diploid cell of the next sporophyte generation. The zygote divides by mitosis to produce a seed. Subsequent seed dispersal ensures survival of the species.

Note
Make sure you know the difference between pollination and fertilisation. Many students don't!

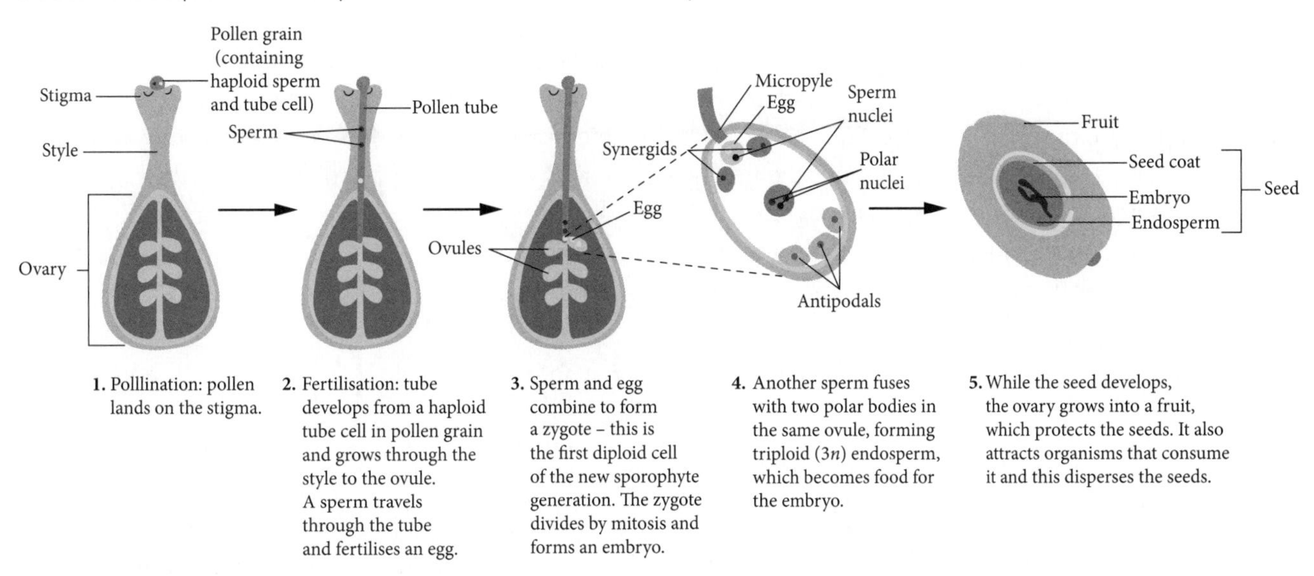

FIGURE 1.6 Pollination and fertilisation in angiosperms

Fungi: budding, spores

Fungi can be unicellular (e.g. yeast) or multicellular (e.g. mushrooms). Their life cycles consist of two stages: vegetative (growth) and reproductive. They are **haploid-dominant** organisms, which means they spend most of their life in the vegetative stage as haploid cells. Most species are capable of asexual and sexual reproduction; however, some unicellular species only reproduce asexually.

Unicellular fungi reproduce asexually by budding or fission, which allows the population size to quickly increase in favourable conditions. When nutrients are limiting, or conditions are unfavourable, they mate by **conjugation** to produce a diploid cell. The diploid cell then undergoes sporulation, a form of meiosis, to produce haploid **spores**, which then return to the haploid stage of the life cycle.

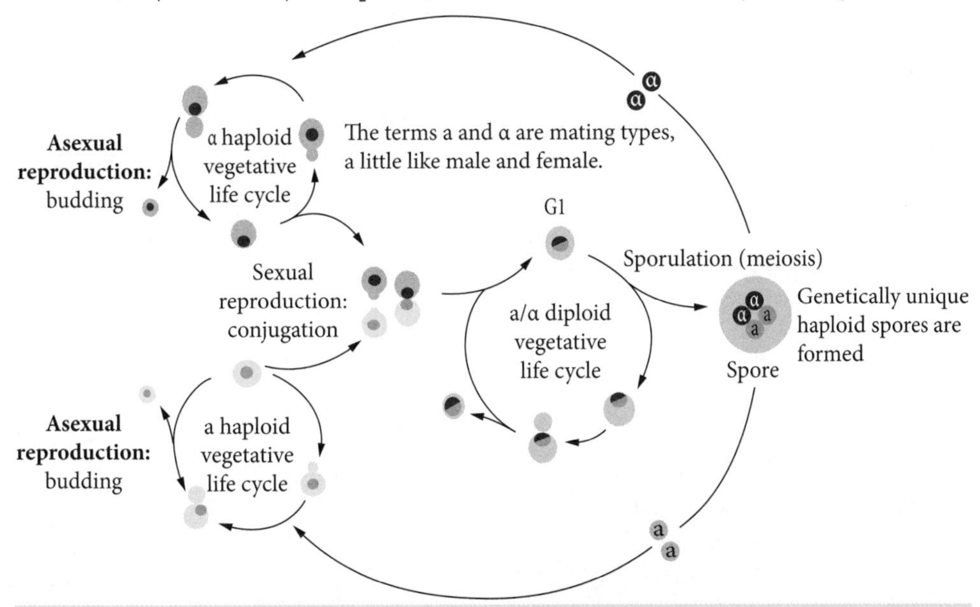

FIGURE 1.7 Asexual and sexual reproduction in unicellular fungi

Multicellular fungi spend most of their life in a vegetative state as thread-like structures called **hyphae**. Collectively, hyphae form the multicellular haploid **mycelium**, which plays an important role in both asexual and sexual reproduction, with both modes involving the production of microscopic spores. The spores are dispersed from the parent fungus in order to colonise new environments and ensure continuity of the species.

The mycelium has specialised structures that produce genetically identical haploid spores (mitospores) via mitosis. The spores are dispersed and upon germination each will produce a new hypha, which will continue to divide via mitosis, producing a new fungal colony genetically identical to the original mycelium.

If the mycelium of one fungus comes in contact with the mycelium of another, the cytoplasm of the haploid cells fuse (**plasmogamy**). Initially the nuclei do not combine, and the cells remain **heterokaryotic**. The mycelium, now containing heterokaryotic tissue, forms a fruiting body (e.g. a typical mushroom). Once the fruiting body has matured, the two haploid nuclei in the heterokaryotic cells fuse (**karyogamy**) to form a diploid zygote. The zygote then undergoes meiosis to produce genetically unique spores (meiospores). The production and dispersion of these spores ensures that genetic variation is maintained within the species.

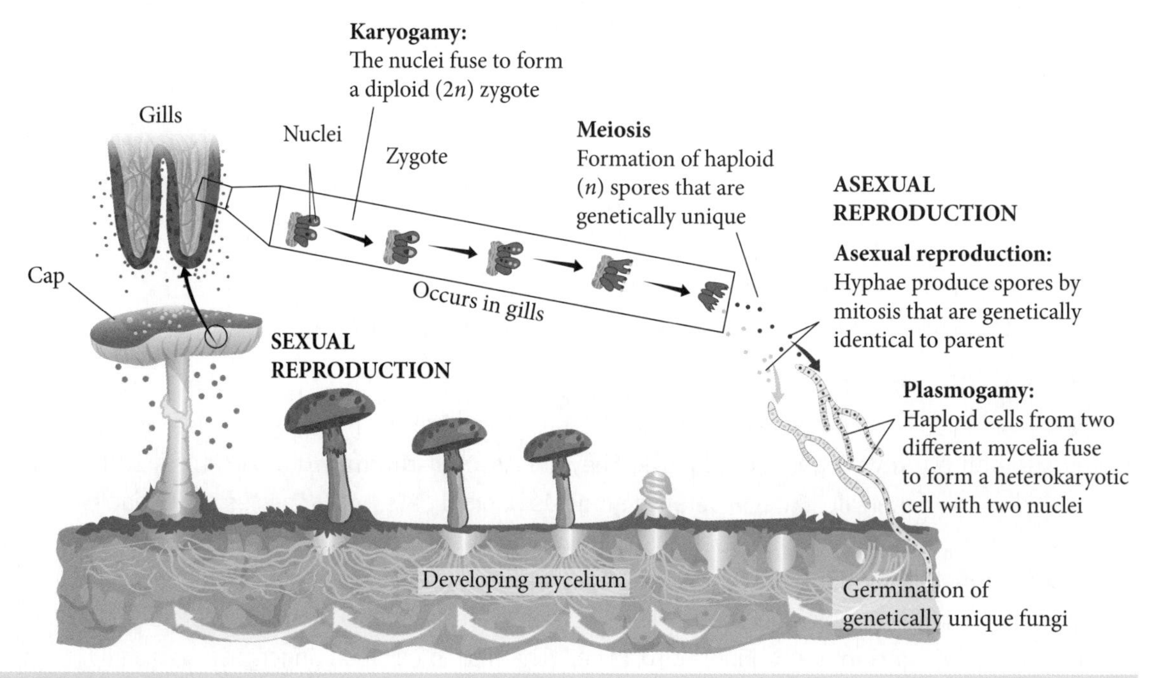

FIGURE 1.8 Asexual and sexual reproduction in multicellular fungi

Bacteria: binary fission

Bacteria have one circular chromosome located in the nucleoid, a region of the cytoplasm that controls cell activity and replication. In addition to the single chromosome, bacteria can contain small pieces of circular **deoxyribonucleic acid (DNA)**, called **plasmids**. Although plasmids are not essential for the survival of bacteria, they can carry genes that play an important role in maintaining genetic variation and continuity of species (e.g. genes for antibiotic resistance).

Bacteria can only reproduce through a process of asexual reproduction called **binary fission**, in which a bacterium increases in size, replicates its DNA, then divides into two identical daughter cells. Individuals in a population of bacteria are therefore clones of each other.

Despite reproducing via binary fission, bacteria can still acquire genetic variation through three mechanisms: conjugation, **transformation** and **transduction**. The most common of these is conjugation, the direct transfer of DNA from one bacterium to another via a cytoplasmic bridge.

Note
Conjugation in bacteria is not considered to be sexual reproduction.

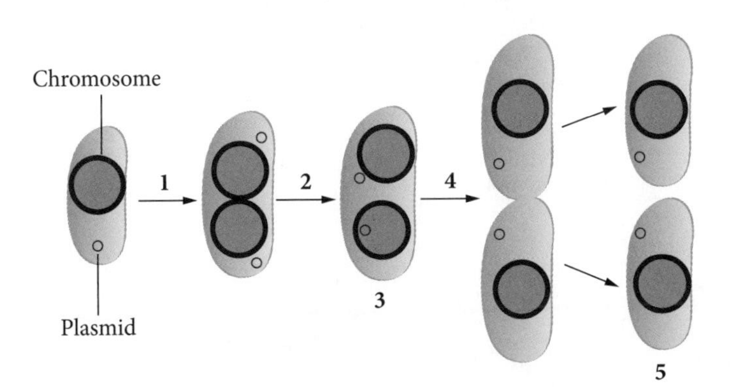

1. Single circular chromosome and plasmids replicate independently of each other.
2. Cell elongates by building more cell wall.
3. Replicated chromosomes and plasmids begin to separate, each moving towards a separate pole as the cell elongates further.
4. Cleavage furrow begins to form and cell wall forms in the cleavage furrow.
5. Two identical daughter cells are produced.

FIGURE 1.9 Binary fission in bacteria

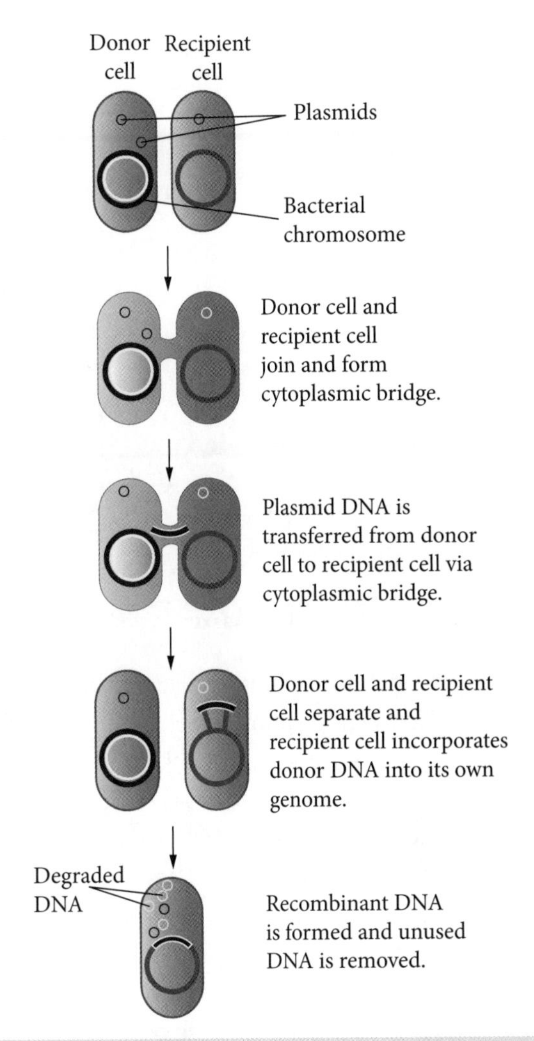

FIGURE 1.10 Genetic variation in bacteria can be achieved by conjugation.

Protists: binary fission, budding

The kingdom Protista consists of mainly unicellular eukaryotic organisms that are not classified as animals, plants or fungi (e.g. *Paramecium*). Most use asexual reproduction as their primary mode of reproduction; however, some species also reproduce sexually.

Methods of asexual reproduction in **protists** include binary fission, multiple fission and budding. Binary fission in protists involves a mitotic-like cell division where the chromosomes are first duplicated, then the **nucleus** divides, and this is then followed by cell division, giving rise to two separate independent organisms.

Multiple fission is like binary fission; however, there are multiple rounds of chromosome duplication and cell division, giving rise to many independent organisms.

Budding occurs when a small outgrowth from the parent protist cell forms. The chromosomes again duplicate, followed by a nuclear split. The daughter nucleus and a portion of cytoplasm migrate to the bud before it splits, giving rise to a smaller independent organism, which continues to grow until it reaches a comparable size to the parent protist cell.

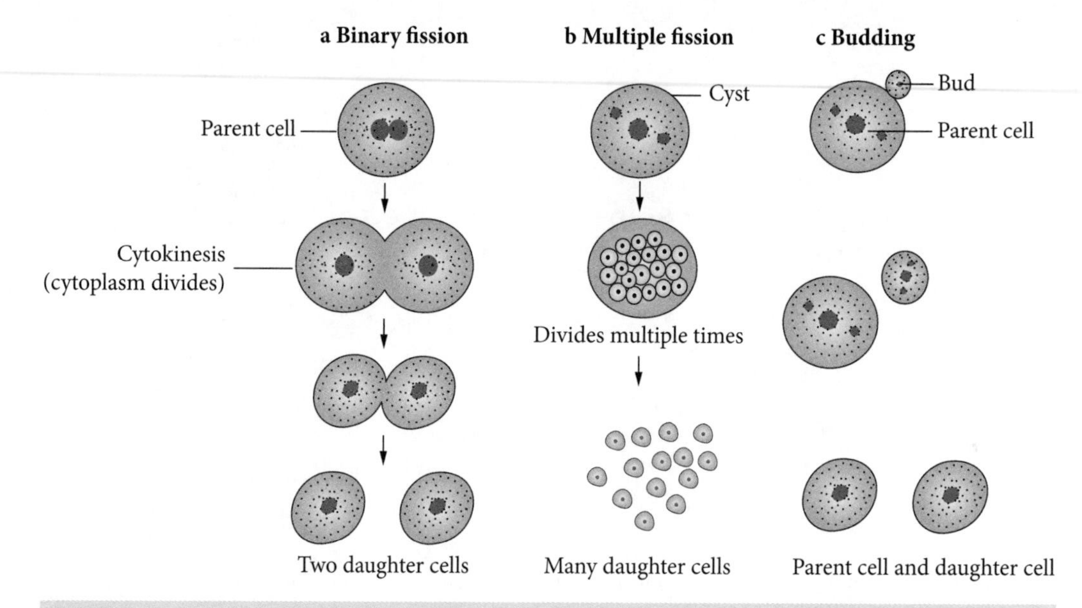

FIGURE 1.11 Asexual reproduction in protists: a binary fission, b multiple fission and c budding

In unfavourable conditions, protists can use sexual reproduction to introduce genetic variation into the population, which ensures continuity of species. Methods of sexual reproduction in protists include **syngamy** or conjugation. In one form of syngamy, an adult diploid cell undergoes meiosis to produce four genetically unique haploid gametes. Gametes from different individuals permanently fuse to produce a new diploid cell line.

Some protists (e.g. *Paramecium*) have two types of nuclei: a micronucleus and a macronucleus. The smaller diploid micronucleus is the reproductive nucleus, while the larger macronucleus contains more DNA and controls the metabolic activity of the cell. These protists reproduce sexually via conjugation, which involves temporary fusion of cells and exchange of genetic information.

> **Note**
> The methods of introducing genetic variation in bacteria, and multiple fission and sexual reproduction in protists, address the 'including but not limited to' component of the syllabus dot point.

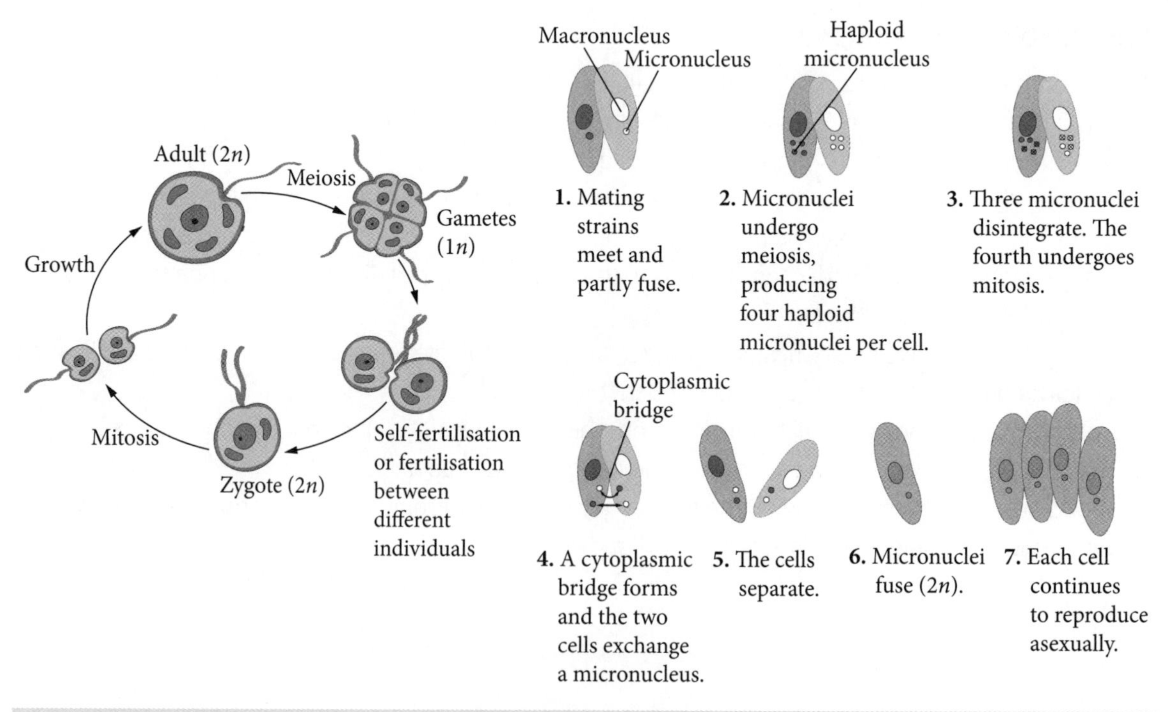

FIGURE 1.12 Sexual reproduction in protists: a syngamy and b conjugation

1.1.2 Analysing the features of fertilisation, implantation and hormonal control of pregnancy and birth in mammals

Note

Remember: 'analyse' requires you to identify the different components and the relationships between them, and from that to draw out and relate the implications.

In all mammals, sexual reproduction involves internal fertilisation and results in genetically unique offspring. The male and female reproductive organs produce haploid gametes, the sperm and eggs.

For internal fertilisation to occur, the penis is inserted into the vagina, where seminal fluid, typically containing hundreds of millions of sperm, is deposited upon ejaculation. Sperm move through the cervix and the uterus, and a small number reach the fallopian tubes. Here, one sperm fertilises an egg, forming a zygote, the first diploid cell of a new individual.

In placental mammals, the developing zygote travels back down the fallopian tube to the uterus, where it implants into the **endometrium**. After **implantation**, several key developmental stages occur as the cell mass develops into an embryo and then a **foetus** prior to birth.

CASE STUDY 1

HUMANS

After fertilisation, the zygote goes through several mitotic cell divisions to form a morula. This occurs as the cell mass travels from the fallopian tube to the uterus. Once the morula reaches the uterus, cell division continues, giving rise to a **blastocyst**, the first structure with different cell types. Conception is considered the point at which the blastocyst implants into the endometrium. Following implantation of the blastocyst, rapid cell differentiation and specialisation occur, giving rise to the embryo (followed by the foetus stage) and placenta. Embryonic tissues give rise to the umbilical cord, which attaches to the placenta, allowing the exchange of blood, nutrients and waste material with the mother.

The embryonic period (from fertilisation to 8 weeks) is the stage in which the neural tube (later the central nervous system) and other major organ development occurs. The foetal period extends from 8 weeks after fertilisation until birth, and is characterised by more human-like features and the continued specialisation, growth and development of all organ systems required for independent function. The average gestation period, measured from the time of the mother's last menstrual period to the time of birth, is 40 weeks in humans.

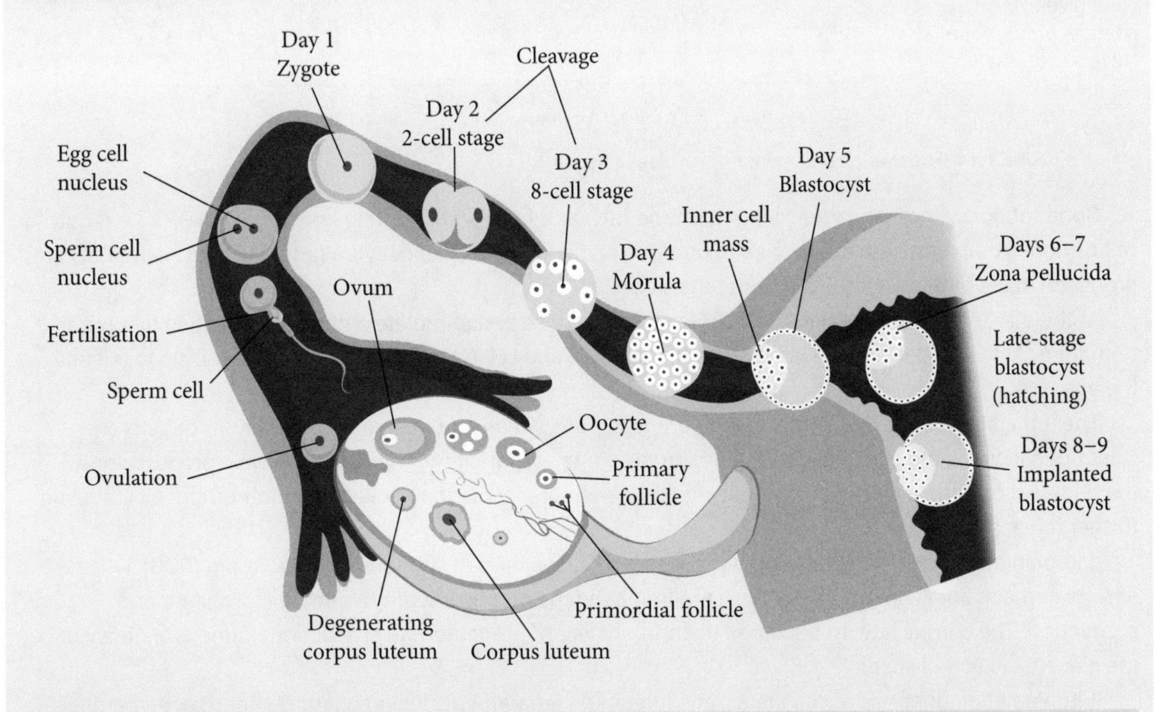

FIGURE 1.13 Ovulation, fertilisation and implantation

The production of gametes, **pregnancy** and birth are all highly regulated by specific **hormones**.

Ovarian and menstrual cycles

The ovarian cycle is responsible for:

1 the maturation of a follicle and the egg contained within it
2 the release of the egg during ovulation
3 the subsequent production of the **corpus luteum**.

The menstrual cycle is responsible for the preparation and maintenance of the uterine wall in readiness for implantation of the zygote. If fertilisation does not occur, or if implantation is unsuccessful, the endometrium and egg will be shed during the menstrual period.

The ovarian and menstrual cycles work in conjunction with each other and on average last for 28 days.

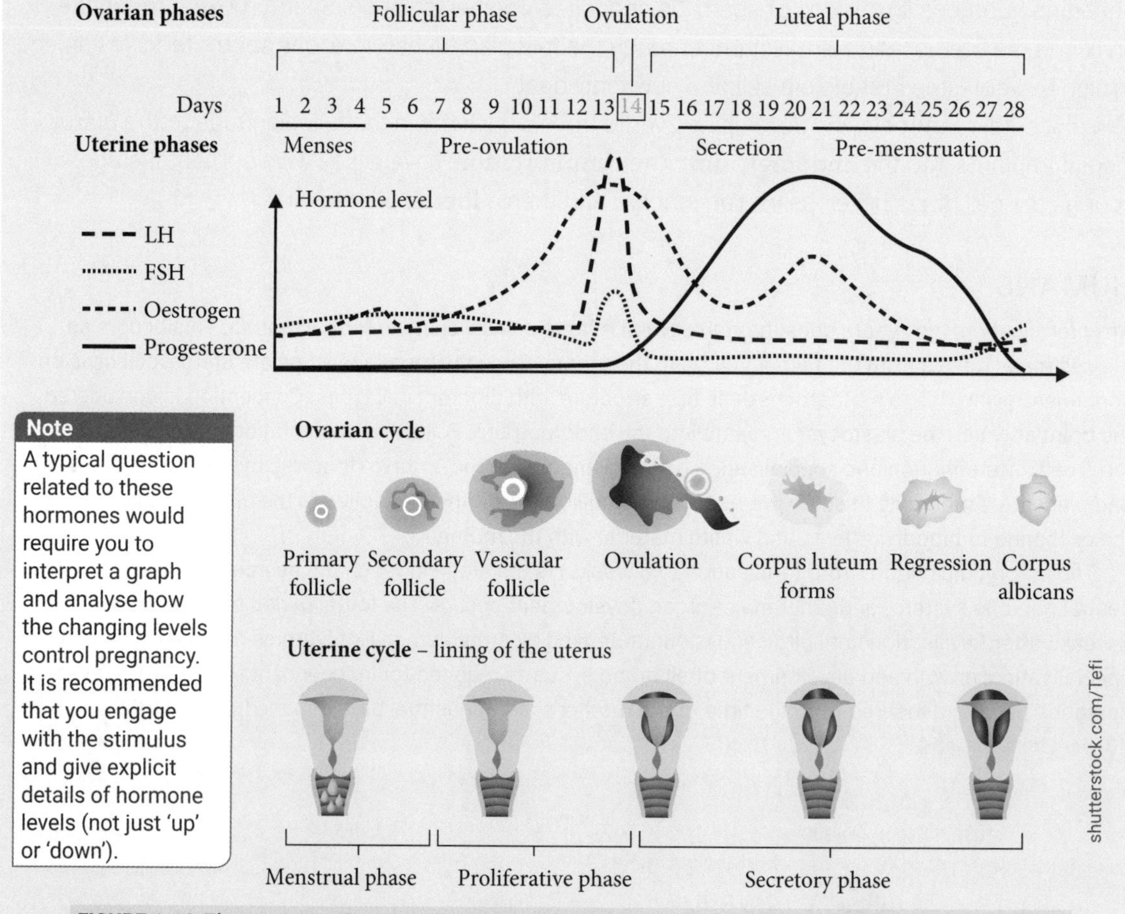

Note

A typical question related to these hormones would require you to interpret a graph and analyse how the changing levels control pregnancy. It is recommended that you engage with the stimulus and give explicit details of hormone levels (not just 'up' or 'down').

FIGURE 1.14 The ovarian and menstrual cycles in humans

Gonadotropins, including luteinising hormone (LH) and follicle-stimulating hormone (FSH), are a family of hormones that stimulate the ovaries to produce and release mature oocytes (eggs). Both LH and FSH are produced in the pituitary gland and target the **ovary**.

At the start of a new ovarian cycle, FSH levels slowly rise and stimulate a new follicle within the ovary to mature. At the 14-day point, a large spike in both FSH and LH induces the now mature follicle to release the egg.

The remaining follicle tissue forms the corpus luteum, which produces oestrogen and progesterone. Oestrogen induces the thickening of the endometrium to prepare it for implantation, while progesterone, the primary hormone involved in maintaining pregnancy, preserves the thickened endometrium by inhibiting further release of LH and FSH.

If implantation occurs, the developing placenta releases human chorionic gonadotropin (hCG), which assists in maintaining pregnancy by further stimulating the corpus luteum to continue progesterone production. The corpus luteum becomes the primary site of progesterone production during pregnancy once the placenta is established.

If implantation does not occur, the corpus luteum degenerates, reducing oestrogen and progesterone levels, which in turn leads to the shedding of the endometrium during the menstrual period. At this point, the ovarian and menstrual cycles commence again.

At 40 weeks, the fully grown foetus places increased pressure and strain on the uterine wall. This stimulates increased production of oestrogen from the placenta, which in turn begins to inhibit progesterone production. The falling levels of progesterone trigger muscular contractions of the uterus.

The increased oestrogen levels and commencement of contractions also stimulate the production of oxytocin from both the placenta and the pituitary gland of the mother. Increased levels of oxytocin induce greater and more frequent uterine contractions and the dilation of the cervix. This moves the foetus down and out of the birth canal during the birthing process. The commencement of the birthing process is an example of a positive feedback mechanism, where the production of oxytocin induces contractions, which in turn leads to more oxytocin production.

Other hormones involved in the birthing process include endorphins and adrenaline. Levels of both these hormones increase as birth draws near. Endorphins, which are produced in the pituitary gland, have a calming effect, and offer some resistance to pain during the birth. The adrenaline increase is linked to heightened awareness so that the female has time to find a 'safe' place prior to birthing.

1.1.3 Impact of scientific knowledge on the manipulation of plant and animal reproduction

Note

The focus of this syllabus dot point is the impact of scientific knowledge on the manipulation of plant and animal reproduction in agriculture, *not* the reproductive technologies themselves. This evaluation requires you to address the practices that existed prior to a specific example of scientific knowledge, how the accumulation of this knowledge furthered reproductive techniques in agriculture, and any possible negative consequences. Based on the evidence provided, you must also provide a judgement.

Scientific research has improved our understanding of genetics and inheritance, and has led to the development of technologies that have enabled humans to manipulate plant and animal reproduction in agriculture. The impact of new scientific knowledge is summarised in Table 1.5.

TABLE 1.5 The impact of new scientific knowledge

Scientific knowledge	Before science knowledge was acquired	Impact of new scientific knowledge
Physical traits are passed from one generation to the next	Animals and plants were harvested from the wild	Development of agriculture: • **Selective breeding** has created animals and plants with desired characteristics (e.g. increased milk production in dairy cattle)
Laws of heredity: Gregor Mendel (1860s) proposed that offspring inherit one 'factor' from each parent	The cause of phenotypes in plants and animals was unknown	• Desirable traits can be selected in a predictable way • Improved quality and yields
Chromosomal theory of inheritance: Sutton and Boveri (early 1900s) proposed that chromosomes exist in pairs, inherited from each parent	The physical 'factors' inherited from parents were unknown	• Provided the basis for better understanding and control of outcomes of selective breeding, **artificial pollination** and **artificial insemination**
Structure of DNA discovered: Watson and Crick (1953)	Little understanding of DNA replication and how the genetic code is 'read'	• Understanding of transcription and translation, and how mutations cause disease and allow evolution • Basis of **cloning**
DNA sequencing invented (early 1970s)	Not possible to 'see' a DNA sequence	• Helped development of **recombinant DNA** technology • Identification of alleles responsible for traits

››

TABLE 1.5 cont.

Scientific knowledge	Before science knowledge was acquired	Impact of new scientific knowledge
Development of recombinant DNA technology (1970s)	DNA could not be transferred between species	• Genes can be transferred between species to create **transgenic organism**. Widespread in plant agriculture
Whole-animal cloning (1996)	Animals could only be bred through fusion of egg and sperm	• Genetically identical animals could be cloned • Potential to produce animals with a specific trait to be used for human therapy
Discovery of CRISPR-Cas9 Charpentier and Doudna (2012)	Genetic manipulation was time consuming and expensive	• Cheap, fast and accurate editing of specific genes to make **genetically modified organisms (GMOs)**

1.2 Cell replication: How important is it for genetic material to be replicated exactly?

Cell replication is the process by which a single parent cell copies its DNA and divides into two or more daughter cells. This section explores the two types of cell replication, **mitosis** and **meiosis**. Mitosis begins with the first cell division after fertilisation and gives rise to all the cell types (except gametes), tissues, organs and bones of the body. Mitosis produces cells to grow, repair damaged tissues and constantly replenish others, including old skin cells and the lining of our intestines. During meiosis, sexually reproducing organisms produce gametes, the reproductive cells of the organism. New genetic variation is generated during this process and is transmitted to the next generation during reproduction.

1.2.1 Modelling the processes involved in cell replication

Mitosis and meiosis

Models are essential for scientific research, because they simplify complex structures and concepts that cannot be seen by eye. They also provide predictive power and can shape the direction of future research. Models also have limitations. For example, those used to demonstrate cell replication highlight the key structures and stages of each process, but do not usually show detail at the molecular level.

Types of models that could be used to demonstrate knowledge and understanding of mitosis and meiosis include:

- diagrams
- physical replicas
- computer simulations.

Whatever the type of model, key features to include when modelling mitosis and meiosis are shown in Figure 1.15 and Table 1.6.

Note

This is an 'including but not limited to' dot point. You must be familiar with modelling mitosis, meiosis and DNA replication, and be able to apply this skill to other processes involved in cell replication.

Note

Remembering the names of each stage of mitosis and meiosis is *not* a specific requirement of the syllabus. Your primary focus should be the behaviour of chromosomes during each stage and the outcome of each process.

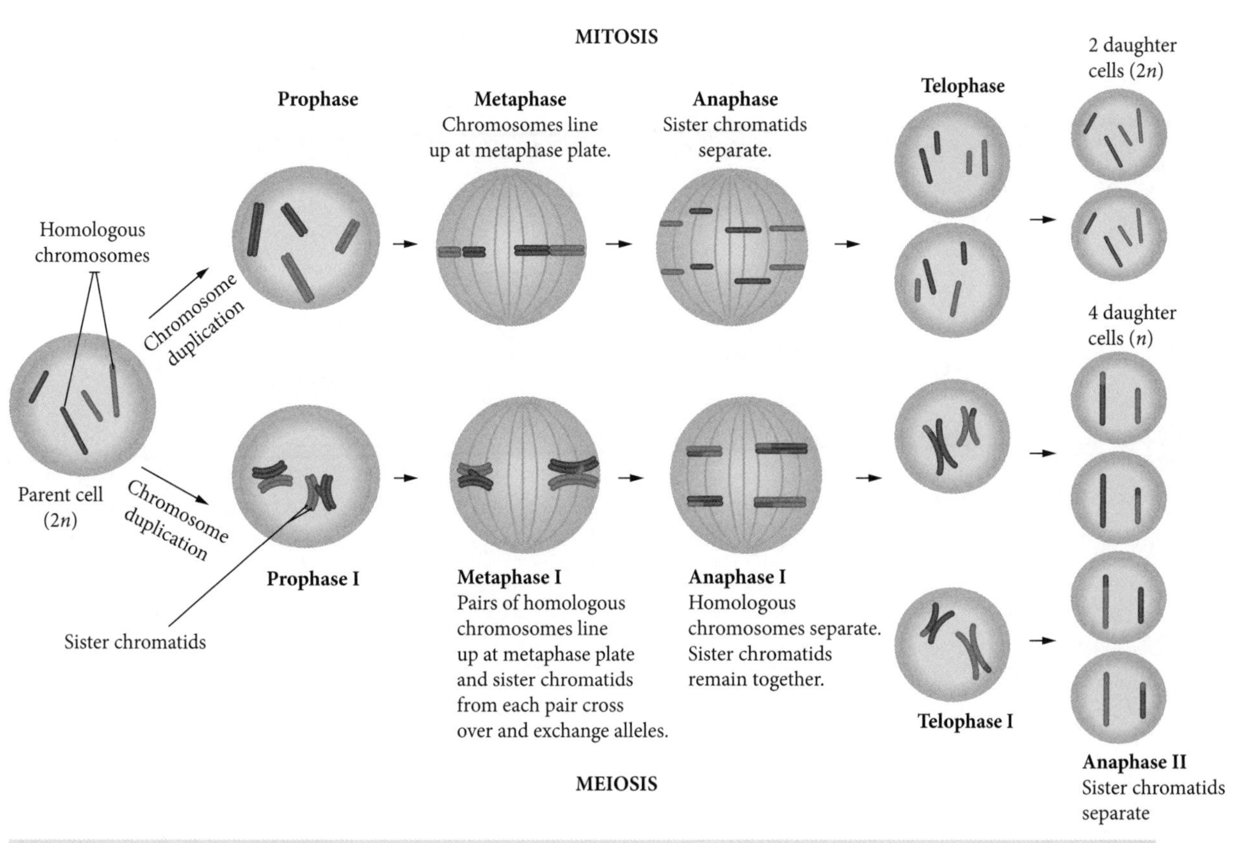

FIGURE 1.15 Mitosis and meiosis

TABLE 1.6 A comparison of mitosis and meiosis

	Mitosis	Meiosis
Parent cell	Diploid (2*n*)	Diploid (2*n*)
Genetic recombination	No exchange of genetic information between homologous chromosomes	Genetic information exchanged during prophase I when crossing over occurs between non-**sister chromatids** of homologous chromosomes
Number of divisions	One	Two (meiosis I and meiosis II)
Number of daughter cells	Two genetically identical daughter cells	Four genetically unique daughter cells
Number of chromosomes	• Same number as parent cell • Diploid (2*n*)	• Half the number of the parent cell • Haploid (*n*)
Location	All tissues	Gonads

DNA replication using the Watson and Crick DNA model, including nucleotide composition, pairing and bonding

Key features of the Watson and Crick model of DNA

In 1953, Watson and Crick proposed that DNA was a double-stranded, helical molecule with two sugar–phosphate backbones consisting of complementary nitrogenous base pairs (adenine (A), thymine (T), cytosine (C) and guanine (G)). Each base combines with a sugar molecule (**deoxyribose**) and a phosphate group to form a **nucleotide**. The sugar and phosphate groups are bound together by **covalent bonds**, which provide strength and stability, while weaker **hydrogen bonds** bind the complementary nitrogenous bases (A-T and C-G).

The sugar–phosphate backbones are 'anti-parallel' because they run in opposite directions; they are sometimes called the forward and reverse, or plus (+) and minus (−), strands. The anti-parallel backbones and complementary base pairings are vital to DNA replication and gene transcription.

There are three main differences between DNA and RNA. RNA is single stranded, the sugar in the backbone is ribose (instead of deoxyribose), and thymine (T) is replaced by another base, called uracil (U).

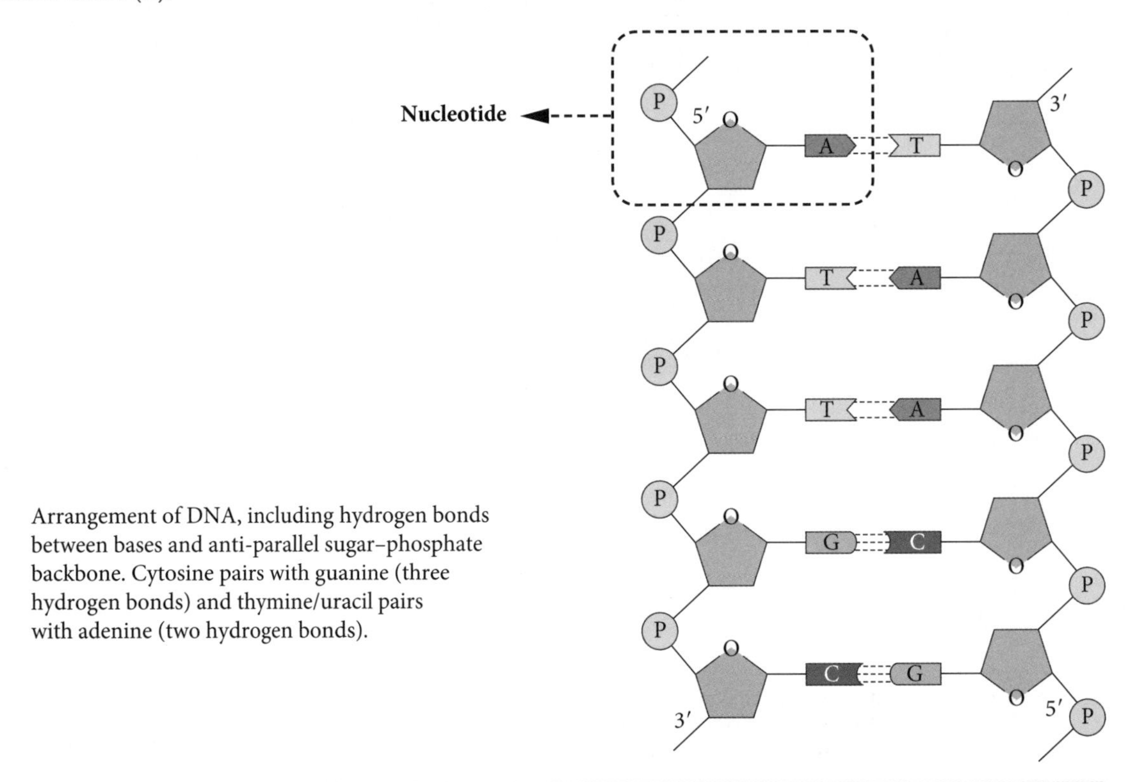

FIGURE 1.16 The structure of DNA

DNA replication explained

DNA replication is the process by which two identical DNA helices are made from a single double-stranded DNA helix molecule. The first stage involves the **enzyme** DNA helicase, which moves along the DNA molecule, breaking the hydrogen bonds between the complementary base pairs. This causes the two parent DNA strands to unwind and separate, creating a replication fork. Each strand acts as a template to synthesise two new identical DNA molecules. In the second stage, **primer** molecules initiate a second enzyme, **DNA polymerase**, to begin the replication process of each of the two parent DNA strands. The enzyme facilitates the addition of new complementary nucleotides in the 5' to 3' direction. This addition is continuous in the leading strand and fragmented in the lagging strand. Each of the two resulting DNA molecules consists of a strand from the original molecule and a newly formed strand. This is known as semi-conservative replication.

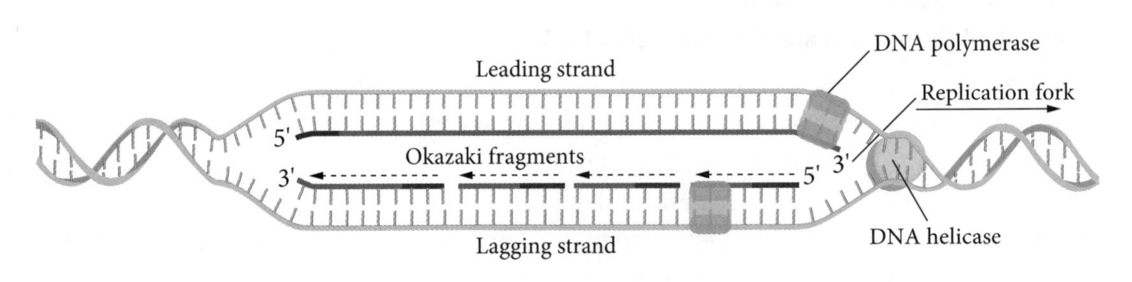

FIGURE 1.17 DNA replication

1.2.2 Assess the effect of the cell replication processes on the continuity of species

During mitosis, it is essential for cell replication to result in two daughter cells with identical genetic information. This is because the genetic information contains **genes** that ultimately code for **proteins**. Proteins play important structural, enzymatic and hormonal roles within the body. If the genetic sequence is altered during cell replication, the structure, and therefore function, of the resultant proteins may be affected.

> **Note**
> You should know the purpose, process and outcomes of both mitosis and meiosis. You will then be well placed to 'assess' the importance of each process to the continuity of species. This requires a value judgement.

Meiosis, which occurs only in sexually reproducing organisms, is responsible for the production of genetically unique haploid gametes. It is essential that all chromosomes are correctly replicated during the initial stages of meiosis. However, the subsequent exchange of alleles between **homologous chromosomes** results in genetic variation, which is essential to the survival of the species, and in the longer term may lead to speciation.

1.3 DNA and polypeptide synthesis: Why is polypeptide synthesis important?

Nucleic acids, in the form of DNA and RNA, are common to all life. DNA contains the genetic code, which is used as a template for polypeptide synthesis. This involves two key processes, **transcription** and **translation**, each involving different forms of RNA. The end-product of polypeptide synthesis is proteins, including:

- enzymes that carry out biochemical reactions inside cells
- structural proteins that maintain the shape and stability of cells
- signalling proteins that relay messages within a cell
- hormones that communicate between cells.

This section explores the importance of polypeptide synthesis for the survival and replication of cells.

1.3.1 Constructing appropriate representations to model and compare DNA in eukaryotes and prokaryotes

The three domains of life are Bacteria, Archaea and Eukarya. Bacteria and Archaea are **prokaryotic** organisms and are unicellular. The Eukarya can be unicellular eukaryotes (kingdom Protista and some of kingdom Fungi) or multicellular eukaryotes (kingdoms Animalia, Plantae and Fungi). Table 1.7 compares the features of prokaryotic and eukaryotic cells.

TABLE 1.7 A general comparison of prokaryotic cells and eukaryotic cells

	Prokaryotic cells	Eukaryotic cells
Size	• 0.5–5.0 µm (micrometres)	• 10–100 µm (micrometres)
Outer cell structures	• Cell membrane • Cell wall always present	• Cell membrane • Cell wall depends on the kingdom
Internal structures	• Cytosol • No membrane-bound organelles • Nucleoid	• Cytoplasm • Membrane-bound organelles • Nucleus
Ribosomes	• Present, smaller, 70S • Free floating in cytosol	• Present, larger, 80S • Free floating in cytoplasm or bound to endoplasmic reticulum

DNA in eukaryotes and prokaryotes

Nucleic acids (DNA and RNA) are common to both prokaryotes and eukaryotes. The building blocks of DNA (nucleotides) and their function (to produce proteins) are the same in all organisms, but there are key differences between the two types of cells in the location, arrangement and replication of DNA.

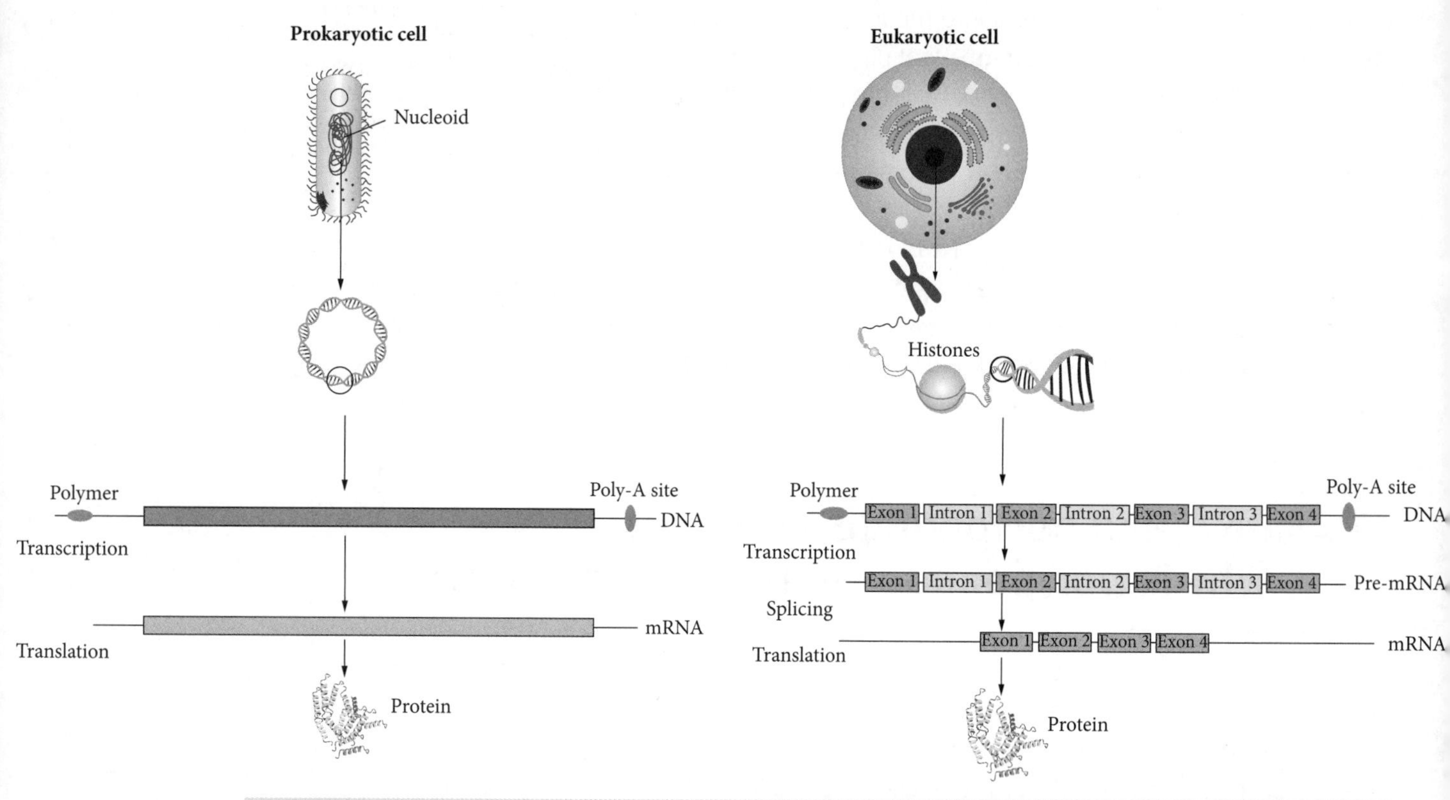

FIGURE 1.18 The genetic material and mechanism of protein production of a prokaryotic cell and a eukaryotic cell

TABLE 1.8 A comparison of genetic material (DNA) in prokaryotes and eukaryotes

	Prokaryotic cells	Eukaryotic cells
Location of chromosomal DNA	• Nucleoid	• Nucleus
Chromosome quantity	• One circular chromosome	• Multiple linear chromosomes; number depends on the species
DNA packaging	• DNA packaged into a single circular chromosome • No histone proteins	• DNA packaged into linear chromosomes • DNA tightly wrapped around histone proteins to form nucleosomes
Additional DNA	• Plasmids (small, circular DNA)	• No plasmids • Mitochondrial and chloroplast (mtDNA, cpDNA)
DNA replication	• Single replication origin (oriC) • 2000 base pairs per second • Chromosomal and plasmid DNA replicate independently	• Multiple replication origins • 100 base pairs per second • Chromosomal, mitochondrial and chloroplast DNA replicate independently
Coding DNA	• Fewer genes than eukaryotes (thousands) • No **introns** (non-coding DNA) • Coding DNA organised into operons (gene clusters)	• More genes than prokaryotes (tens of thousands) • **Exons** (coding DNA) separated by introns, allowing for gene **splicing** and multiple protein products per gene

A notable difference between the two cell types is the presence of non-chromosomal, plasmid DNA in prokaryotes. Plasmids are small, circular, double-stranded DNA molecules that reside in the cytoplasm and act independently of the larger, single, circular chromosome. Plasmids are not essential for the survival of a bacterium; however, they can carry genetic material, such as antibiotic-resistance genes, that are advantageous for survival in certain environments, such as hospitals.

The eukaryotic cells of animals and plants have an additional small genome in their mitochondria, known as the mitochondrial genome (mtDNA), and plants also have a genome in their chloroplasts (cpDNA). Both genomes are small, circular, double-stranded molecules that lack introns. They exist as multiples copies within each organelle and replicate independently of the chromosomal DNA. The mtDNA genome contains genes that are essential for energy metabolism, while genes in the cpDNA are required for photosynthesis in plants. The presence and independent function of mtDNA and cpDNA is evidence that supports the endosymbiotic theory of the origin of eukaryotic life.

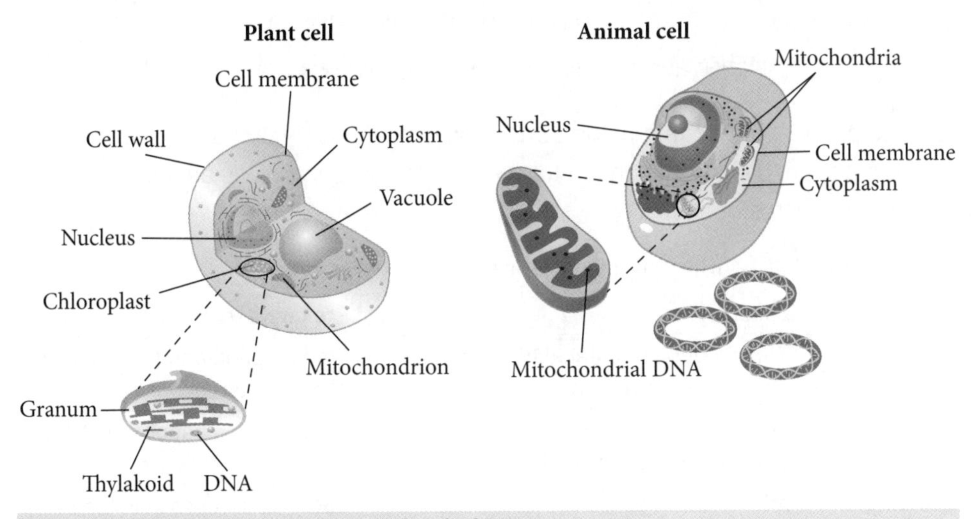

FIGURE 1.19 Chloroplast DNA and mitochondrial DNA

1.3.2 Modelling the process of polypeptide synthesis

Transcription and translation

Note
This is an 'including' syllabus dot point. This means you may be asked a question about any of the topics in this section.

A gene is the basic unit of heredity. In eukaryotes, a typical gene consists of a **promoter**, followed by alternating exons (coding DNA) and introns (non-coding DNA). **Transcription factors** bind to the promoter of a gene to activate **gene expression**, which is the process of converting the information in a gene into a functional protein product.

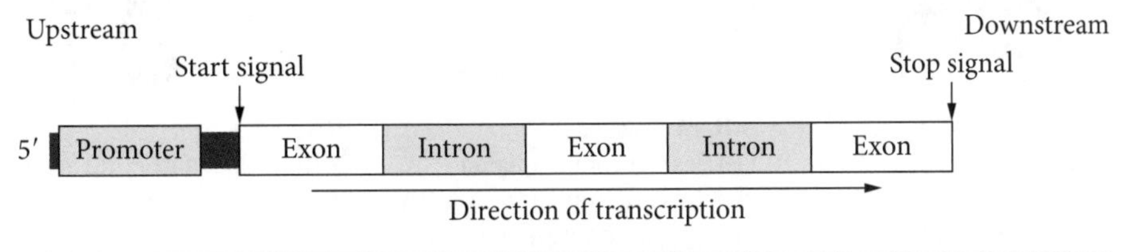

FIGURE 1.20 The basic structure of a gene in a eukaryote

The first stage in gene expression is transcription, in which the genetic code is copied into a single-stranded RNA molecule, called **messenger RNA (mRNA)**. The mRNA is then used as a template to synthesise a polypeptide, or string of **amino acids**, in the process of translation, which involves **ribosomes**. Amino acids are the building blocks, or monomers, of polypeptides. They are simple organic compounds containing a carboxyl group (—COOH), an amino group (—NH_2) and an 'R group', which varies between each of the 20 different amino acids required to make proteins.

An amino acid group (NH_2)

A carboxylic acid group

$$H_2N-CH(R)-COOH$$

The R group will change depending on the amino acid.

FIGURE 1.21 The basic structure of an amino acid

The basic chemistry of transcription and translation is the same in prokaryotes and eukaryotes, but there are some important differences. In the following sections, eukaryotes are used as an example to explore the details of each step.

TABLE 1.9 A comparison of transcription and translation in eukaryotes and prokaryotes

	Eukaryotes	Prokaryotes
Timing	Transcription and translation do not occur at the same time.	Transcription and translation can occur simultaneously.
Location	Transcription happens in the nucleus; translation happens in the cytoplasm.	Transcription happens in the cytoplasm.
mRNA processing	mRNA is processed in the nucleus.	mRNA is processed in the cytoplasm.
mRNA modification	mRNA undergoes modification before translation, including splicing, capping and polyadenylation.	These do not occur.

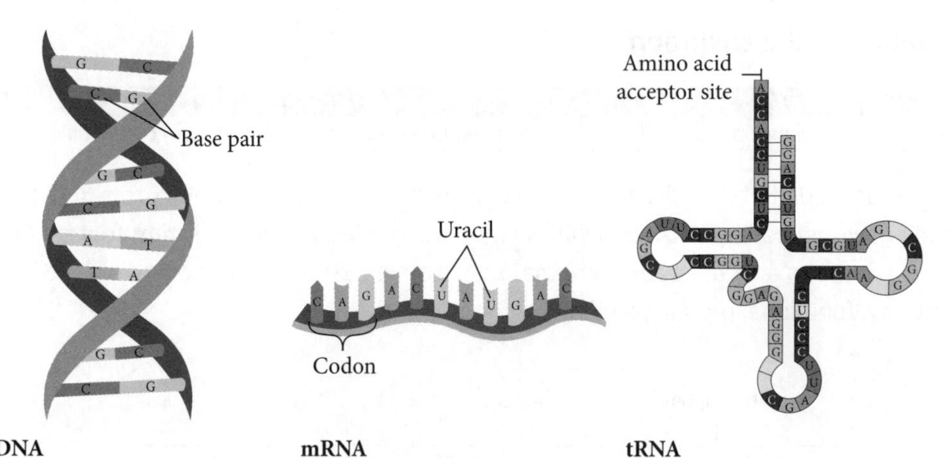

DNA
Contains the genetic code. The mRNA makes a copy of the template strand.

mRNA
Carries the complementary genetic code copied from DNA during transcription to the ribosomes for protein synthesis.

tRNA
Delivers specific amino acids during protein synthesis.

FIGURE 1.22 Key molecules involved in polypeptide synthesis

Note

The *trans-* prefix means 'move'. So 'transcription' means 'move the scribe (writing) from DNA to mRNA'. The language of nucleotides stays the same. 'Translation' means 'move the language from nucleotides to amino acids'.

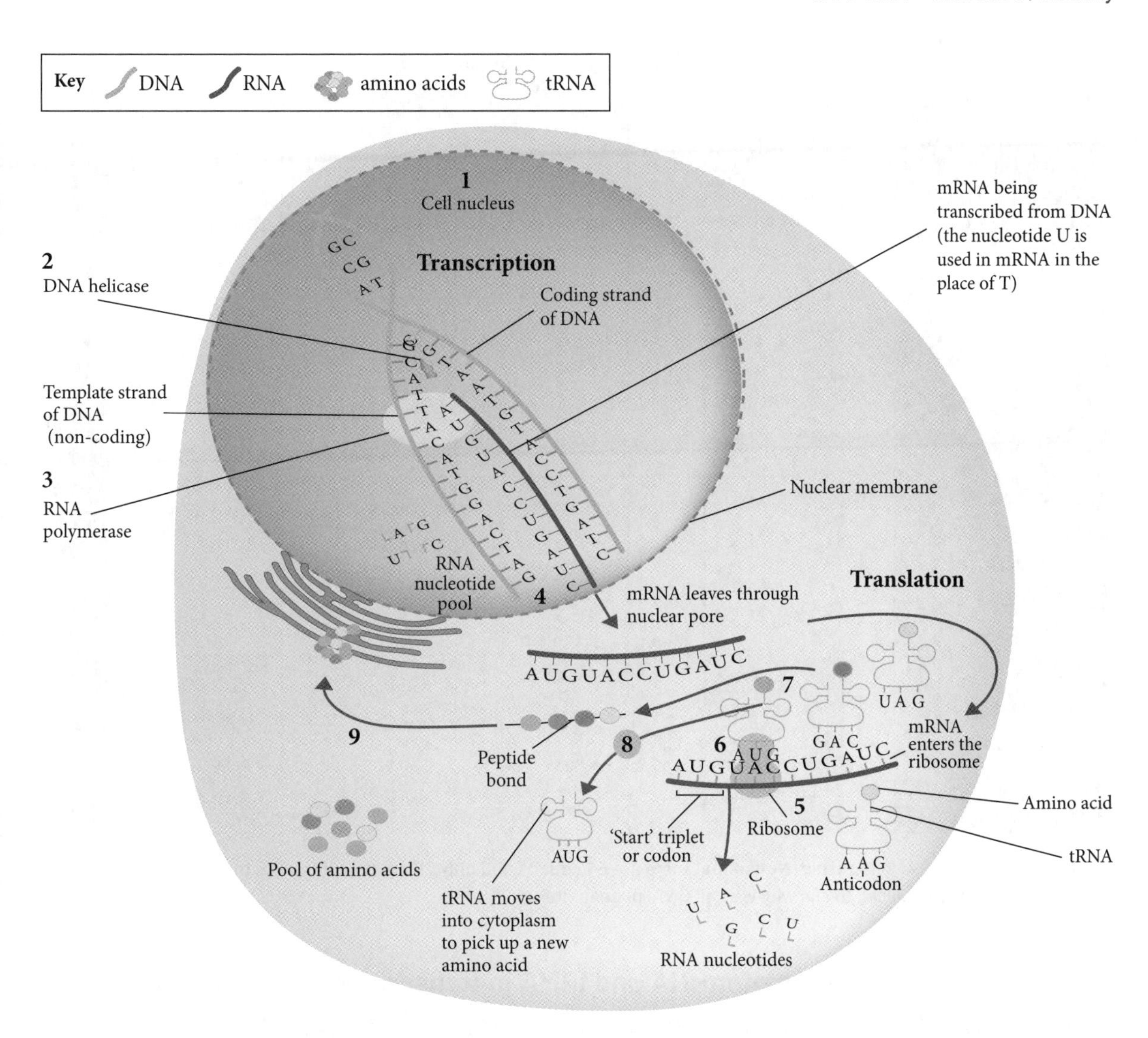

When modelling polypeptide synthesis, it is essential to incorporate the following processes and structures:

Transcription (nucleus)

1. A cell is stimulated to produce a specific protein.
2. DNA helicase unwinds and unzips the gene responsible for coding the protein. This exposes the bases of the coding strand and complementary template strand of the gene.
3. RNA polymerase binds to the promoter at the start of the gene on the template strand and runs the length of the gene, adding free nucleotides as it proceeds, until it reaches a stop codon. Both exons and introns are transcribed into a single-stranded pre-mRNA sequence. The pre-mRNA contains uracil (U) in place of thymine (T).

RNA processing (nucleus)

4. The introns are removed from the pre-mRNA and the exons are bound back together to form the final mRNA, in a process called splicing. Alternate splicing, which can also remove exons, produces combinations of proteins from the same gene. The final mRNA sequence exits the nucleus through a nuclear pore into the cytoplasm.

Translation (ribosomes)

5. The mRNA transfers to the ribosome, which consists of a small subunit and a large subunit of ribosomal RNA (rRNA). The mRNA passes through the ribosome, and the nucleotides are read in sets of three (codons), always starting with a start codon (AUG).
6. The ribosome facilitates a second type of RNA molecule, called transfer RNA (tRNA) to bind with the codon via a complementary anticodon sequence.
7. The other end of each tRNA is bound by a specific amino acid, which will be incorporated into the growing polypeptide chain. The amino acids are linked by a peptide bond.
8. Each tRNA molecule can be recycled after it has delivered its amino acid.
9. When a stop codon is reached, the polypeptide chain terminates and the polypeptide is folded into a specific 3D shape to form a functional protein. The mRNA can then be used again to produce more of the same protein.

FIGURE 1.23 Polypeptide synthesis

Ala = alanine
Arg = arginine
Asn = asparagine
Asp = aspartic acid
Cys = cysteine
Gln = glutamine
Glu = glutamic acid
Gly = glycine
His = histidine
Ile = isoleucine
Leu = leucine
Lys = lysine
Met = methionine
Phe = phenylalanine
Pro = proline
Ser = serine
Thr = threonine
Trp = tryptophan
Tyr = tyrosine
Val = valine

Second base

First base	U	C	A	G	Third base
U	UUU, UUC } Phe UUA, UUG } Leu	UCU, UCC, UCA, UCG } Ser	UAU, UAC } Tyr UAA Stop UAG Stop	UGU, UGC } Cys UGA Stop UGG Trp	U C A G
C	CUU, CUC, CUA, CUG } Leu	CCU, CCC, CCA, CCG } Pro	CAU, CAC } His CAA, CAG } Gln	CGU, CGC, CGA, CGG } Arg	U C A G
A	AUU, AUC, AUA } Ile AUG Met/Start	ACU, ACC, ACA, ACG } Thr	AAU, AAC } Asn AAA, AAG } Lys	AGU, AGC } Ser AGA, AGG } Arg	U C A G
G	GUU, GUC, GUA, GUG } Val	GCU, GCC, GCA, GCG } Ala	GAU, GAC } Asp GAA, GAG } Glu	GGU, GGC, GGA, GGG } Gly	U C A G

FIGURE 1.24 An mRNA codon table. Notice that there are 64 codons and only 20 amino acids. This means multiple codons encode the same amino acid; this is why we call DNA 'degenerate'.

Assessing the importance of mRNA and tRNA in transcription and translation

> **Note**
> Assessing the importance of mRNA and tRNA requires a detailed knowledge of the role that each plays in transcription and translation, followed by a value judgement. Both are very important. Let's explore why!

Transcription and translation are essential processes required to synthesise a protein from a gene. The genetic code must be accurately copied into the mRNA sequence and this sequence must be interpreted correctly by **transfer RNA (tRNA)** to build a polypeptide that will become a functional protein. If mRNA and tRNA did not exist or function correctly, proteins could not be made in a cell and life would not exist.

Neither mRNA nor tRNA can detect a gene mutation. As a result, a mutation could potentially produce an incorrect mRNA sequence during transcription and code for an incorrect amino acid during translation, resulting in an abnormal phenotype. This further highlights the importance of both molecules in polypeptide synthesis.

The example in Figure 1.25 illustrates how the DNA code is correctly copied by mRNA and tRNA to produce a polypeptide, and how protein production can be affected if a mutation is present in a gene. The third **codon** on the template strand codes for the amino acid tyrosine. The corresponding mRNA codon and tRNA **anticodon** are UAC and AUG respectively. If a mutation occurred at the third base of this triplet on the template strand and changed the guanine to cytosine (AT**C**), the consequences would likely be significant. The resulting mRNA codon would be UAG, which would induce a stop in polypeptide synthesis. The protein would not be produced.

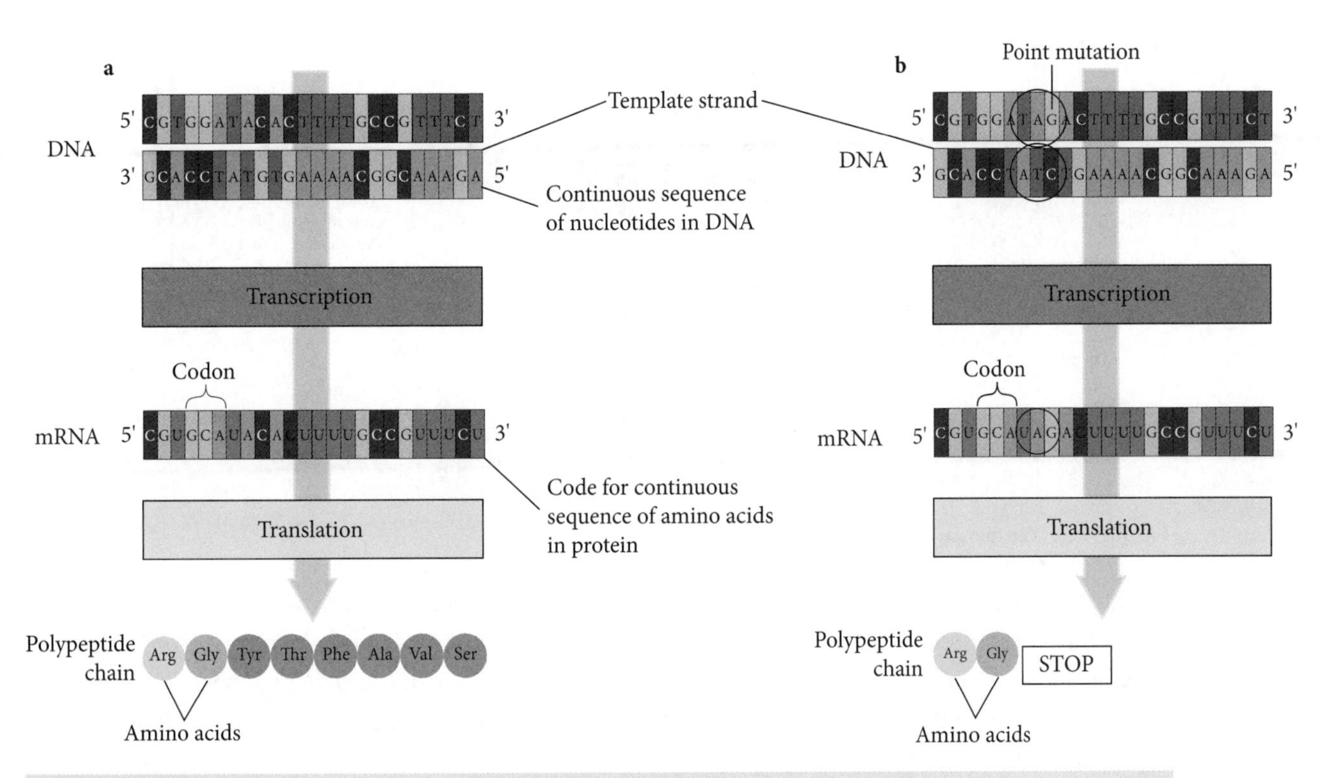

FIGURE 1.25 Polypeptide synthesis, showing a the correct process and b the impact of a mutation

Analysing the function and importance of polypeptide synthesis

Note

Analysing the importance of polypeptide synthesis requires you to identify the different components involved, identify the relationship between these components, and explore the implications of the different components not functioning correctly. (This was mostly covered in section 1.3.2.)

Polypeptide synthesis is important because it requires the accurate translation of the DNA code into a functional protein. These proteins have important structural, enzymatic and hormonal roles within cells (see section 1.3.3). Polypeptide synthesis is also important in the context of disease. The process will faithfully incorporate a genetic mutation into a polypeptide, even if it has the potential to produce a non-functional protein (see section 2.1.2 in Module 6 for the effects of mutations on the phenotype).

Assessing how genes and environment affect phenotypic expression

It is important to revise some biological terms and concepts before assessing how genes and environment affect phenotypic expression. The genomes of eukaryotic organisms are organised into chromosomes. For example, the human genome consists of 46 chromosomes and approximately 25 000 genes spread across them. A set of chromosomes includes 22 **autosomal** pairs and 1 pair of sex chromosomes (XX in females and XY in males).

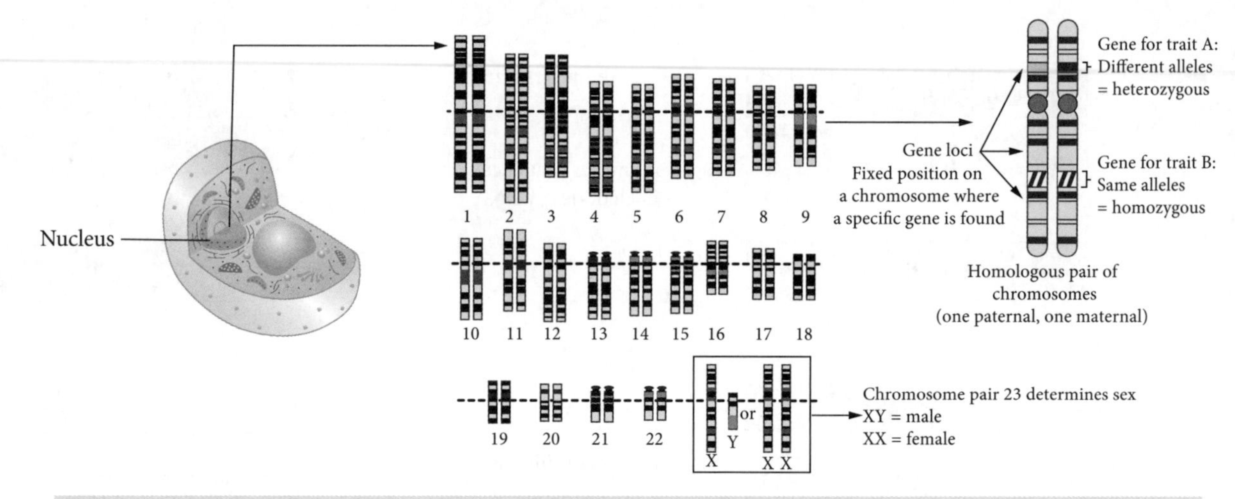

FIGURE 1.26 The human karyotype

Each pair of chromosomes, known as a homologous pair, consists of a paternal chromosome and a maternal chromosome, inherited at fertilisation. The position of a gene on a chromosome is called a **locus**, and at every locus there are two copies (**alleles**) of a gene. If the DNA sequence of the two alleles is identical, the individual is **homozygous**. If the sequence varies between the two alleles, the individual is **heterozygous**.

The full set of genes an organism has is broadly defined as its genotype, but the term is more typically used to describe the two alleles at a single locus in relation to a particular trait or phenotype. The phenotype is the physical expression of the genotype at the protein level, and ultimately the physically observable characteristics of the organism.

The environment contributes to the phenotype by influencing gene expression. The environment includes all factors an organism is exposed to throughout its life; for example, the period of embryonic development in the internal environment of an egg or the uterus, and the external environment such as diet, temperature and light.

Genes can be broadly categorised as constitutive or regulated. **Constitutive genes** are always expressed and are less influenced by the environment. For example, genes that express tRNA and rRNA are always active because they are always needed for translation. In contrast, **regulated genes** are expressed or silenced, depending on stimuli from the environment. Regulated genes are inhibited by **epigenetic** marks on the DNA. These are natural chemical modifications on the DNA that block the activation of transcription.

DNA methylation is a chemical modification of cytosine bases by a methyl group ($-CH_3$). When a region that regulates gene expression (a promoter region) is methylated, the gene cannot be expressed.

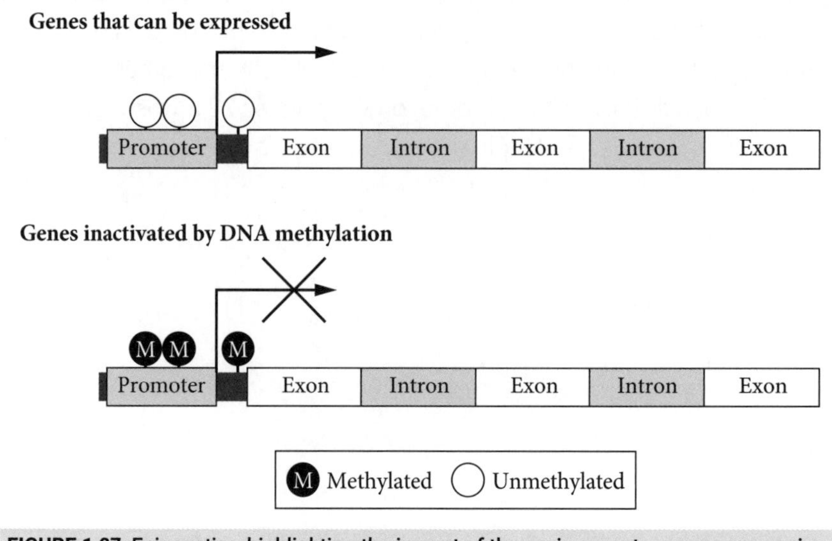

FIGURE 1.27 Epigenetics, highlighting the impact of the environment on gene expression

CASE STUDY 2

EFFECT OF TEMPERATURE ON REPTILE GENDER

In several species of reptiles (e.g. crocodiles and some turtles) the gender of the offspring is determined by the temperature during the egg incubation period. During embryonic development, there is a thermosensitive period (TSP) that influences the expression of the *Sox9* gene, which is involved in sex determination. For example, saltwater crocodiles *(Crocodylus porosus)* have an egg incubation period of 80–90 days. The TSP is between days 25 and 50; during this TSP, eggs kept at 32°C or above will hatch as male offspring, while those kept at 31°C or below will hatch as females.

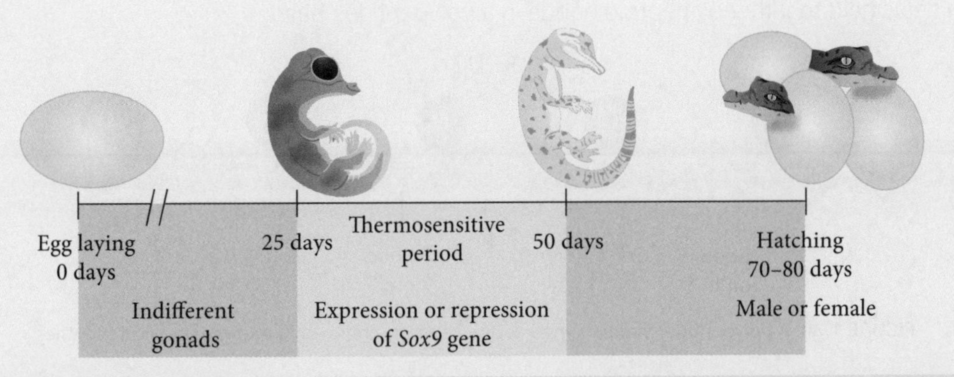

FIGURE 1.28 The effect of temperature on sex determination during embryonic development in the saltwater crocodile

CASE STUDY 3

THE HONEYBEE CASTE SYSTEM

In a honeybee hive, the queen and worker bees are all diploid females and have the same genome. However, they have very different physiologies, behaviours and physical characteristics. Workers collect nectar, pollen and water for the colony, defend against invaders and maintain the hive. The queen lays eggs and releases pheromones to control and unify the hive.

The difference between the queen and the workers is caused by diet. All bee larvae are initially fed royal jelly by nurse workers. Most larvae are weaned off royal jelly to become workers. Larvae destined to be queens are maintained on a rich diet of royal jelly. Chemicals in royal jelly affect which genes are silenced by an epigenetic mark called DNA methylation. The different set of genes expressed creates the queen phenotype.

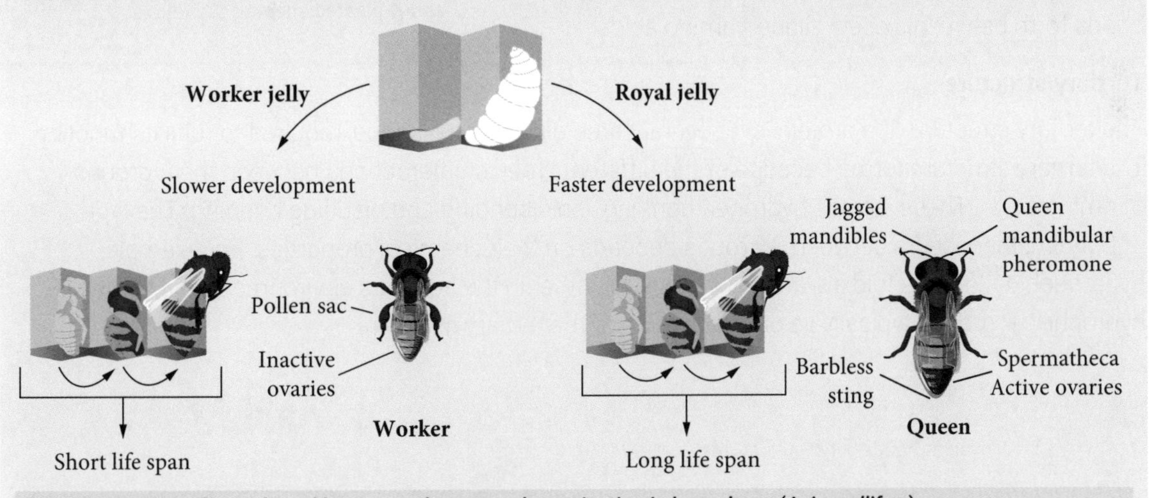

FIGURE 1.29 The effect of royal honey on phenotype determination in honeybees (*Apis mellifera*)

1.3.3 Investigating the structure and function of proteins in living things

There are four levels of protein structure: primary, secondary, tertiary and quaternary.

Primary structure

The **primary structure** is the most basic structure of a protein. It is simply the unique sequence of amino acids held together by peptide bonds in a polypeptide chain.

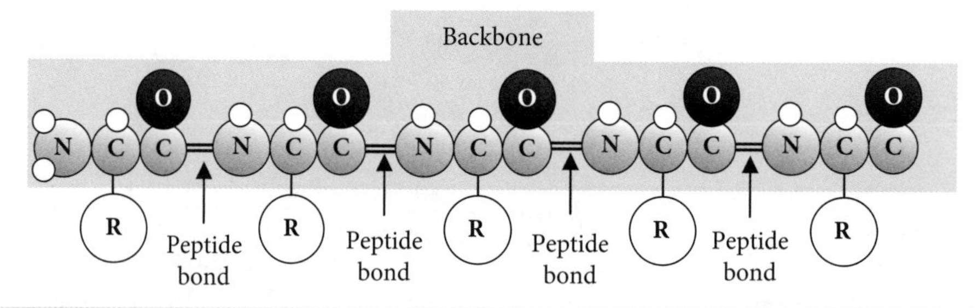

FIGURE 1.30 A polypeptide chain polymer forms as many amino acid monomers bond together.

Secondary structure

The secondary structure of a protein is the pattern of folding that occurs due to interactions between amino acids. Specifically, hydrogen bonds form between oxygen atoms and hydrogen atoms within the backbone, causing the polypeptide chain to coil or pleat. The two most common types of secondary structures are the α-helix and the β-pleated sheet. An α-helix forms when the polypeptide backbone coils around a helical axis in a clockwise direction. In β-pleated sheets, segments of the polypeptide chain line up next to each other, and hydrogen bonds form between closely aligned amino acids.

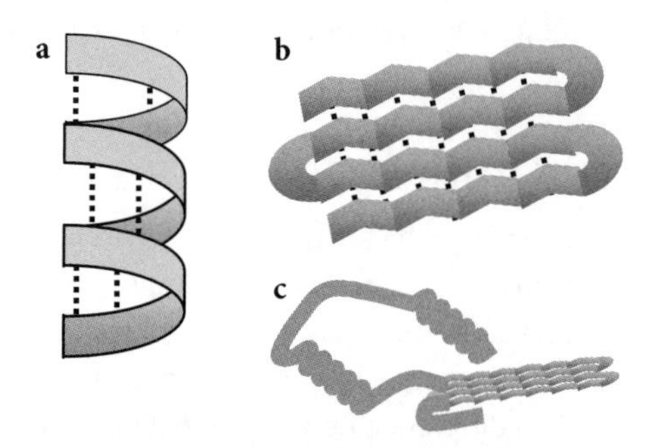

FIGURE 1.31 The secondary structure of proteins: a an α-helix, b a β-pleated sheet and c a protein showing both α-helices and β-pleated sheets

Tertiary structure

The tertiary structure of a protein is its overall three-dimensional shape, required to fulfil its function. The tertiary structure forms because of the different types of interactions between the R groups of amino acids. These include hydrogen bonding, ionic bonding and disulfide bonding. The type of bonding that occurs between R groups depends on their chemical properties. For example, hydrophobic R groups fold inwards towards each other in the aqueous environment of a cell, while hydrophilic R groups typically lie on the outside of the tertiary structure.

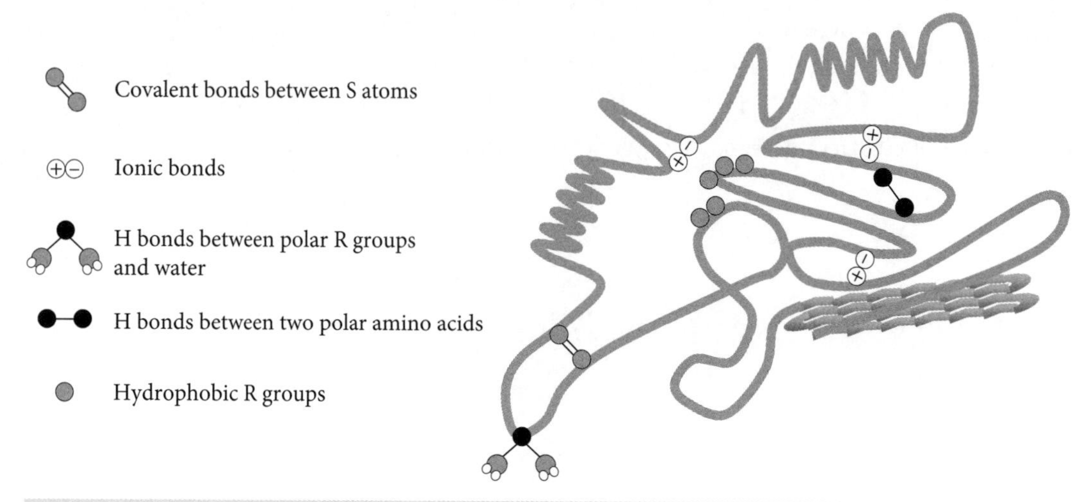

FIGURE 1.32 The tertiary structure of proteins: three-dimensional folding

Quaternary structure

Some proteins are formed from more than one polypeptide chain. The quaternary structure is the way in which the different subunits are packaged together to form the overall structure of the protein. For example, haemoglobin, the protein in red blood cells that binds and transports oxygen around the body, is made of four polypeptide subunits: two α-globin and two β-globin molecules. The α-globin and β-globin subunits are coded for by different genes, the haemoglobin alpha gene (HBA) and the haemoglobin beta gene (HBB) respectively.

Antibodies are another common example of proteins made from multiple polypeptides. Each antibody is made from four **polypeptides**, two heavy chains and two light chains. They join to create a Y-shaped molecule that has a highly variable region (HVR). The HVR is different for each antibody and provides the specificity required to bind different antigens.

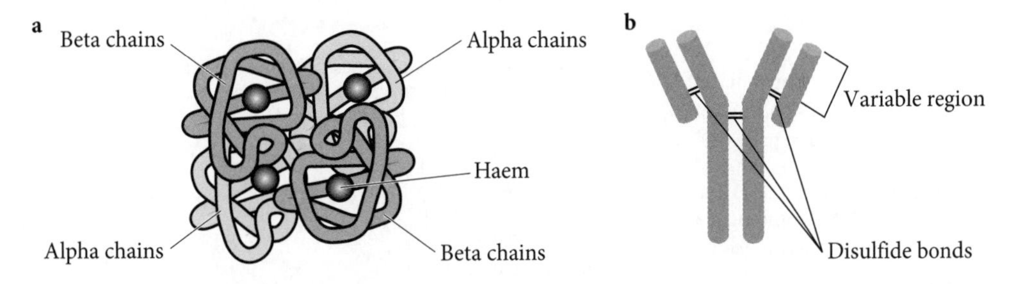

FIGURE 1.33 a A haemoglobin protein, with four polypeptide chains forming its 3D structure. b An antibody molecule held together by disulfide bonds between cystine amino acids on the polypeptide chains

Protein function

Proteins are complex macromolecules that play important structural, functional and regulatory roles within an organism. The total collection of proteins that can be produced by an organism is referred to as its **proteome**. In eukaryotic organisms, the number of proteins produced is significantly higher than the number of genes. Humans have an estimated 25 000 genes and 200 000 different proteins. This protein diversity can be explained by **alternative splicing**.

Proteins can be broadly classified as either globular or fibrous, and further subdivided based on their specific function. Globular proteins have an irregular amino acid sequence and a spherical shape determined by the tertiary, or quaternary, structure. They are functional molecules and include:

- enzymes (e.g. lactase, which breaks down milk)
- antibodies (e.g. immunoglobulin G)
- transport proteins (e.g. haemoglobin, cytochrome C)
- hormones (e.g. insulin and growth hormone).

In contrast, fibrous proteins are thin, elongated molecules made from repetitive amino acid sequences that give rise to a secondary structure only. They are all structural molecules and include actin and myosin, which are involved in muscle contractions, and collagen, the most abundant protein in the human body, which gives skin its elasticity.

Fibrous proteins are also less susceptible to environmental changes, such as temperature and pH, because of their molecular structure and chemical properties.

1.4 Genetic variation: How can the genetic similarities and differences within and between species be compared?

A **pedigree** is a diagrammatic model that can be used to track the inheritance of a genetic trait through generations. An understanding of Mendelian inheritance allows farmers to combine pedigrees with **Punnett squares** to predict the likely genotypes, and therefore phenotypes, of offspring from a plant or livestock cross. Doctors can also use this information to predict the likelihood of a person inheriting a genetic mutation if their family has a history of a particular disease. This can also help people to make an informed decision about having children, based on the risk of passing on a heritable condition, or enable a person to better monitor their own health and take preventative action as required.

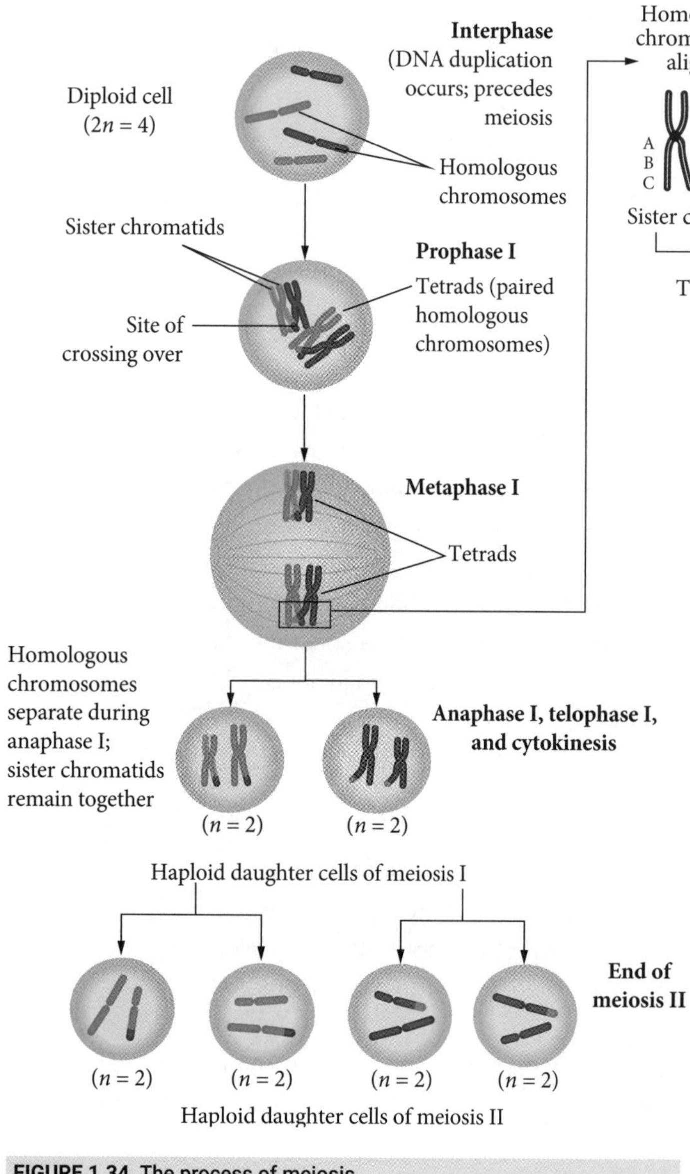

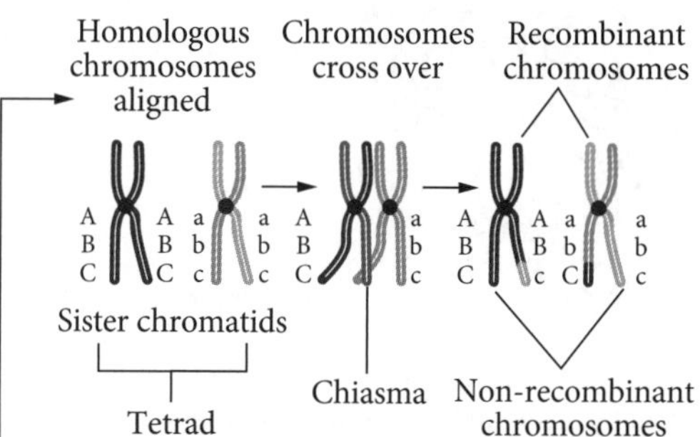

FIGURE 1.34 The process of meiosis

TABLE 1.10 How genotype variation is produced by meiosis, fertilisation and mutation

Process	How genotype variation is produced
Meiosis Crossing over	• Alleles are exchanged between non-sister chromatids on homologous chromosomes • This results in new combinations of alleles on a chromosome • This produces new variation of genotypes upon fertilisation
Independent assortment	• Genes and alleles of one trait are inherited independently of the genes and alleles of another trait • This produces new variation of genotypes at loci across chromosomes upon fertilisation
Random segregation	• Alleles of the same gene separate randomly and equally into daughter cells • This produces new variation of genotypes at loci across chromosomes upon fertilisation
Fertilisation	• The union of a male and a female gamete • A gamete has a genetically unique haploid genome • Every offspring will have a new, unique genotype at loci across the genome, created by the fusion of the two haploid genomes
Mutation	• A change in the DNA sequence • A mutation can occur during meiosis to produce a new allele in a gamete • The mutant allele will combine with an allele in another gamete at fertilisation to produce a new genotype at a gene locus

1.4.1 Predicting variations in the genotype of offspring by modelling meiosis

Note

Meiosis is such an important biological process that you will always be assessed on it in the HSC, with a particular focus on how variation is introduced via crossing over, independent assortment and random segregation. Know how to model your understanding by drawing and interpreting images of the process.

Knowledge of how chromosomes behave during meiosis can be used to predict the variation in the genotypes of offspring. Genotype variation is produced by crossing over of homologous chromosomes, independent assortment and random segregation during meiosis, fertilisation and mutations.

Let's model how these processes create genotype variation in a hypothetical organism that has a diploid chromosome number of 6 ($2n = 6$, or 3 pairs of chromosomes). Each pair of chromosomes has two homologues, one maternal and one paternal. The maternal chromosome is represented by an uppercase letter and the paternal by a lowercase letter:

Pair 1: X and x

Pair 2: Y and y

Pair 3: Z and z

Independent assortment of these three chromosome pairs can produce eight possible gamete combinations: XYZ, XYz, Xyz, XyZ, xYZ, xYz, xyZ and xyz. This can be modelled mathematically by 2^n where n = the number of chromosomes per gamete (or number of homologous pairs). Therefore $2^3 = 8$. Remember, though, that this same process occurs in both parents, so the actual number of variants that can occur in the zygote is $2^n \times 2^n = (2^n)^2$. In this example the number of zygote variants = 64.

Applying the same process to humans, each egg and sperm has 2^{23} possible combinations, and the resulting diploid has $(2^{23})^2$ possible variants. Although these numbers are incredibly large, they have not yet considered the genetic recombination that occurs during crossing over, or the random nature of fertilisation. To do this would increase the variation significantly.

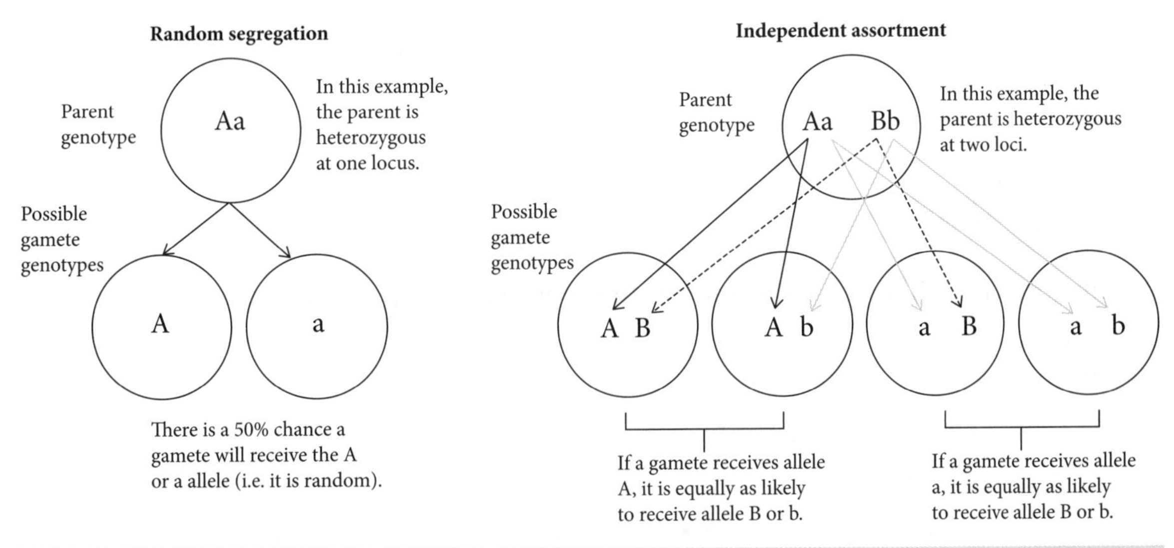

FIGURE 1.35 Random segregation and independent assortment during meiosis. Note: these two diagrams model the possible genotype outcome of random segregation and independent assortment. They are not illustrating the actual process of meiosis.

1.4.2 Modelling the formation of new combinations of genotypes produced during meiosis

Note

This syllabus outcome is an 'including but not limited to' statement. You need to know all the inheritance pattern examples explicitly mentioned in the dot point.

Recall that, in diploid organisms, chromosomes exist as homologous pairs, each having a copy (allele) of a gene at every locus. The genotype is the combination of two alleles for a particular gene, while the phenotype is the physical expression of the genotype. Each homologous pair of chromosomes consists of a **paternal** and a **maternal** chromosome inherited at fertilisation.

When both alleles for a particular characteristic are the same (that is, they have the same DNA sequence), the organism is homozygous for that trait, while an individual with two different alleles is heterozygous. Typically, but importantly not always, one allele is more likely to be expressed in the phenotype of a heterozygous individual. This allele is referred to as the **dominant allele**. The other allele is **recessive** and can only be expressed in the phenotype of a homozygous recessive individual. Each allele is allocated a letter, with the uppercase representing the dominant allele and lowercase the recessive allele.

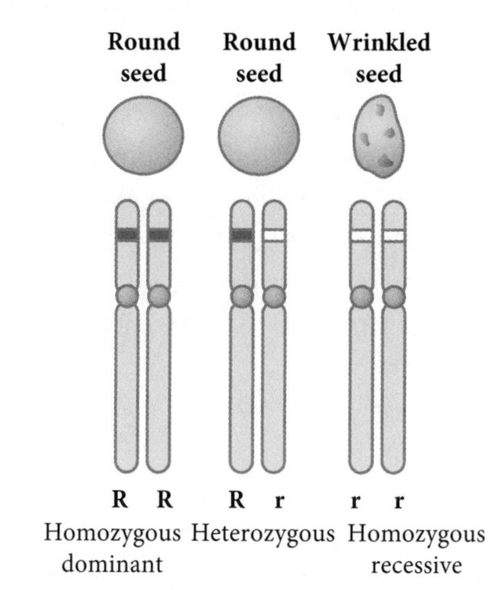

FIGURE 1.36 A comparison of genotype and phenotype for seed shape in pea plants. The round seed allele (R) is dominant over the recessive wrinkled seed allele (r).

For example, in pea plants, the round seed allele (R) is dominant over the recessive wrinkled seed allele (r). Therefore, both RR and Rr plants would have identical 'round seed' phenotypes. In contrast, all plants with wrinkled seeds must be homozygous recessive (rr). The genotypes can be summarised as:

- RR = homozygous dominant
- rr = homozygous recessive
- Rr = heterozygous.

Constructing and interpreting information and data from pedigrees and Punnett squares

Punnett squares

Punnett squares can be used to determine:

- the likely genotypes and phenotypes of offspring from a cross where the parental genotypes are known
- the parental genotype and phenotype based on the observed phenotype of offspring
- which allele is dominant and which is recessive, by comparing phenotypic outcomes.

For example, consider breeding experiments with pea plants where the gene for flower colour has two alleles: the dominant purple allele and the recessive white allele.

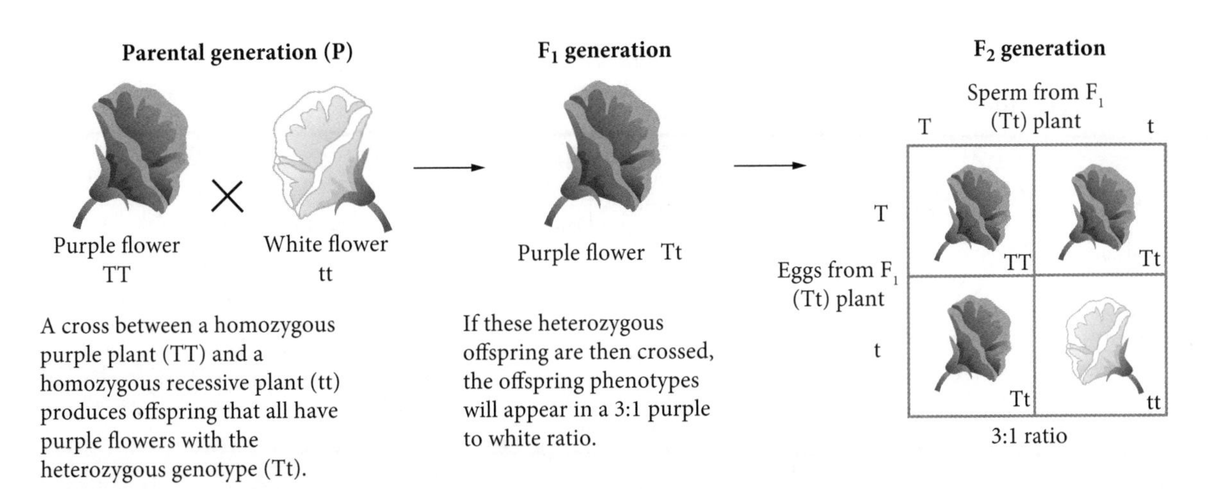

FIGURE 1.37 Using a Punnett square to determine genotype and phenotype ratios of offspring

If the genotype of a purple-flowered plant was not known (TT or Tt), a test cross could be performed with a white-flowered plant (tt). If any of the offspring were white (tt), the purple-flowered parent plant must be heterozygous. In this cross, you would expect to observe a Tt : tt genotype ratio of 1 Tt : 1 tt, and a purple : white phenotype ratio of 1 : 1.

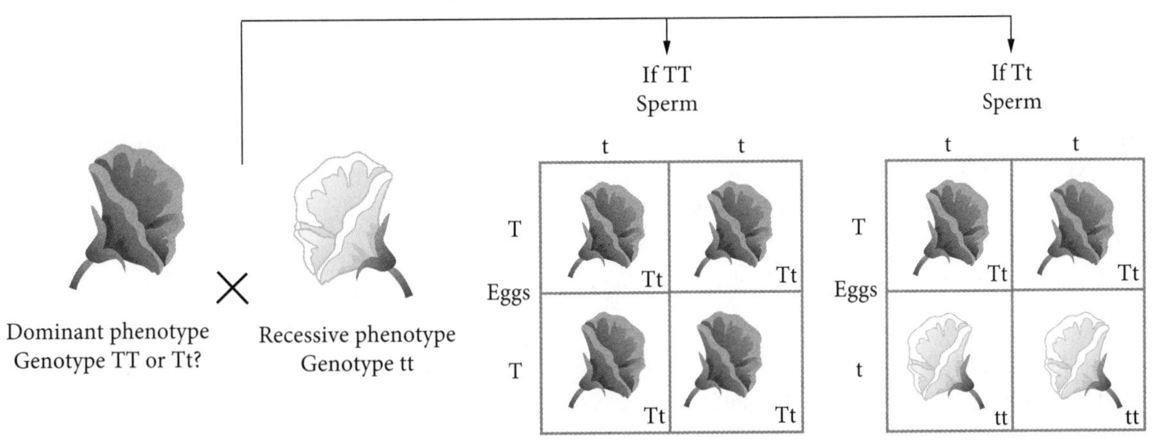

FIGURE 1.38 Using a Punnett square to determine parental genotypes and phenotypes

If all offspring from a single cross were purple, the purple-flowered parent plant is likely to be homozygous dominant. However, this can only be determined with confidence after multiple repeats of the cross.

> **Note**
> A single test cross would not produce statistically significant results. Remember, which allele is passed on at fertilisation depends on chance.

Pedigrees

A pedigree is a visual chart that shows family lineages, relationships and the phenotypic expression of a chosen trait. Pedigrees have characteristic patterns that can be used to:

- determine the type of inheritance shown in the lineage
- determine the genotypes of individuals in the lineage
- predict the likelihood of offspring having the trait.

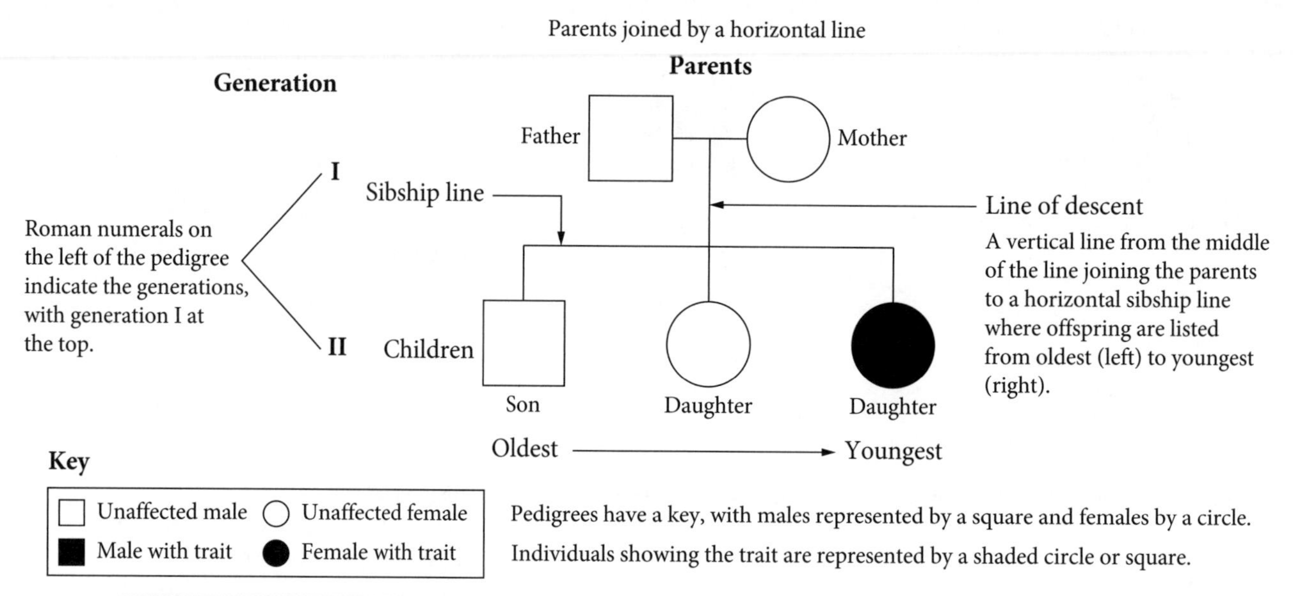

FIGURE 1.39 Setting up a pedigree

Autosomal recessive

Autosomal inheritance relates to genes that are on any chromosome except the sex chromosomes. There is an equal likelihood that a trait encoded by an autosomal gene will be observed in the phenotypes of each sex. Examples of autosomal recessive diseases in humans are cystic fibrosis and sickle cell anaemia.

> **Note**
> When two parents do not express a trait and they have any offspring with the trait, then both parents must be heterozygous, and the trait must be recessive.

For an autosomal recessive trait, both parents of an affected person must carry at least one copy of the recessive allele (Tt or tt). If they are heterozygous (Tt), they will express the dominant allele in their phenotype and can be referred to as carriers of the recessive allele. It is for this reason that recessive traits are not typically seen in every generation.

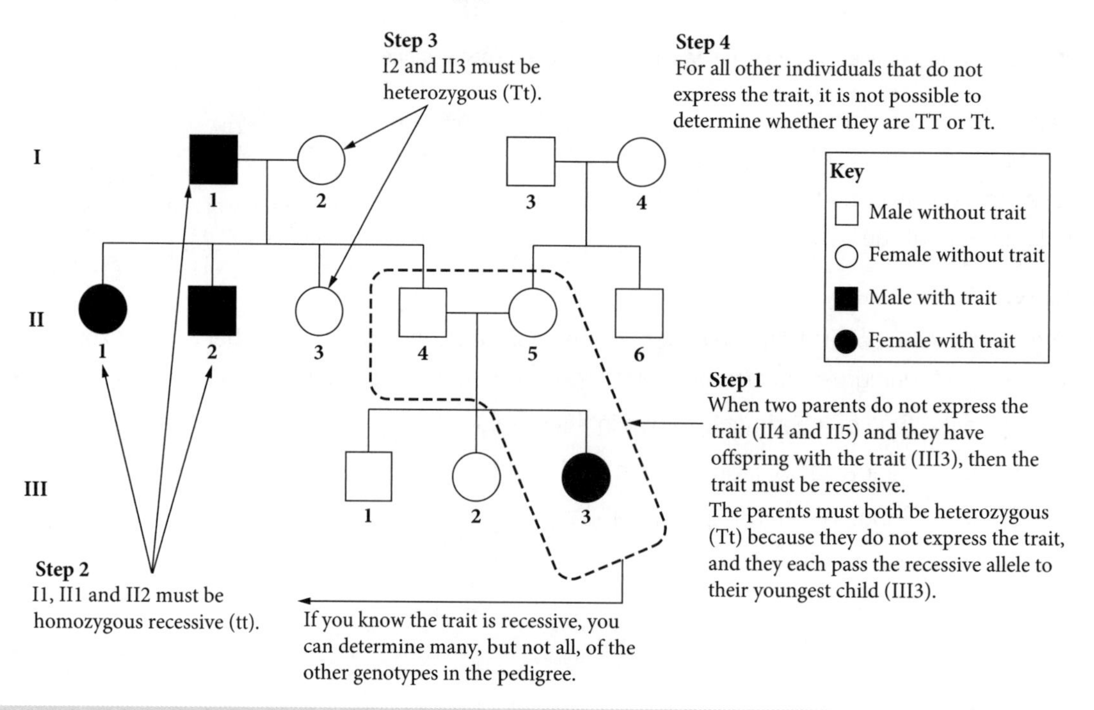

FIGURE 1.40 Determining autosomal recessive traits from a pedigree

Autosomal dominant

Unlike autosomal recessive traits, autosomal dominant traits rarely skip a generation. Affected offspring always have at least one affected parent and at least one copy of the dominant allele. Huntington's disease is an example of an autosomal dominant disease in humans.

Note
When two parents express a trait and they have any offspring without the trait, then both parents must be heterozygous, and the trait must be dominant.

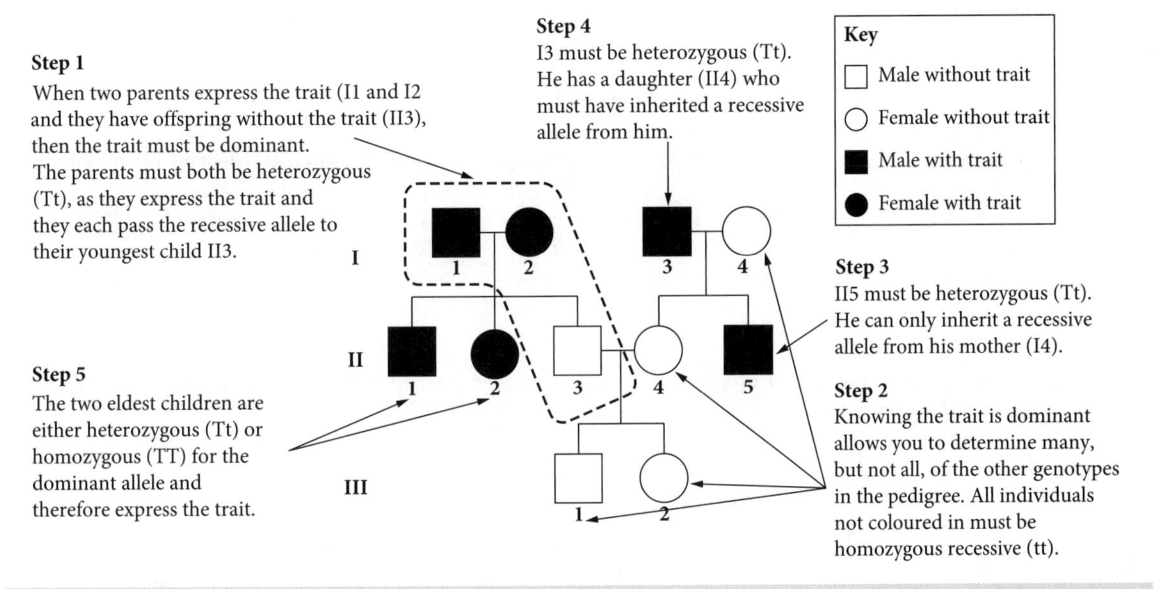

FIGURE 1.41 Determining autosomal dominant traits from a pedigree

Sex-linkage

Sex-linkage refers to genes on the sex chromosomes. While the focus is on humans, it is important to understand that the nature of sex determination varies between species. The XY sex determination system is found in humans, most mammals, and some insects and plants. Females have two copies of the X chromosome and males have one X and one Y chromosome. The male is said to be hemizygous. The sex of the offspring is therefore determined by the father, as each sperm contains either an X or a Y chromosome, while the mother's egg contains only an X chromosome.

Note
Always read a question carefully. You might assume that a sex-linkage question will be about the XY sex determination system in humans. However, you could easily be asked a question that contains stimulus information about another species. For example, birds have a ZW sex determination system. The male bird has two Z chromosomes, and the female has one Z and one W. Female birds would therefore be more likely to exhibit a sex-linked trait.

Sex-linked traits are represented by a superscript letter above the relevant chromosome, usually the X. If the trait is dominant, an uppercase letter is used; for a recessive trait, a lowercase letter is used. Assuming a trait is X-linked recessive, there are three possible genotypes in females: normal female (X^TX^T), carrier female (X^TX^t) and female with the trait (X^tX^t). In contrast, there are only two possible outcomes for males: normal male (X^TY) or male with the trait (X^tY). A single recessive gene in males has the same phenotypic effect as a single dominant allele.

Drosophila melanogaster (fruit fly) is a common animal model used in inheritance studies. The gene for eye colour is sex-linked, with the red eye allele dominant over the white eye allele. Crosses typically produce offspring phenotype ratios different from those expected for autosomal traits.

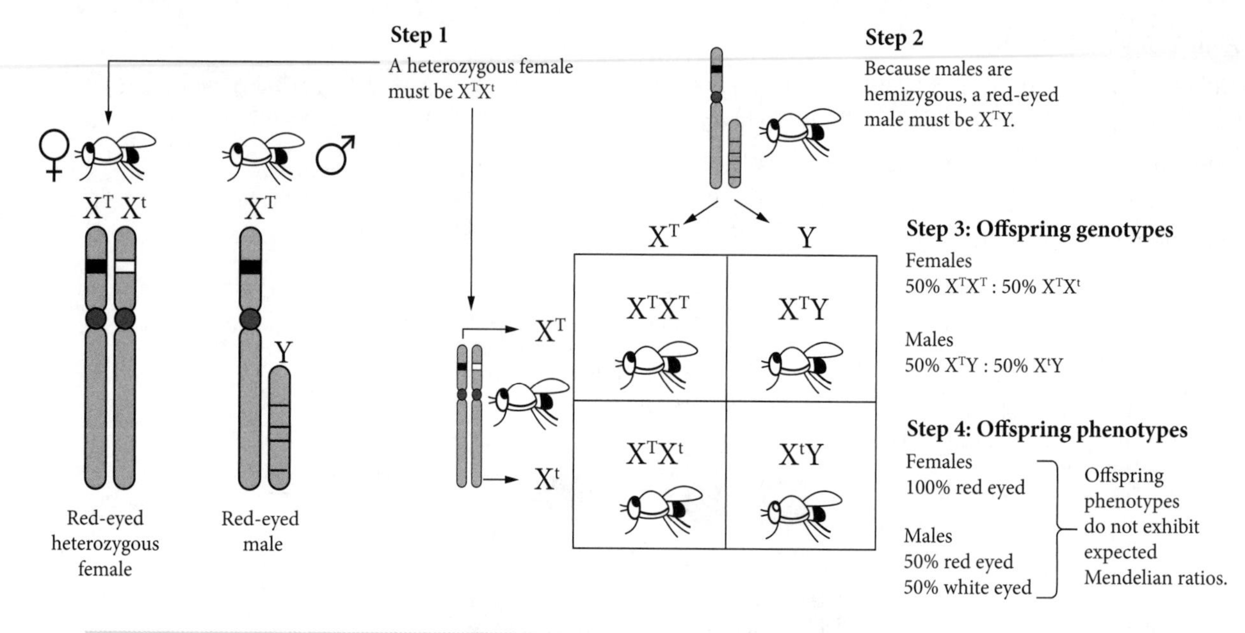

FIGURE 1.42 A sex-linked Punnett square of a heterozygous female × red-eyed male. These crosses produce an offspring phenotype ratio different from the usual 3:1.

Using a pedigree to determine sex-linkage

While it is easy to use a pedigree to rule out sex-linkage, it is more problematic to say with certainty that a trait is sex-linked. Certain features of sex-linkage can resemble cases of autosomal inheritance. Certain key features *must* be present in a pedigree for it to be X-linked.

- Males are more likely to be affected than females.
- If a mother has the trait, her sons will have the trait.
- If a daughter has the trait, the father must also have it, and the mother either has it or is a heterozygous carrier.

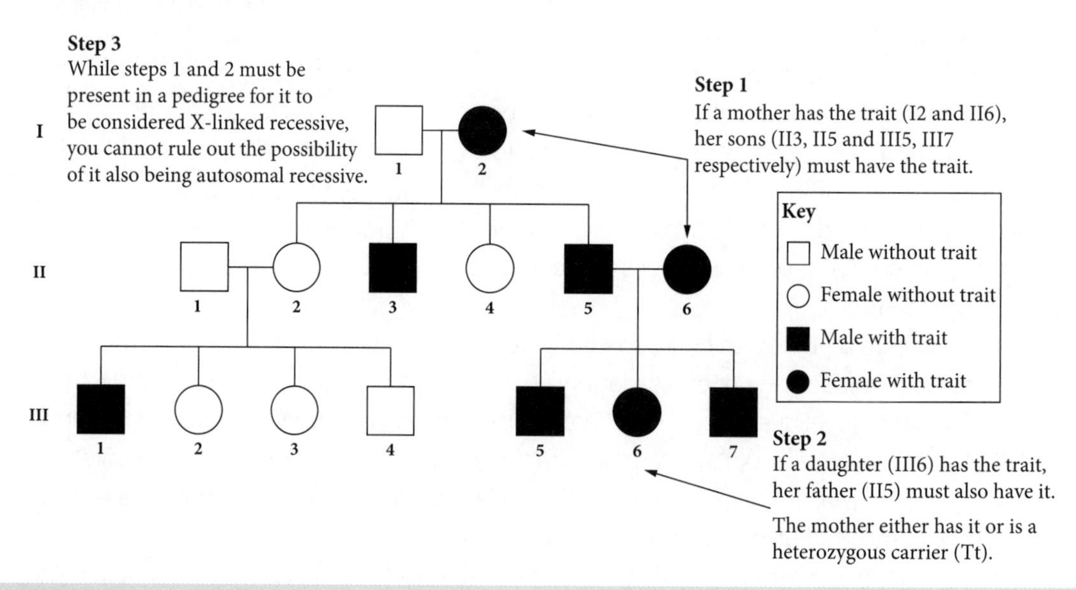

FIGURE 1.43 Using a pedigree to determine sex-linkage

Ornithine transcarbamylase (OTC) deficiency is an X-linked recessive condition. The condition is caused by a mutation in the OTC gene on the X chromosome, which normally produces an enzyme involved in the breakdown and removal of ammonia, a toxic breakdown product of protein. The disease results in the accumulation of ammonia in the blood, which travels to the brain, causing coma, brain damage and death in childhood.

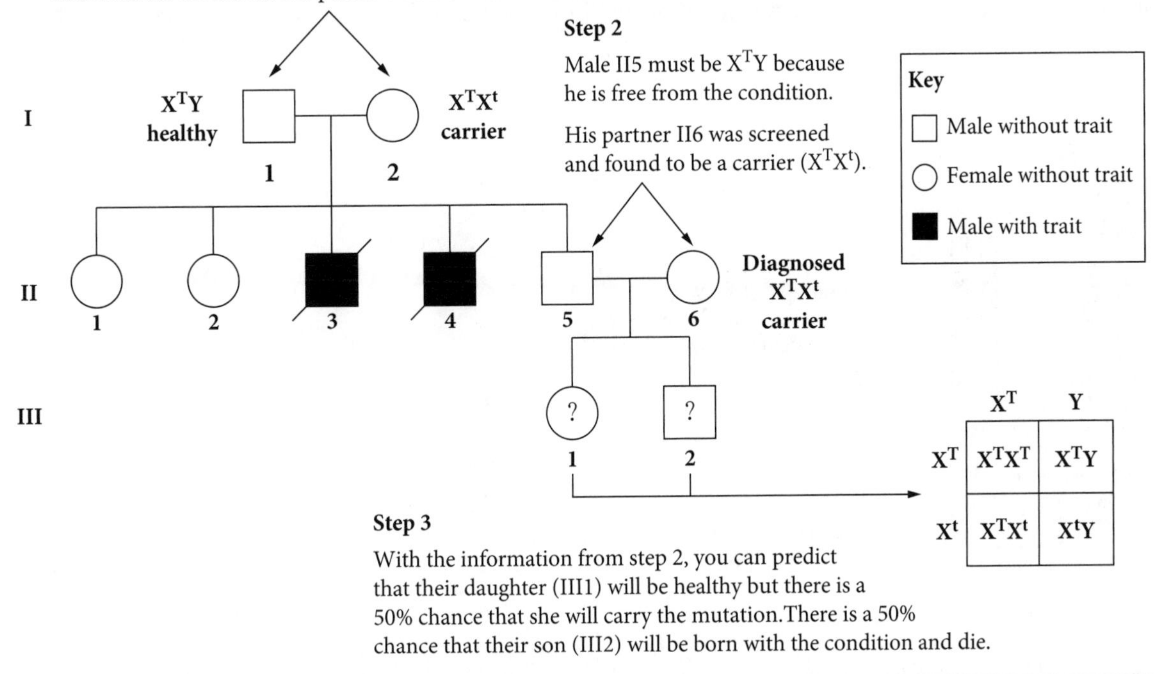

FIGURE 1.44 An example of a sex-linked trait: OTC deficiency is an X-linked recessive condition.

Codominance

Codominance occurs when both alleles of a gene are fully expressed in the phenotype of a heterozygote. In other words, one allele is not dominant over the other and there is no recessive allele. Each allele is represented by a different uppercase letter. The outcome of this is that two codominant alleles result in three possible phenotypes.

Two common examples are ABO blood group (which is also an example of **multiple allele inheritance** due to the presence of the third O allele) in humans and red-white hair colour in cattle.

In the ABO blood groups, the A and B refer to different sugar molecules on the surface of red blood cells, while the O indicates that neither A nor B is present. The A and B alleles (I^A and I^B) are codominant, and they are both dominant over the third, recessive O allele (i). If an individual inherits the A allele (I^A) from one parent and the B allele (I^B) from the other, they will have AB blood group (I^AI^B) and express both sugars on the surface of their red blood cells.

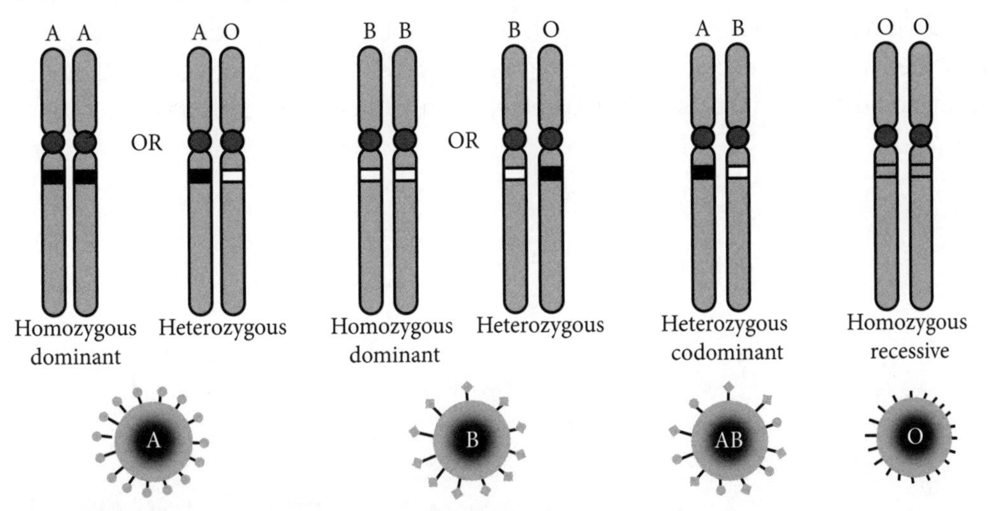

FIGURE 1.45 The ABO blood groups are an example of codominance.

In cattle, if an individual with two red alleles (RR) is mated with an individual with two white alleles (WW), all F_1 offspring will be roan (RW), where both red and white hairs are present. The genotype of an individual can therefore be easily determined based on its phenotype. An F_1 heterozygous cross will produce an offspring phenotypic ratio different from the usual 3 : 1 expected from an autosomal dominant–recessive trait.

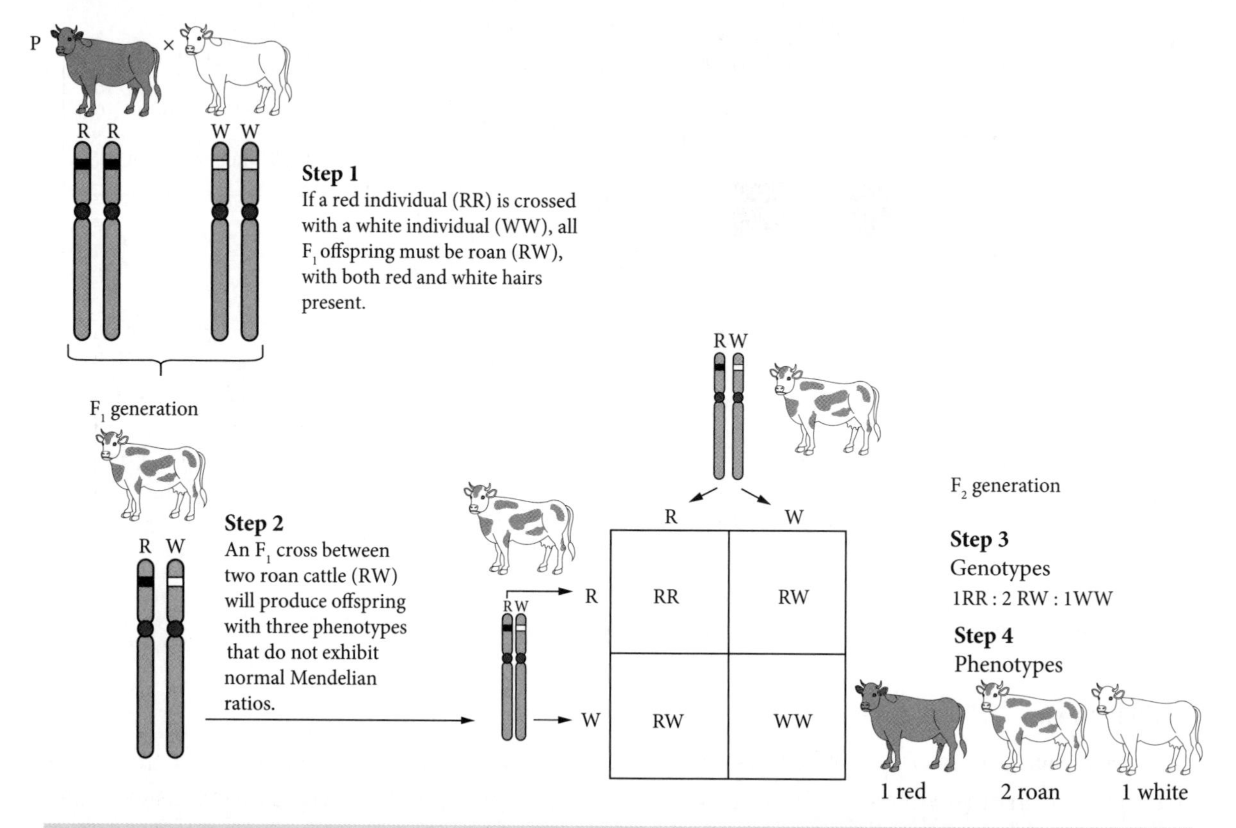

FIGURE 1.46 A codominant Punnett square. The F_1 heterozygous cross produces an offspring phenotype ratio different from the usual 3 : 1.

Incomplete dominance

Like codominance, **incomplete dominance** can produce three phenotypes from a set of two alleles. However, incomplete dominance occurs when neither allele of a gene pair is expressed fully in the phenotype of a heterozygote. Instead, there is a third phenotype with blended characteristics. For example, flower colour in snapdragons (*Antirrhinum majus*) exhibits incomplete dominance. When a homozygous red-flowering plant (RR) is crossed with a homozygous white-flowering plant (WW), all F_1 offspring will have pink flowers (RW). As with codominant traits, the genotype of an individual can be determined from its phenotype. An F_1 heterozygous cross will also produce an offspring phenotypic ratio different from the usual 3 : 1 expected from an autosomal dominant–recessive trait.

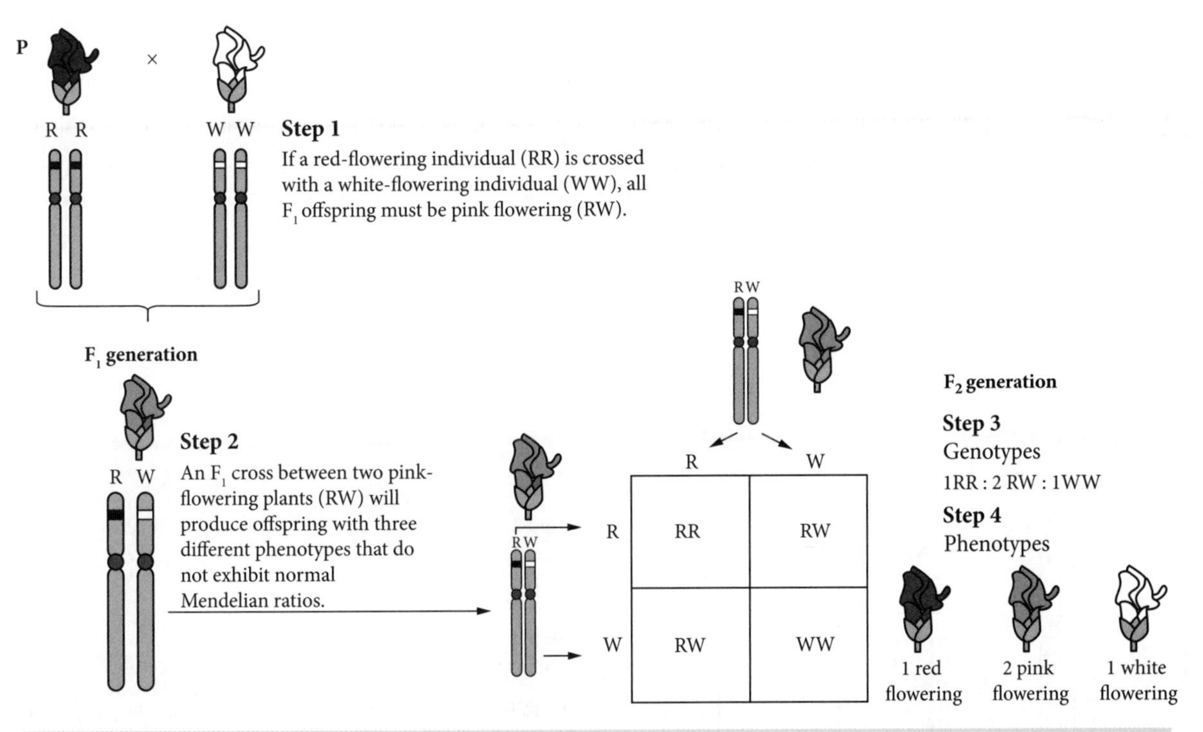

FIGURE 1.47 Incomplete dominance Punnett square. The F_1 heterozygous cross produces an offspring phenotype ratio different from the usual 3 : 1.

Using a pedigree to determine codominance (and incomplete dominance)

Both codominance and incomplete dominance are relatively easy to determine from pedigrees. To distinguish codominance from incomplete dominance, further details would be needed. Three key features must be present in a pedigree to demonstrate either codominance or incomplete dominance.

- Three different phenotypes are present, typically indicated by different colours, one for each phenotype.
- If both parents are heterozygous, offspring can have any one of the three phenotypes.
- If each parent is homozygous for the different alleles, then all offspring will be heterozygous and show the intermediate phenotype.

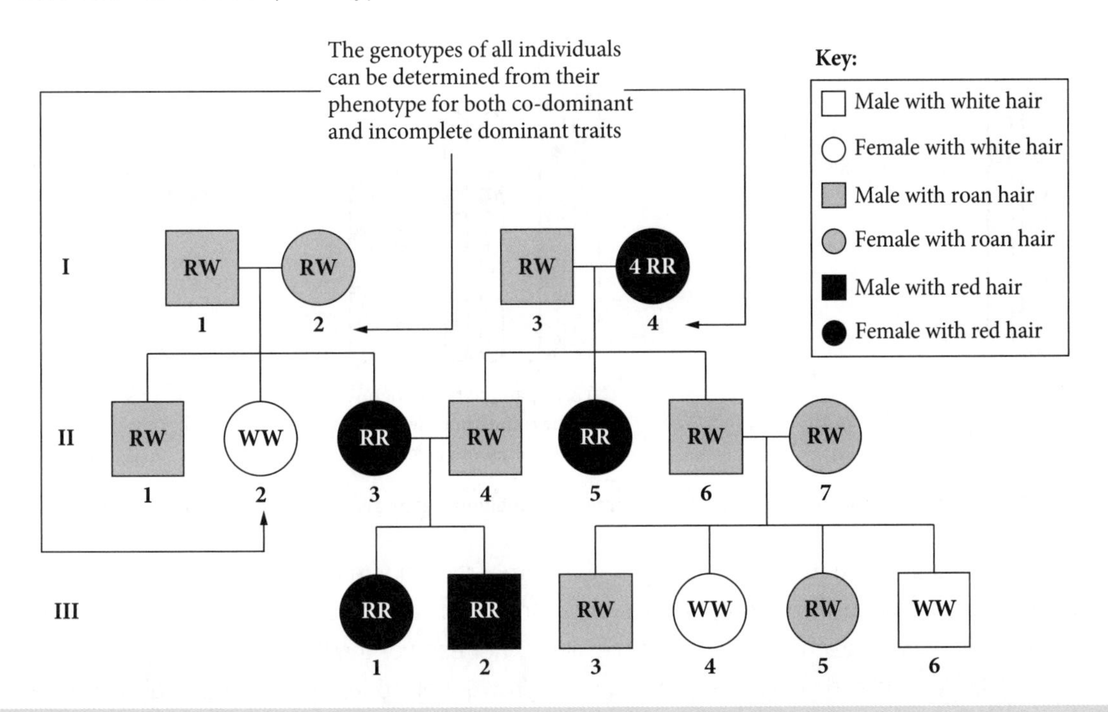

FIGURE 1.48 Using a pedigree to determine codominance (or incomplete dominance)

Multiple alleles

Multiple allele inheritance occurs when there are three or more possible alleles for a gene. While the genotype of an individual will only ever have two alleles at a locus, there is a higher number of possible genotypes (and hence phenotypes) at the locus.

For example, the ABO blood group, discussed earlier in codominance inheritance, is also an example of multiple allele inheritance. The three alleles, I^A, I^B and i, can give rise to six genotypes, I^AI^A, I^Ai, ii, I^Bi, I^BI^B and I^AI^B, which in turn are expressed as one of four phenotypes (blood types): A, B, AB or O.

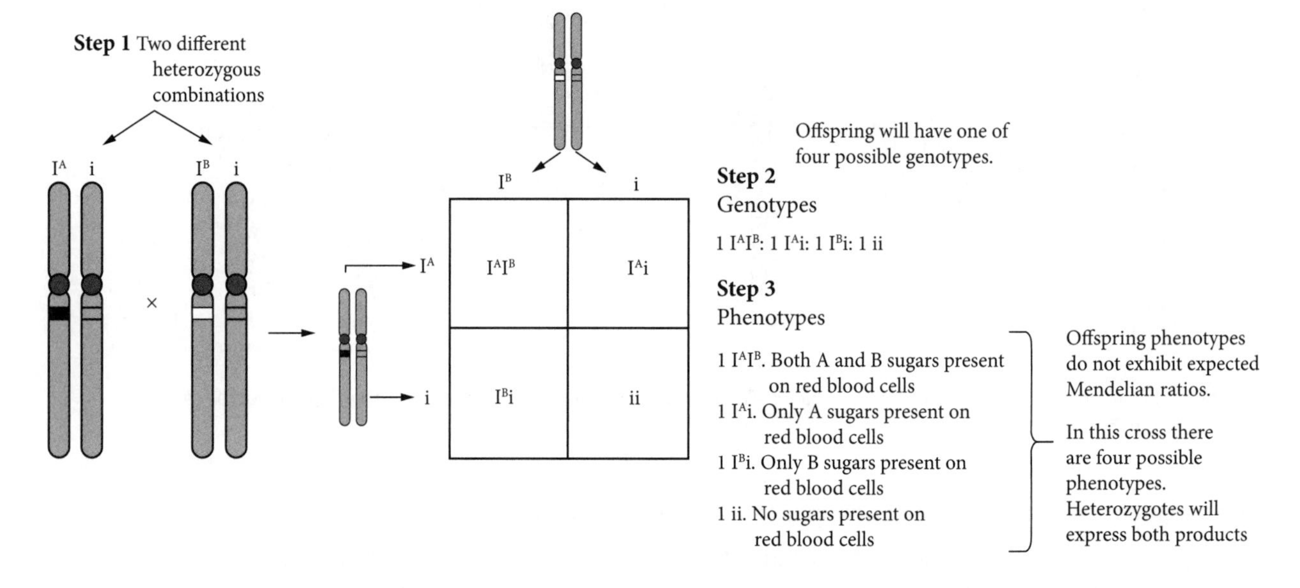

FIGURE 1.49 A Punnett square showing multiple allele inheritance

Polygenic inheritance

Polygenic traits are determined by more than one gene and by the interactions among the different alleles of these genes. The genes are often situated on different chromosomes (e.g. skin colour, eye colour and height).

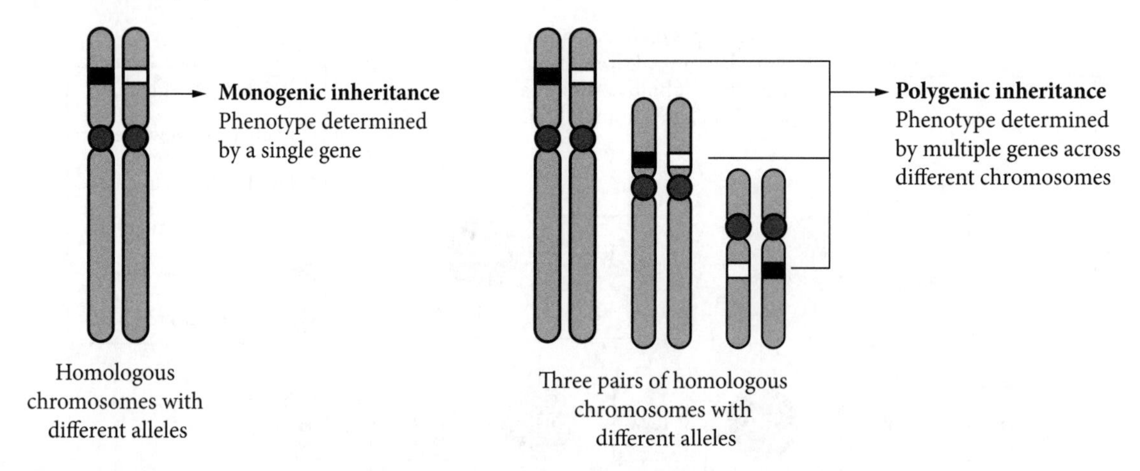

FIGURE 1.50 Polygenic inheritance. Phenotype is determined by multiple genes across different chromosomes.

The different combinations of alleles that result from fertilisation have an additive effect on the phenotype of an individual. For example, consider a simplified case study, where height is determined by three different genes, each with two possible alleles. For each of the three genes, one allele is a contributing allele as it leads to further production of growth hormone, while the other allele is referred to as the non-contributing allele as it does not play a role in the variation. The greater the number of contributing alleles, the taller the person.

If two heterozygotes for each of the three genes (AaBbCc × AaBbCc) had a child, there are seven different height variations (phenotypes) possible for the child, with 64 possible allele combinations in a genotype ratio of 1 : 6 : 15 : 20 : 15 : 6 : 1.

	ABC	ABc	AbC	aBC	Abc	aBc	abC	abc
ABC	6	5	5	5	4	4	4	3
ABc	5	4	4	4	3	3	3	2
AbC	5	4	4	4	3	3	3	2
aBC	5	4	4	4	3	3	3	2
Abc	4	3	3	3	2	2	2	1
aBc	4	3	3	3	2	2	2	1
abC	4	3	3	3	2	2	2	1
abc	3	2	2	2	1	1	1	0

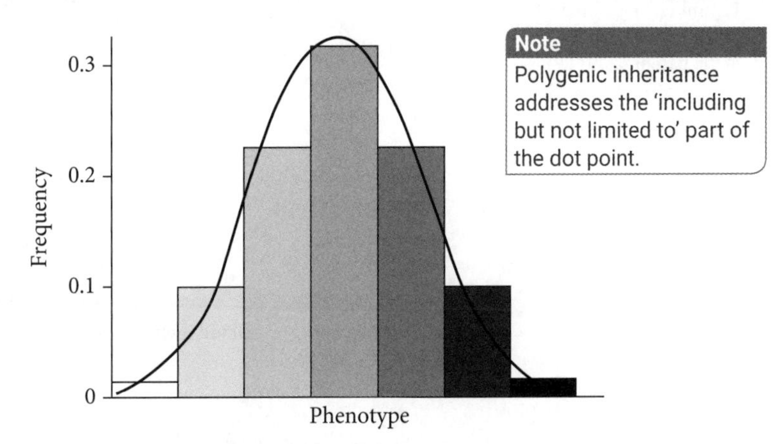

Note
Polygenic inheritance addresses the 'including but not limited to' part of the dot point.

FIGURE 1.51 A polygenic inheritance Punnett square and phenotype frequency

1.4.3 Collecting, recording and presenting data to represent frequencies of characteristics in a population, in order to identify trends, patterns, relationships and limitations in data

Examining frequency data

This section provides examples of how frequency data can be used to identify trends, patterns and relationships within a population. A *trend* is the general tendency of a set of data to move in a certain direction. A *pattern* is when a set of data repeats in a predictable way. A *relationship* exists between two or more variables when changing one variable causes a change in the others.

Note
The focus is on the skills associated with the collection and analysis of the data, not on remembering the specific facts and figures associated with each case study.

Patterns

MENDELIAN INHERITANCE

CASE STUDY 4

One of the best examples of where patterns can be observed in frequency data is Gregor Mendel's pea breeding experiments. Mendel carried out an estimated 30 000 crosses of pea plants over a seven-year period, investigating seven characteristics, each having two different 'factors' (alleles). Through a series of test crosses, Mendel first created 'pure-bred' (homozygous dominant and homozygous recessive) plants for each characteristic.

He then crossed these 'pure-bred' plants (P generation). From the resulting phenotypes of the offspring (F_1 generation), he was able to determine which 'factor' was dominant and which was recessive. For example, from his cross between 'pure-bred' tall and short plants he established that the tall 'factor' masked the short 'factor'.

For all seven characteristics, Mendel observed a specific pattern. The offspring of an F_1 cross always produced offspring (F_2 generation) in a 3 : 1 phenotype ratio. This occurs because the homozygous dominant (TT) and heterozygous (Tt) individuals have the same phenotype, due to the recessive allele remaining hidden in the heterozygotes. The 25% of offspring that express the recessive allele in their phenotype have inherited one copy from each parent.

The patterns observed in Mendel's data are further highlighted with his crosses between plants that were 'pure-bred' (homozygous) for two separate 'factors' (alleles).

››

››

Mendel crossed two pure-bred plants that differed only in a single trait.

The initial 'pure-bred' plants are referred to as the parent plants (P).

Pure-bred tall — Pure-bred short

P: TT, tt

Segregation

Gametes: T, T, t, t

Fertilisation

F_1: Tt, Tt, Tt, Tt

All tall hybrid offspring

All offspring (F_1) had the same physical appearance (phenotype) as one of the parent plants. He concluded that one 'factor' was dominant over the other and that the recessive 'factor' was masked in the F_1 generation (Tt).

He then performed a cross between two of the F_1 generation (Tt × Tt).

F_1: Tt, Tt

Segregation

Gametes: T, t, T, t

Fertilisation

F_2: TT, Tt, Tt, tt

3 tall offspring : 1 short offspring

The phenotypes of the offspring (F_2) were always in a 3 : 1 ratio, strengthening the conclusion that one 'factor' was dominant over the other.

FIGURE 1.52 The cross between two pure-breeding plants for seed colour and subsequent F_1 cross, producing Mendel's classic 3 : 1 phenotype ratio

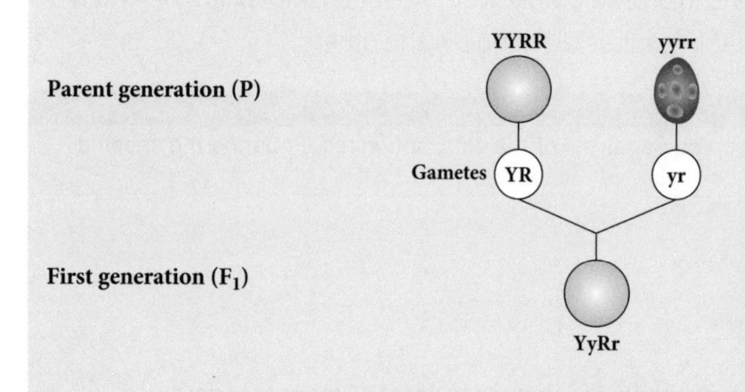

Mendel crossed plants that were 'pure bred' (homozygous) for two separate 'factors' (alleles): seed colour (yellow YY and green yy) and seed shape (round RR and wrinkled rr).

First generation (F_1)

YyRr

All offspring (F_1) had the same physical appearance (phenotype) as one of the parent plants. He concluded that the yellow and round 'factors' were dominant over the green and wrinkled 'factors', which were masked in the F_1 generation. The offspring produced were hybrid for both traits (dihybrid YyRr).

Second generation (F_2)

He then performed a cross between two of the F_1 generation (YyRr × YyRr).

He predicted that the two separate 'factors' would be inherited independently of each other and be observed in the 3 : 1 ratio.

Sperm

Egg	YR 1/4	Yr 1/4	yR 1/4	yr 1/4
YR 1/4	YYRR	YYRr	YyRR	YyRr
Yr 1/4	YYRr	YYrr	YyRr	Yyrr
yR 1/4	YyRR	YyRr	yyRR	yyRr
yr 1/4	YyRr	Yyrr	yyRr	yyrr

9/16 Yellow round

3/16 Green round

3/16 Yellow wrinkled

1/16 Green wrinkled

The phenotypes of the offspring (F_2) were always in a 9 : 3 : 3 : 1 ratio.

When examined separately, each 'factor' appears in the 3 : 1 phenotypic ratio.

3 yellow : 1 green

3 round : 1 wrinkled

FIGURE 1.53 Identifying patterns in data using genotype frequencies: cross between pea plants that were pure-bred for two separate 'factors' and subsequent F_1 cross

››

»

Mendel's dihybrid cross experiments always produced offspring in a 9 : 3 : 3 : 1 phenotype ratio. For example, for seed shape and colour the F_2 generation consisted of:

- 315 plants with round yellow seeds
- 108 plants with round green seeds
- 101 plants with wrinkled yellow seeds
- 32 plants with wrinkled green seeds.

When each trait is examined separately, the resultant phenotypes are close to his predicted 3 : 1 ratio:

- 423 plants with round seeds (315 + 108) and 133 with wrinkled seeds (101 + 32) = 3.2 : 1
- 416 plants with yellow seeds (315 + 101) and 140 with green seeds (108 + 32) = 2.97 : 1

The identification of a pattern can be used to predict the outcome of crosses in different contexts. However, it is important to note that there are limitations in the data of inheritance studies. For example, predictable patterns of inheritance cannot be observed in more complex polygenic traits that are encoded by multiple genes.

Analysing single nucleotide polymorphisms

A **single nucleotide polymorphism (SNP)** (pronounced 'snip') is a point mutation that is present in more than 1% of the population. Just like the alleles of a gene, each individual has two copies of each SNP, one on each homologous chromosome.

Single nucleotide polymorphisms are the most common type of genetic variation between individuals. The human genome contains a SNP every 1000 nucleotides, which means there are about 4–5 million SNPs in a single person's genome. These SNPs are biological markers that help scientists to identify alleles associated with specific diseases. They can also be used to determine ancestry, by comparing the frequency of every SNP in a person's genome with a database of SNPs from geographical populations around the world.

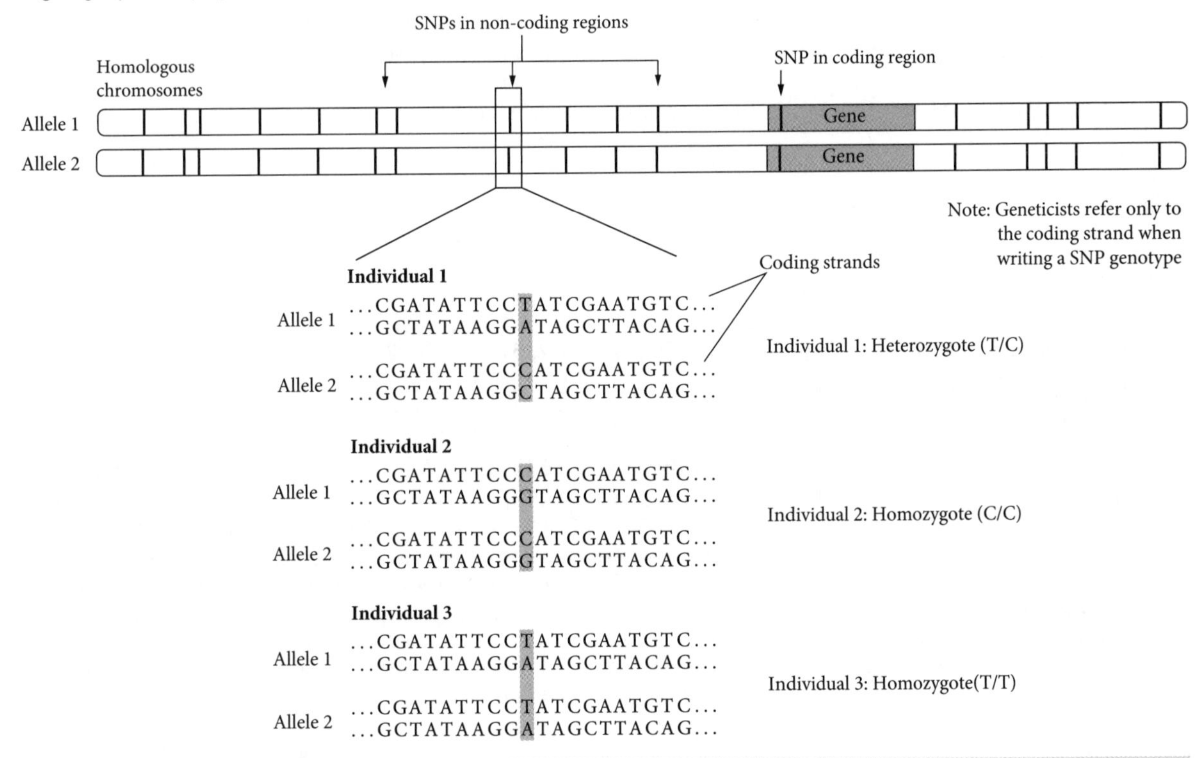

FIGURE 1.54 Single nucleotide polymorphism

Relationships

CASE STUDY 5

LACTOSE TOLERANCE IN HUMANS

The study of lactose tolerance in humans is an example of using frequency data to identify a relationship between a specific SNP and the ability to consume dairy products. In humans, the enzyme lactase is produced in babies to break down lactose, the main carbohydrate in breast milk. In our ancestors, the LCT gene that expresses lactase was 'switched off' after a child was weaned. This trait is called lactase non-persistence (LNP; encoded by the LNP allele) and gives rise to the lactose intolerance phenotype.

Approximately 10 000 years ago, in northern European populations, a mutation occurred in the LCT gene that coincided with the domestication of livestock. The mutation stopped the LCT gene from being 'switched off' and allowed individuals to continue consuming milk and other dairy products into adulthood. This lactase persistence (LP) allele is dominant over the LNP allele and provides an adaptive advantage due to the opportunity it provides for increased nutritional uptake.

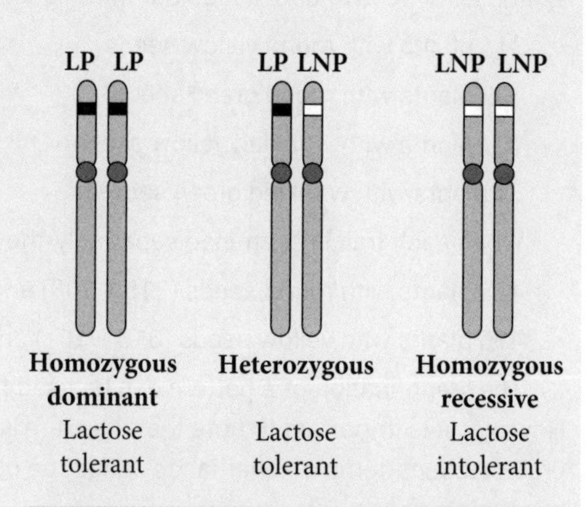

FIGURE 1.55 Possible genotypes for lactose tolerance in humans

The mutation is present in approximately 30% of the world's population, mainly in people of European descent, and is less common in areas of the world where dairy products aren't consumed.

Table 1.11 shows the frequency of the LNP and LP alleles in Caucasian people in the United Kingdom and in Indigenous Mexicans. The LP has a population frequency of 75% in the UK, where dairy products are widely consumed. In contrast, the frequency of the LP allele is only 0.0028% in Indigenous Mexicans, who represent the ancestral population in this region before the arrival of Spanish people and domestication of livestock.

A limitation of the LNP study is that it can be difficult to tell whether the SNP causes a condition or simply co-segregates with a gene it is sitting next to. Also, this study makes assumptions about the ethnic background of the people, and it relies on the people in the study knowing their family history.

TABLE 1.11 The frequency of LNP and LP alleles in Caucasians in the UK and in Indigenous Mexicans

	United Kingdom*			Indigenous peoples of Mexico**		
	Number of individuals	**LP allele**	**LNP allele**	**Number of individuals**	**LP allele**	**LNP allele**
Homozygous dominant (LP : LP)	1881	3762	0	0	0	0
Heterozygous (LP : LNP)	1236	1236	1236	2	2	2
Homozygous recessive (LNP : LNP)	227	0	454	179	0	358
Total (frequency)		**4998 (0.747)**	**1690 (0.253)**		**2 (0.0055)**	**360 (0.994)**

*Data source: Davey Smith G. et al. (2009) 'Lactase persistence-related genetic variant: population substructure and health outcomes', *European Journal of Human Genetics*, 17: 357–67, accessed at https://www.nature.com/articles/ejhg2008156

**Data source: Ojeda-Granados et al. (2016) 'Association of lactase persistence genotypes with high intake of dairy saturated fat and high prevalence of lactase non-persistence among the Mexican population', *Journal of Nutrigenetics and Nutrigenomics*, 9: 83–94, accessed at https://www.researchgate.net/publication/304712810_Association_of_Lactase_Persistence_Genotypes_with_High_Intake_of_Dairy_Saturated_Fat_and_High_Prevalence_of_Lactase_Non-Persistence_among_the_Mexican_Population

Trends

Scientists can also identify trends in data. In the example shown in Figure 1.56, scientists plotted the average size of eggs in a wide range of species against the number of offspring produced. They discovered a trend in which the size of eggs declines as the number of eggs, and therefore offspring produced, increases. This information can be used to provide insights into the physiology and evolution of different species.

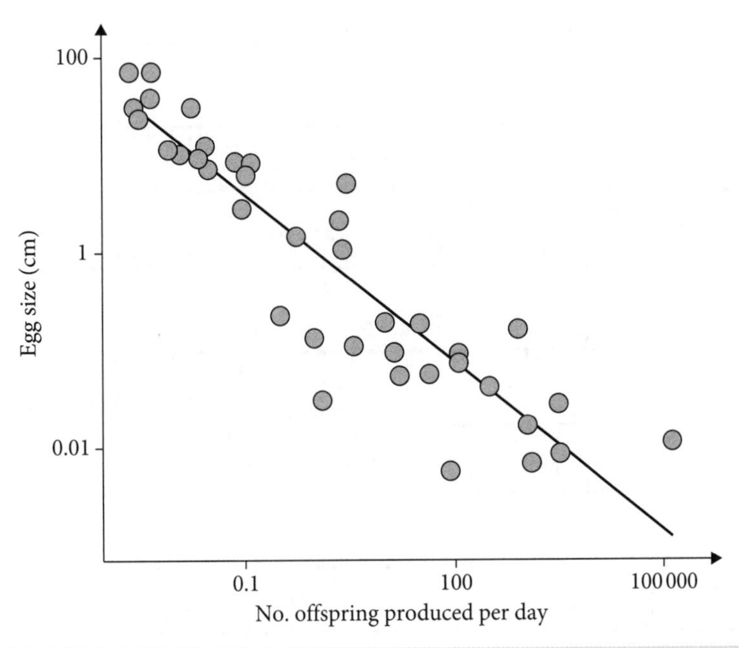

FIGURE 1.56 The relationship between egg size and number of offspring produced per day

1.5 Inheritance patterns in a population: Can population genetic patterns be predicted with any accuracy?

The structure of DNA was discovered in the early 1950s. In the relatively short time since, technologies have been developed that can quickly and accurately sequence DNA. The data can be used in large-scale collaborative bioinformatics studies in various fields of population genetics; for example, to determine the mutations responsible for diseases, the evolutionary relationships between species, and as a conservation management tool to measure genetic variation in endangered plants and animals.

1.5.1 The use of technologies to determine inheritance patterns in a population

DNA sequencing and profiling

DNA sequencing is a laboratory technique used to determine the exact nucleotide sequence of a DNA segment, or of a whole genome. **DNA profiling** generates a specific DNA pattern, or profile, of an individual at specific loci that can then be matched to a known control. The **polymerase chain reaction (PCR)** and **electrophoresis** are essential to both techniques.

Polymerase chain reaction

The first step in any genetic experiment is the extraction of DNA from cells. Because the amount of DNA extracted from a sample is often not enough for applications in the laboratory, a PCR is first used to make millions of copies of the region of DNA that is of interest. The sample can then be used for gene cloning, DNA sequencing or DNA profiling.

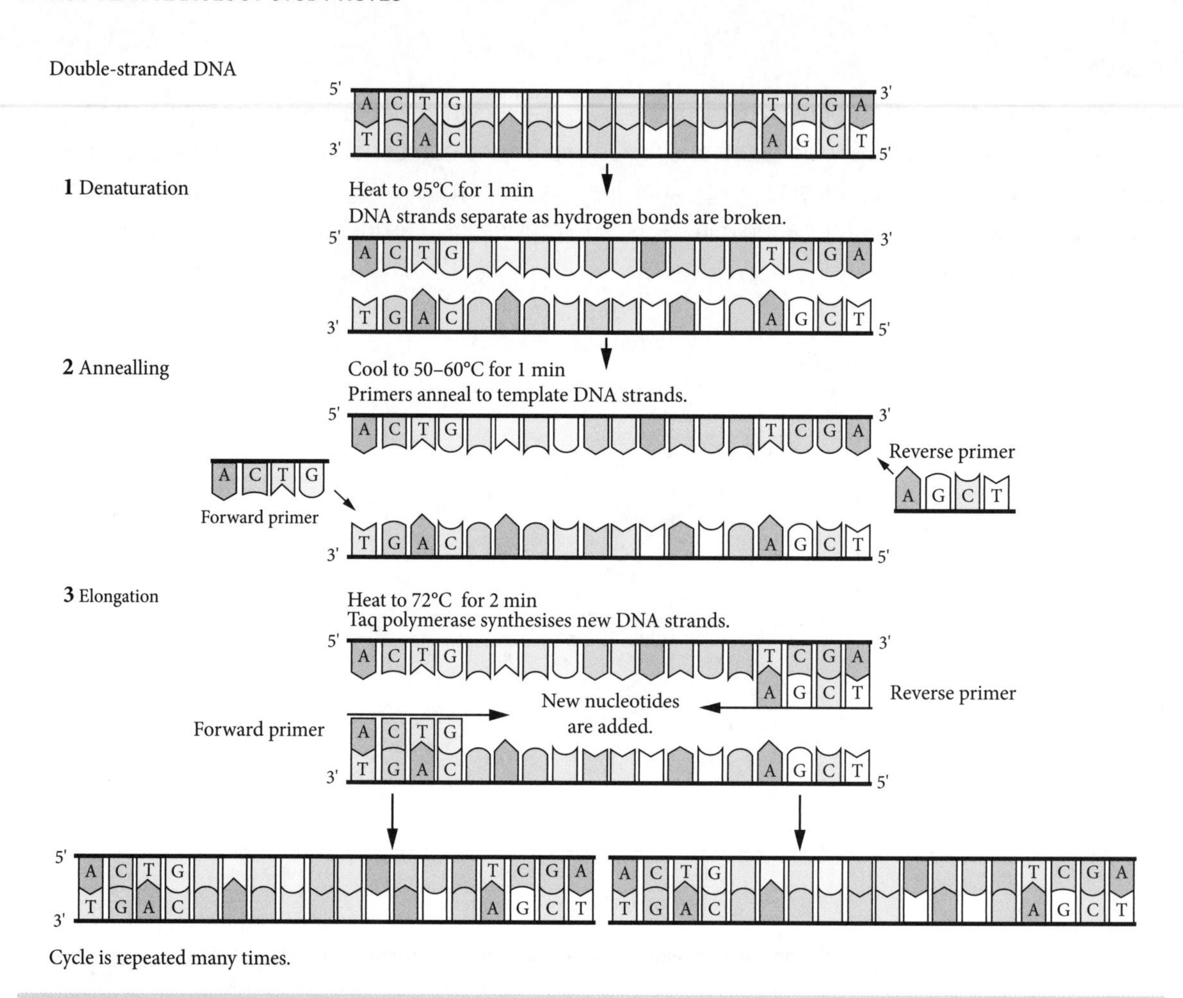

FIGURE 1.57 The polymerase chain reaction

A PCR consists of a DNA sample, primers (short single-stranded pieces of DNA, usually 18–22 bases long) that are complementary to either end of the DNA region to be amplified, free nucleotides called deoxynucleotide trisphosphates (dNTPs; A, T, C and G), Taq polymerase, a buffer to maintain a stable pH, and magnesium chloride ($MgCl_2$), which Taq requires to be an active enzyme. Taq is a DNA polymerase capable of functioning at high temperatures during PCR. The enzyme was originally isolated from *Thermus aquaticus* (Taq), a species of bacterium that lives in hot springs.

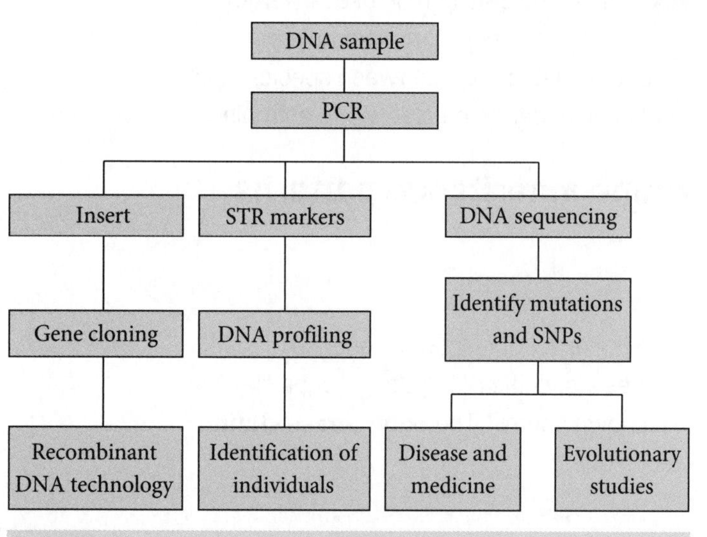

FIGURE 1.58 Applications of the polymerase chain reaction

The ingredients are placed into a PCR tube, which is then inserted into a thermocycler machine. A thermocycler uses variations in temperature to control the DNA synthesis process via three steps, which are repeated 30–40 times. After each cycle, the quantity of DNA is doubled. This results in the exponential amplification of a specific region of DNA, formerly 100–1000 bases in length, to a large quantity.

The product of a PCR can be used as an insert to clone a gene (see Module 6, section 2.3.4), used as a marker in DNA profiling studies, or sequenced to identify a SNP or genetic mutations.

Gel electrophoresis

Gel electrophoresis is a technique that separates segments of DNA based on their size (length). The simplest type of gel electrophoresis is called agarose gel electrophoresis. Liquid agarose, which contains a dye that binds to DNA, is poured into a casting chamber, and sets as a gel. The gel is submerged in a buffer solution and the DNA samples are loaded into wells at the negative electrode end. An electric charge is applied, the buffer conducts electricity, and the negatively charged DNA molecules are pulled through the pores in the gel towards the positive terminal. The shorter fragments migrate faster and further than the larger fragments of DNA. The dye in the gel binds to the DNA as it migrates through the gel, which is then illuminated with blue or UV light. Electrophoresis is routinely used to confirm that the correctly sized PCR product has been amplified, by comparing it to a size standard or marker, or it can be used as a diagnostic tool.

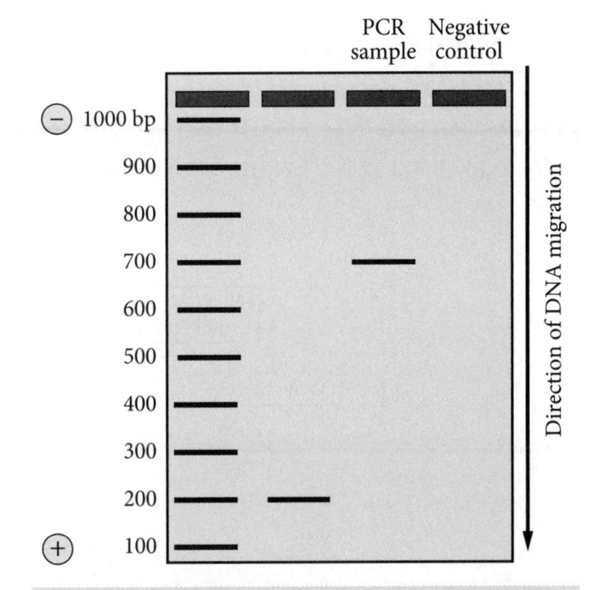

FIGURE 1.59 Using gel electrophoresis to confirm the size of PCR products

For example, cystic fibrosis (CF) is an autosomal recessive disease most commonly caused by the F508del mutation in the CFTR gene. The mutation results in deletion of three base pairs (bp) and the removal of a single amino acid from the CFTR protein (1480 to 1479). Primers are used to amplify the target DNA region during PCR before running a gel electrophoresis. Individuals who have CF will have a single band appear that is three bp shorter than a normal healthy control. CF is discussed in more detail in Module 6 (section 2.1.3).

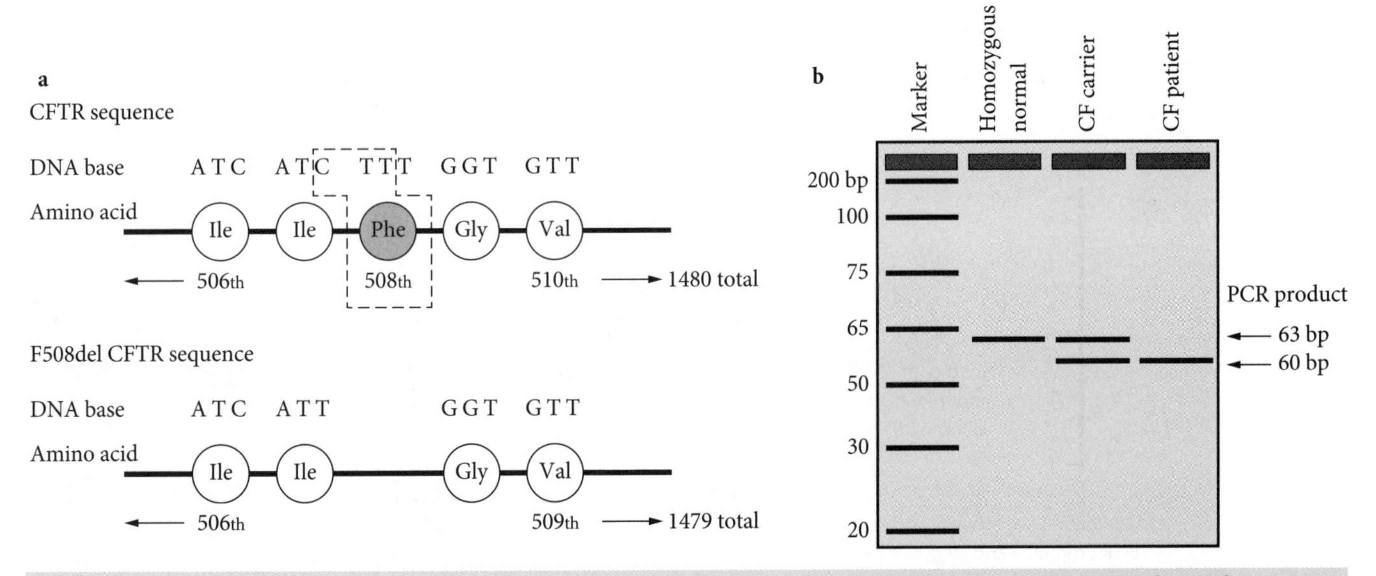

FIGURE 1.60 Using gel electrophoresis to diagnosis CF. a PCR product for the CF patient will be 3 bp shorter than a normal control. b Gel electrophoresis to visualise and diagnose CF

Capillary electrophoresis is a more specialised type of gel electrophoresis, in which DNA migrates through small capillaries containing a gel polymer. The DNA is first labelled with different-coloured fluorescent dyes before it is loaded into the polymer. The coloured dyes can be attached to the primer used in a PCR, which 'tags' the PCR product, or the dye can be attached to nucleotides that are incorporated into a piece of DNA during a sequencing reaction. Capillary electrophoresis has a very high resolution of one base pair. The dye on the DNA is excited by a laser as it passes a small window in the machine, and this releases a colour that is detected by a receiver and interpreted by software as a signal. It is used to visualise the products of **short tandem repeats (STRs)** in DNA profiling or the products of a DNA sequencing reaction.

9780170465267

DNA sequencing

Recall that DNA sequencing is the process of determining the exact order of nucleotides in a segment of DNA, such as a PCR product, or an entire genome. The main method used is single-molecule DNA sequencing, called Sanger sequencing.

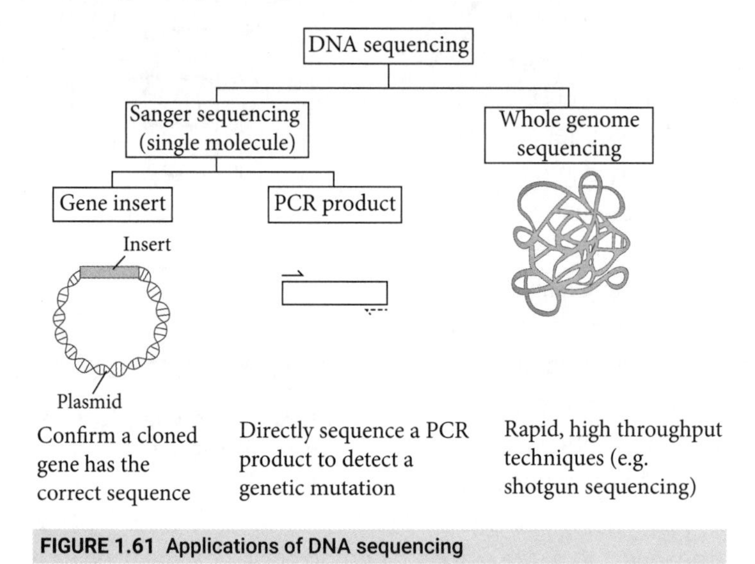

FIGURE 1.61 Applications of DNA sequencing

Sanger sequencing is similar to PCR but there are some important differences. The first difference is that a sequencing reaction has only one primer, which means there is no exponential amplification. The second difference is that, in addition to the four dNTPs, the reaction contains four modified versions of the nucleotides, called dideoxynucleotide triphosphates (ddNTPs), or **chain-terminating nucleotides**. Each ddNTP is modified with a fluorescent dye (C = blue, G = black, A = green and T = red) and they each terminate the reaction when they are incorporated into the growing DNA strand. Each piece of DNA emits a colour as it passes the laser in a capillary electrophoresis machine.

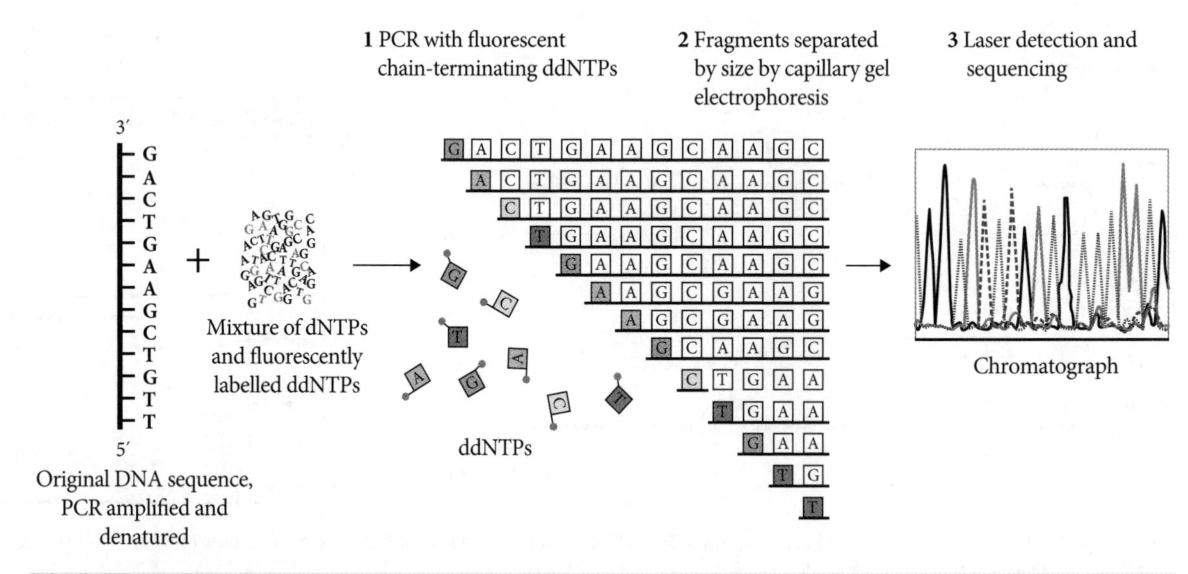

FIGURE 1.62 Sanger sequencing and detection with capillary electrophoresis

DNA sequencing can be used to study the inheritance of disease-causing mutations in a family or population, or to create a phylogenetic tree to show the evolutionary relationship between species. Phylogenetic trees are created by sequencing a gene, such as cytochrome B (*CYTB*), and comparing the sequence between species. DNA is like a 'molecular clock', which means genetic change accumulates at a relatively constant rate over time. The more closely related two species are, the more similar their genetic sequences. Distantly related species therefore have a greater number of genetic differences than closely related species.

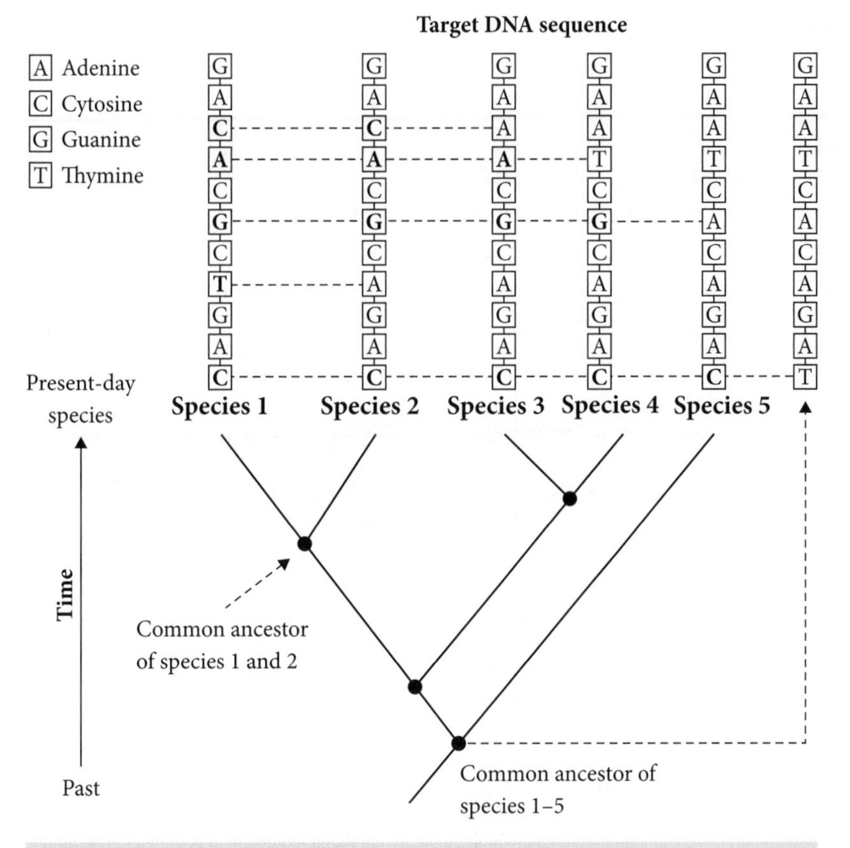

FIGURE 1.63 The use of DNA sequencing to create a phylogenetic tree

DNA profiling

Recall that DNA profiling is used to generate a specific DNA pattern of an individual that can then be matched to a known control. The technique is used in paternity testing, or to identify the perpetrator of a crime by matching a DNA profile generated from bodily fluids left at the crime scene with the DNA profile of a suspect.

DNA profiling relies on highly polymorphic genetic markers, such as short tandem repeats (STRs). Also known as microsatellites, STRs are loci that consist of mono-, di-, tri- or tetra-nucleotide repeats located in non-coding regions of the genome. The number of repeats can vary between alleles in the same individual and between individuals. The repeats are flanked by unique DNA sequences, and this is where primers bind in a PCR. In practice, STRs are normally several hundred base pairs long.

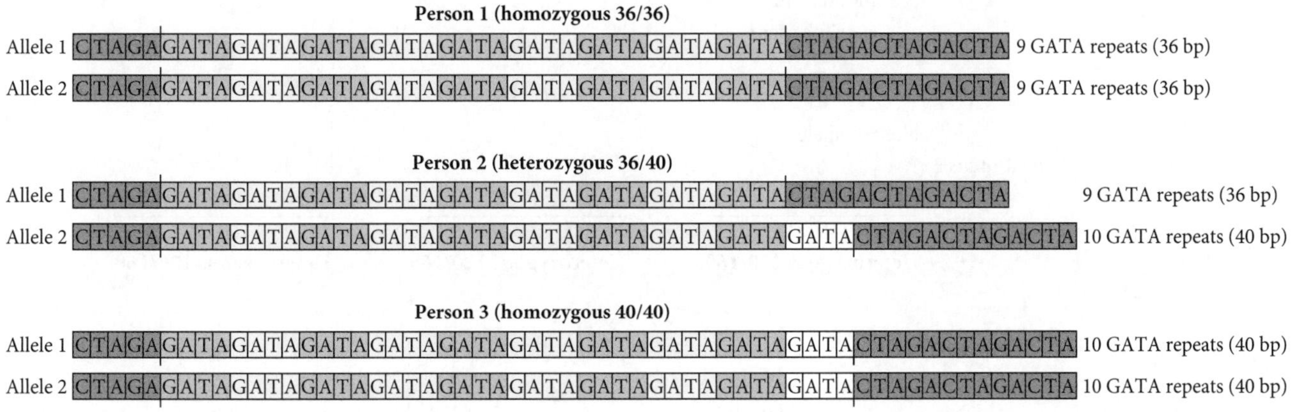

Person 1 (homozygous 36/36)

Allele 1 CTAGAGATAGATAGATAGATAGATAGATAGATAGATAGATACTAGACTAGACTA 9 GATA repeats (36 bp)

Allele 2 CTAGAGATAGATAGATAGATAGATAGATAGATAGATAGATACTAGACTAGACTA 9 GATA repeats (36 bp)

Person 2 (heterozygous 36/40)

Allele 1 CTAGAGATAGATAGATAGATAGATAGATAGATAGATAGATACTAGACTAGACTA 9 GATA repeats (36 bp)

Allele 2 CTAGAGATAGATAGATAGATAGATAGATAGATAGATAGATAGATACTAGACTAGACTA 10 GATA repeats (40 bp)

Person 3 (homozygous 40/40)

Allele 1 CTAGAGATAGATAGATAGATAGATAGATAGATAGATAGATAGATACTAGACTAGACTA 10 GATA repeats (40 bp)

Allele 2 CTAGAGATAGATAGATAGATAGATAGATAGATAGATAGATAGATACTAGACTAGACTA 10 GATA repeats (40 bp)

FIGURE 1.64 The genotypes of three people at an STR

A person with an equal number of repeats on both alleles will have one product in the PCR (homozygous), while a person with a different number of repeats on each allele (heterozygous) will have two products in the PCR. An individual is usually genotyped at 10–15 STR loci to generate a unique DNA profile.

Figure 1.66 shows an example of DNA profiling in a paternity case using a single STR locus. In paternity cases, the mother and father share an allele with a child. In this case there were two potential fathers, but potential father 2 was excluded because he did not share either allele with the child. Potential father 1 is therefore the likely father of the child. It should be noted that when only one STR locus is examined, an allele may be shared by chance. Additional STR loci need to be tested to provide enough statistical power to say with confidence that he is the father.

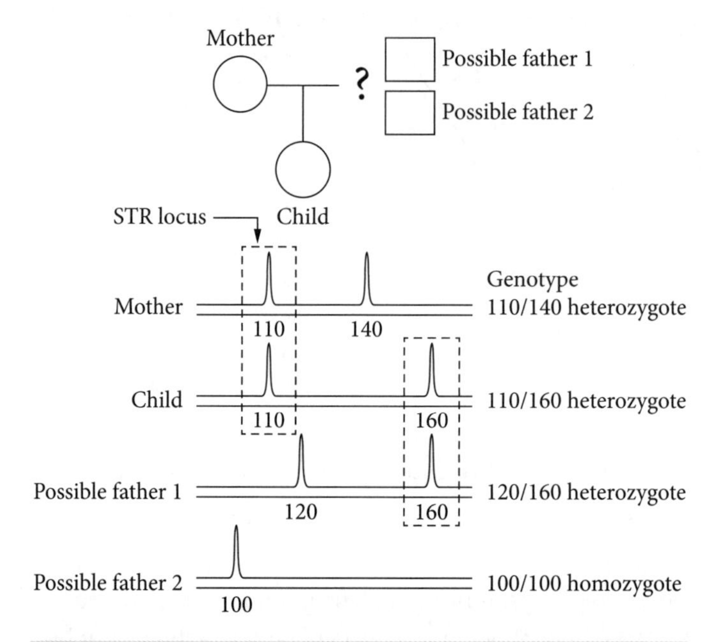

FIGURE 1.65 The use of STR DNA profiling in paternity testing

Figure 1.66 illustrates how DNA profiling can be used to match a suspect to a crime scene. In this case the genotype of a suspect must exactly match the genotype from the DNA sample obtained from a crime scene. Suspect 2 can be excluded based on STR locus 2. Suspect 1 is likely responsible for the crime as both STR loci match. However, once again, additional loci would be required to provide statistical power to state with confidence that the sample belongs to suspect 1.

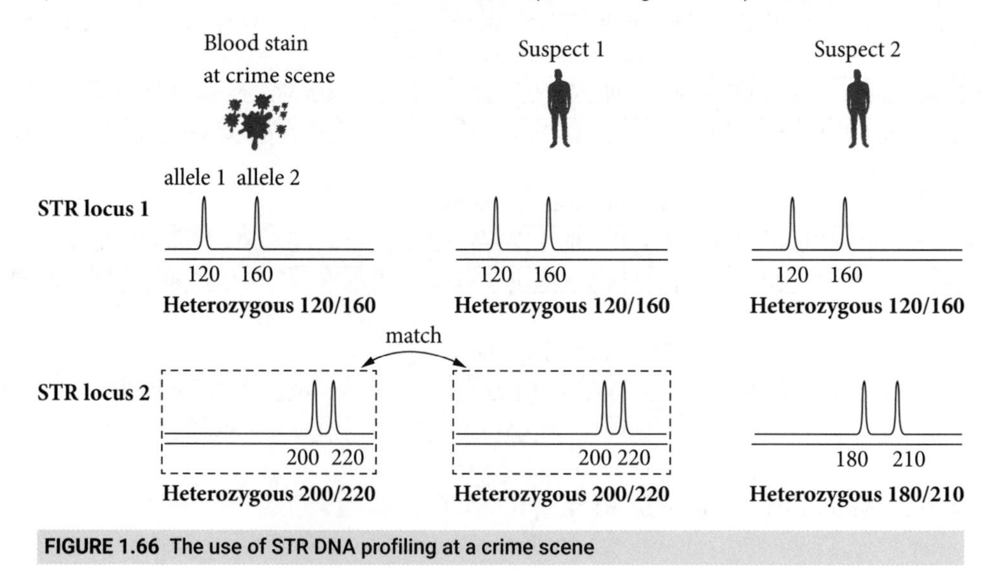

FIGURE 1.66 The use of STR DNA profiling at a crime scene

1.5.2 Identifying trends, patterns and relationships from data analysis of large-scale collaborative projects involving population genetics

Recall that a *trend* is the general tendency of a set of data to move in a certain direction. A *pattern* is when a set of data repeats in a predictable way. A *relationship* exists between two or more variables when changing one variable causes a change in the others.

Note

This syllabus dot point requires only a single example. Therefore, you can choose any one of the three provided or another of your choice. You are *not* required to know the specifics of each one; however, with an appropriate and contextually relevant stimulus you could be asked to analyse data and identify trends, patterns and relationships from any population genetics study.

Conservation management

The cheetah (*Acinonyx jubatus*) is a case study that illustrates the application of population genetics in conservation management. The cheetah is a vulnerable species, with a population of about 7000 individuals, with numbers continuing to decline because of reduced prey and habitat, and conflict with humans.

The use of a variety of genetic markers and techniques (e.g. STR markers, SNPs and whole-genome sequencing) in large-scale collaborative projects (e.g. by ecologists, geneticists) over the past 40 years has shown that cheetahs have a very low level of genetic variation, which makes them more vulnerable to diseases, environmental change and extinction.

Why do cheetahs have such low levels of genetic variation? Cheetahs experienced a population bottleneck (see section 2.1.6) about 10 000–12 000 years ago that resulted in a reduced population and fragmentation of the species into six subspecies across Africa and Iran. The cheetah is now confined to only 9% of its historical range. The bottleneck, combined with reduced gene flow and genetic drift, resulted in a low level of genetic variation in the species. The trend and general relationship between population size and genetic variation in a species are shown in Figure 1.67.

Genetic information is central to the success of conservation management strategies. For example, a shared global stud book, with DNA profiles of each captive cheetah, is used for controlled breeding programs to increase gene flow and maintain genetic diversity.

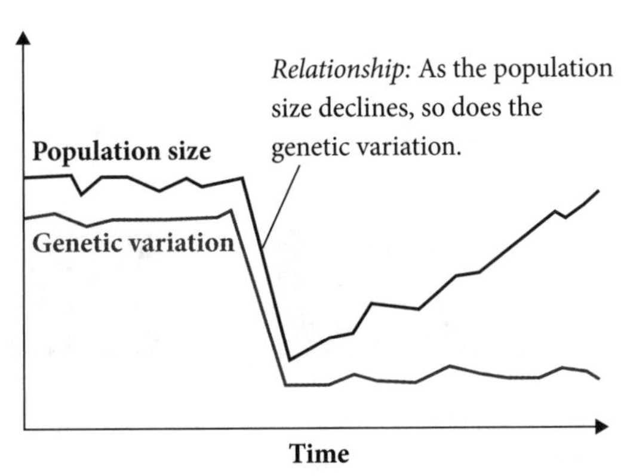

FIGURE 1.67 The trend and relationship between population size and genetic variation

Inheritance of a disease or disorder

Genome-wide association studies (GWAS) are large-scale collaborative projects between geneticists, medical clinicians and bioinformaticians that use genetic data sets to identify a relationship between a SNP and a gene that is associated with an increased risk of developing a disease or disorder. The SNP databases that form the basis of GWAS come from earlier collaborative projects, such as the Human Genome Project and the 1000 Genomes Project.

GWAS use SNPs as markers to identify genes that sit physically close to them on a chromosome. If a SNP is close to a gene, a SNP allele will always be inherited with the same gene allele on a **haplotype**. This is because the combinations will not be broken up by crossing over during meiosis.

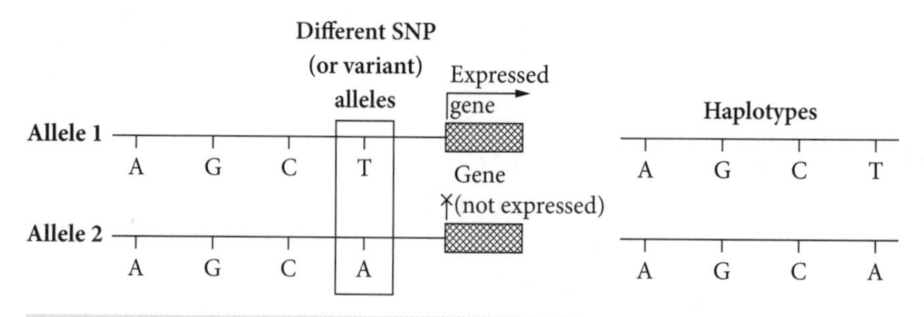

FIGURE 1.68 A haplotype showing four SNPs inherited as a 'block' with a gene

In GWAS, genomic DNA is extracted from a tissue sample from a large number of people with (cases) or without (controls) a disease or disorder. Individuals from each group are genotyped at hundreds of thousands of SNPs across the whole genome. If one allele of a SNP is over-represented in the cases relative to the controls, this suggests that the SNP sits physically close to a gene that is contributing to the phenotype of the disease or disorder. A study of over 38 000 participants identified a SNP allele in the FTO (fat mass and obesity-associated) gene located on chromosome 16. People with this allele had a 1.67-times higher likelihood of developing obesity (see section 4.3.1).

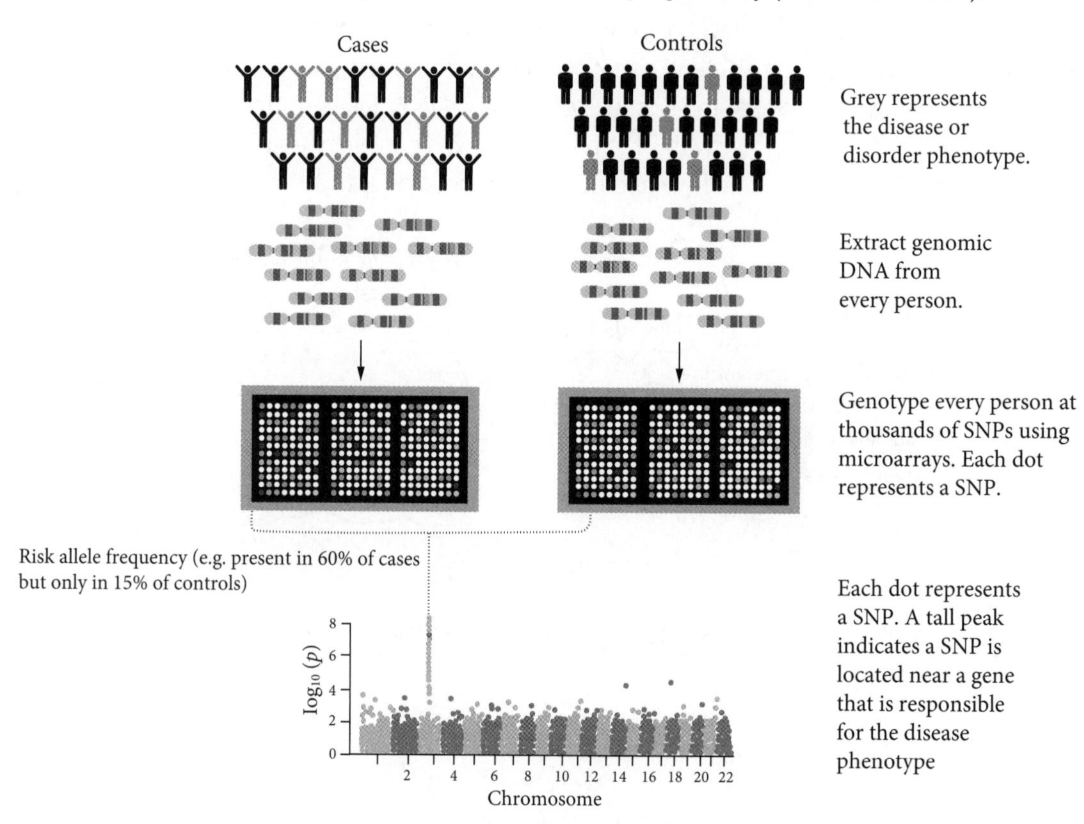

FIGURE 1.69 The use of GWAS to identify a gene associated with an increased risk of a disease or disorder

Human evolution

The predictable pattern of Mendelian inheritance has allowed scientists working in large-scale collaborative studies to use allele frequency data from thousands of genes in humans to identify a relationship between genetic variation and geographic location on Earth. People in Africa have higher levels of genetic variation than people in other parts of the world. This relationship supports the 'Out of Africa' hypothesis of human evolution, with alleles being lost in serial founder effects (see section 2.1.6) as small numbers of people established populations in new areas.

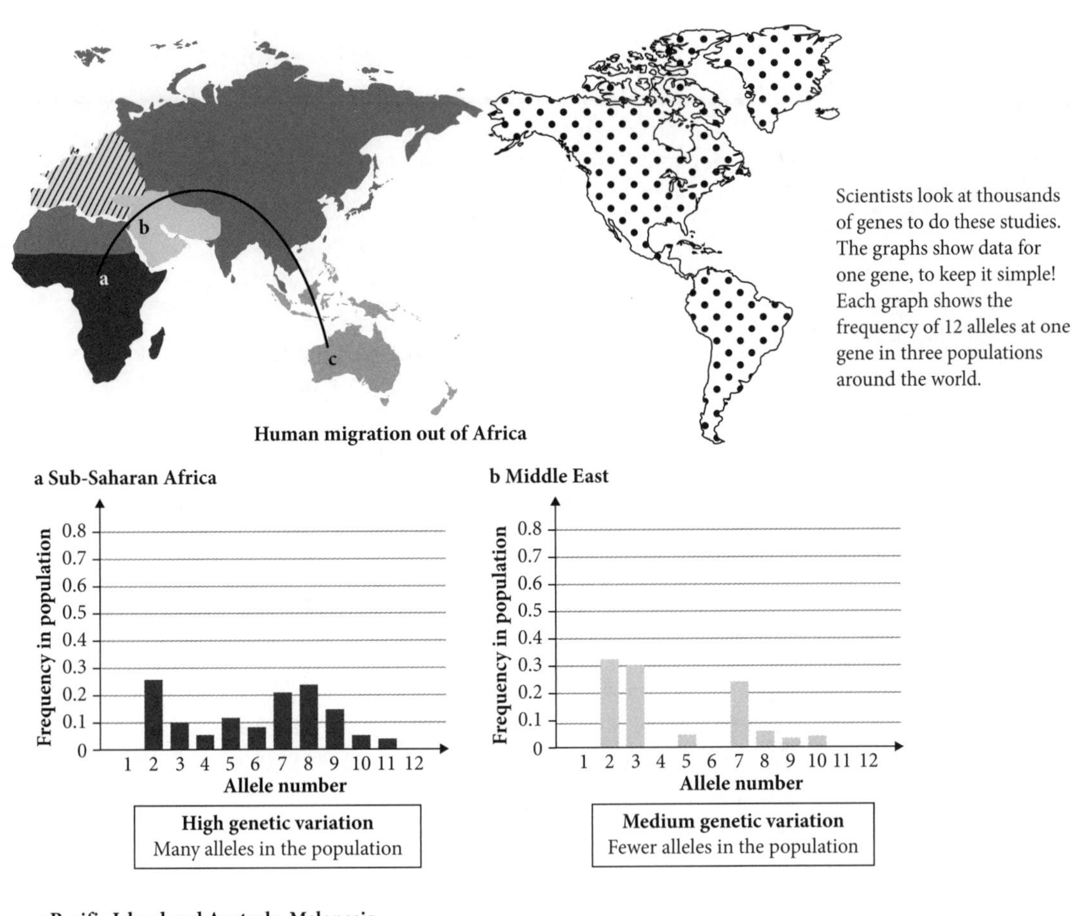

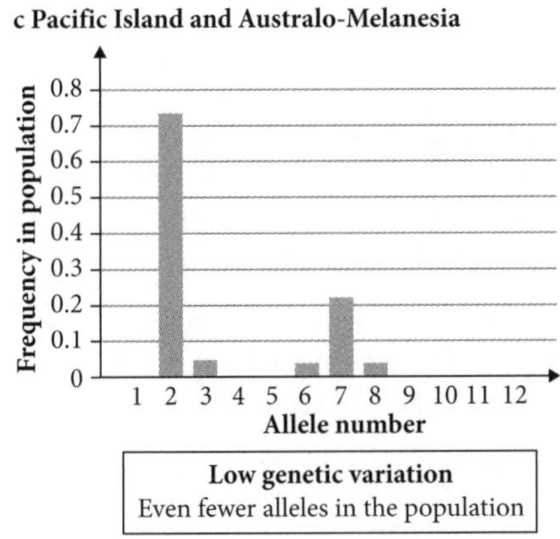

FIGURE 1.70 Patterns of genetic diversity in humans from different regions of the world

Note

You can expect to be asked a question on the advantages and disadvantages of collaborative studies with reference to a specific example.

Glossary

allele A version of a gene; alleles of the same gene have minor differences in their genetic sequence, resulting in variations in the phenotype

alternation of generations Describes an organim's life cycle when it contains both haploid and diploid stages; both stages are multicellular, though one may be dominant over the other

alternative splicing A process that occurs in eukaryotic organisms during gene expression to remove different exon and intron combinations, producing different proteins from the same gene

amino acid A small molecule containing an amino group (NH_2) and a carboxylic acid group (COOH); a building block of proteins

angiosperm A plant that belongs to one of the four major plant groups; has vascular tissue and produces flowers and seed-bearing fruit; heterosporous

anticodon A group of three nucleotides in tRNA that binds to the complementary codon of the mRNA during translation, resulting in the addition of a specific amino acid to the polypeptide chain

artificial insemination A type of assisted reproductive technology that uses any method other than sexual intercourse to introduce semen into the reproductive tract of a female

artificial pollination The mechanical transfer of pollen by humans from the stamen to the stigma of a flower when natural pollinators are absent, or to select for favourable traits

asexual reproduction A type of reproduction that requires only one parent and gives rise to genetically identical offspring

autosomal Describes any chromosome that is not one of the sex chromosomes

binary fission A type of asexual reproduction by unicellular organisms in which the parent cell divides into two approximately equal daughter cells

blastocyst An early embryonic cellular mass that has undergone some cell differentiation prior to implantation in the uterus

bryophyte A plant (e.g. moss) that belongs to one of the four major plant groups; has non-vascular tissue; homosporous

budding A type of asexual reproduction where an outgrowth of a body region separates from the original organism, giving rise to two individuals

chain-terminating nucleotide A nucleotide that has been modified to stop DNA elongation during DNA sequencing; contains a fluorescent marker for detection

https://get.ga/aplus-hsc-bio-u34

A+ DIGITAL FLASHCARDS
Revise this topic's key terms and concepts by scanning the QR code or typing the URL into your browser.

cloning A technique used to make exact genetic copies of genes, cells, tissues or whole organisms

codominance A type of inheritance in which the genotype consists of two alleles of the same gene, equally expressed in the phenotype

codon A group of three nucleotides in mRNA that binds to the complementary anticodon of the tRNA molecule and in doing so codes for a specific amino acid

conjugation The temporary union of two bacteria or protists in order to exchange genetic material

constitutive gene A gene that is always active and typically not influenced by the environment

corpus luteum A hormone-secreting structure that develops in an ovary following the release of an egg; degenerates if pregnancy doesn't occur

covalent bond A chemical bond formed when two atoms share electrons

diploid Describes an autosomal cell that has two sets of chromosomes in homologous pairs, one inherited from each parent; represented as $2n$, where n is the number of chromosomes in the gamete

diploid-dominant The multicellular diploid stage of an organism that dominates, with the only haploid cells being gametes

deoxyribonucleic acid (DNA) A double-stranded nucleic acid that contains the genetic code responsible for the production of proteins during polypeptide synthesis

deoxyribose A five-carbon sugar molecule combined with a phosphate group, forming the backbone of DNA molecules

DNA polymerase An enzyme that catalyses the bonding of nucleotides to form new strands of DNA during cell replication

DNA profiling A laboratory technique in which a specific DNA pattern (profile) of an individual is generated

DNA sequencing A laboratory technique used to determine the exact nucleotide sequence of a segment of DNA or a whole genome

dominant allele An allele that is always expressed in the phenotype; masks the recessive allele in a heterozygous individual

electrophoresis A technique used to separate fragments of DNA, or different proteins, based on their molecular size

embryo An early stage in the development of a vertebrate, between fertilisation of the egg and development of adult characteristics and features (foetus)

endometrium The mucous membrane lining of the uterus; thickens during the menstrual cycle in preparation for implantation of a blastocyst

enzyme A specific protein that acts as a biological catalyst; increases the rate of a biochemical reaction by lowering the amount of energy required for the reaction to proceed

epigenetic Describes the non-genetic influences of gene expression (e.g. the environment)

eukaryotic Describes a complex type of cell that has membrane-bound nucleus and organelles; multicellular (and some single-celled) organisms are made up of this type of cell; includes all animal, plant, fungi and protist cells

exon A region of a gene that codes for an amino acid sequence during polypeptide synthesis

external fertilisation The union of male and female gametes outside the female's body

fertilisation The fusion of an egg cell and a sperm cell, producing a zygote

fission A type of asexual reproduction in which an organism splits along its longitudinal axis, forming two separate organisms

foetus A stage in the development of a vertebrate; follows the embryonic stage and is characterised by the appearance of adult characteristics and features

fragmentation A type of asexual reproduction in which a body part detaches and develops into a new organism and the original organism regenerates the lost body part

fungi One of the four eukaryotic kingdoms; a diverse group categorised by a cell wall, hyphae and heterotrophism

gamete A haploid (n) sex cell capable of fusion with another haploid cell to form a zygote

gametophyte The haploid multicellular stage of a plant's life cycle

gene A segment of DNA that contains the genetic information to code for a specific protein

gene expression The process by which a gene product, usually a protein, is expressed in the phenotype of an organism

genetically modified organism (GMO) An organism whose genome has been altered using genetic engineering techniques in the laboratory

gestation The period of time between conception and birth

haploid Describes a sex cell (gamete), which has one set of chromosomes, half the number (n) of the corresponding diploid ($2n$) number in the same organism

haploid-dominant Describes a life cycle dominated by the unicellular or multicellular haploid stage, with the only diploid cell being the zygote

haplotype The set of alleles that are usually inherited together from a parent

heterokaryotic Describes a double haploid nucleated cell produced during sexual reproduction in fungi

heterosporous Producing spores of two different sizes and sexes

heterozygous Describes a diploid cell that contains different alleles for a particular gene (e.g. Tt)

homologous chromosomes A pair of chromosomes in a diploid organism that carry the same genes, but possibly different alleles, at the same location (locus)

homozygous Describes a diploid cell that contains only one type of allele for a particular gene (e.g. TT or tt)

hormone A regulatory molecule produced by an organism and transported in blood or sap to target cells to stimulate a specific outcome

hydrogen bond A weak chemical bond formed between a hydrogen atom and another electronegative atom when both atoms are already bound via other chemical bonds

hypha A long filament of a fungus, made up of haploid cells; plural *hyphae*.

implantation Attachment and embedding of the blastocyst into the endometrial lining of the uterus

incomplete dominance A type of inheritance in which neither the dominant nor the recessive allele is expressed completely in the heterozygous phenotype; instead, there is an intermediate phenotype

internal fertilisation The union of male and female gametes inside the female's body

intron A non-coding region of a gene that is spliced out prior to translation; alternative splicing of different introns within a gene is important in determining the end gene product

karyogamy The fusion of haploid nuclei within a heterokaryotic cell; occurs during sexual reproduction in fungi

locus The specific point at which a gene is located on a chromosome; plural *loci*

maternal Derived from the mother

meiosis A type of cell division that occurs only in sexually reproducing organisms and results in four genetically unique haploid (*n*) daughter cells

messenger RNA (mRNA) A single-stranded nucleic acid produced during transcription; has the complementary nucleotide sequence to the template strand of DNA (except for thymine, T, which is replaced by uracil, U)

mitosis A type of cell division that results in two genetically identical diploid (2*n*) daughter cells

multiple allele inheritance A type of inheritance where the phenotype for a particular trait can be determined by more than two alleles

mycelium A collection of hyphae in the body of a fungus

nucleic acid A long-chain molecule made from repeating nucleotide monomers; includes DNA and RNA, which are common to all life

nucleotide A molecule consisting of a five-carbon sugar joined to a phosphate group and a nitrogenous base; nucleotides are the monomers (building blocks) of nucleic acids (DNA and RNA)

nucleus A membrane-bound organelle that contains the genetic material of eukaryotes

ovary The female reproductive organ in which an egg is produced

parthenogenesis A type of asexual reproduction where an unfertilised egg develops into a complete individual; the resulting offspring can be haploid or diploid, depending on the species

paternal Derived from the father

pedigree A diagrammatic model used to track the inheritance of genetic traits through generations

plasmid A small, circular DNA structure independent of the chromosome in prokaryotic cells

plasmogamy The fusion of the cytoplasm of haploid cells during sexual reproduction in fungi, producing a heterokaryotic cell (double-nucleated cell)

polymerase chain reaction (PCR) A laboratory technique used to amplify small quantities of DNA

polypeptide An organic molecule made up of a specific sequence of amino acids joined by peptide bonds

pregnancy The period of embryonic and foetal development within the uterus that occurs between fertilisation and birth

primary structure In proteins, the linear sequence of amino acids in a polypeptide

primer A short DNA sequence that binds to the start and end of a target DNA sequence in PCR

prokaryotic Describes a simple type of cell that lacks a membrane-bound nucleus and organelles; includes all archaea and bacteria

promoter A region of DNA near the beginning of a gene that is bound by transcription factors to initiate transcription

protein An organic molecule made up of one or more long chains of amino acids (polypeptides); the specific amino acid sequence and resulting molecular bonds produce a unique 3D structure required for the protein to function

proteome The total collection of proteins that can be produced by an organism

protist A unicellular eukaryotic organism in the kingdom Protista (e.g. protozoans)

pteridophytes (ferns) One of the three groups of vascular plants; reproduce via spores

Punnett square A diagrammatic representation of the possible genotypes of offspring from a parent cross

recessive allele An allele that is only expressed in the phenotype of a homozygous recessive individual (e.g. tt)

recombinant DNA A DNA molecule produced by laboratory methods that combine pieces of DNA from two different organisms

regulated gene A gene that is produced or regulated as required

ribosome A cell organelle in which proteins are synthesised, in prokaryotic and eukaryotic cells

selective breeding The process by which animals or plants are bred based on desired characteristics

sex-linkage A type of inheritance that relates to genes carried on the sex chromosomes

sexual reproduction A type of reproduction in which genetically unique male and female gametes fuse at fertilisation

short tandem repeat (STR) A DNA sequence in introns that contains repeating nucleotide sequences, typically 2–5 base pairs (bp) long; length varies between individuals

single nucleotide polymorphism (SNP) A point mutation that is present in more than 1% of the population

sister chromatids Identical copies of a chromosome, joined at a centromere, formed during replication prior to cell division

splicing The process in which different combinations of introns are removed from pre-mRNA prior to translation in eukaryotic organisms

spore A haploid cell that can develop into a new organism without sexual reproduction; common in plants, algae and fungi

sporophyte The diploid multicellular stage of a plant's life cycle

stigma The receptive tip of the carpel in a flower; a female reproductive organ

style A narrow tube-like structure of the carpel (female reproductive organ in flowers); supports the stigma

syngamy The fusion of two cells or two nuclei to form a new organism

transcription The process in which the base sequence of a DNA template strand is used to produce a complementary single-stranded mRNA sequence

transcription factor A protein that binds to DNA to initiate and control the rate of gene transcription

transduction The process in which a bacteriophage (a virus that infects bacteria) transfers DNA between bacteria; the bacteriophage injects DNA that is a mix of its own genetic material and that of a previously infected bacterium

transfer RNA (tRNA) An RNA molecule that delivers specific amino acids to the ribosome during the translation phase of polypeptide synthesis

transformation The uptake of DNA by bacteria from the environment that is then incorporated into the bacterium's chromosomal or plasmid DNA

transgenic organism A living thing that has had its genome altered through the introduction of a gene or genes from a different species

translation The process in which a polypeptide is formed in a ribosome by using the mRNA base sequence to produce a specific amino acid sequence

zygote The first diploid cell ($2n$) formed by the union of haploid (n) male and female gametes

Exam practice

Multiple-choice questions

Solutions start on page 245.

Reproduction

Question 1

Hydra are small eukaryotic organisms capable of both asexual and sexual reproduction. The image shows a *Hydra* reproducing asexually.

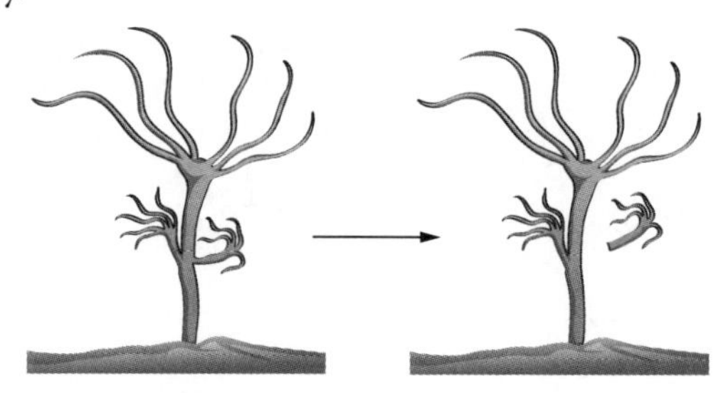

What type of asexual reproduction is shown?

A Budding

B Binary fission

C Fragmentation

D External fertilisation

Question 2

Which of the following is an advantage of asexual reproduction over sexual reproduction?

A Species that utilise asexual reproduction can exist in a wider and more volatile range of environments.

B Increased genetic variation results from recombination of alleles during asexual reproduction.

C Species that utilise asexual reproduction are genetically identical and therefore better suited to stable environments.

D Fertilisation occurs externally to the female and therefore less parental care is required.

Question 3 ©NESA 2020 SI Q3

The following four events occur during reproduction in a placental mammal.

1 Fertilisation

2 Implantation

3 Ovulation

4 Placental formation

In which order do these events occur?

A 2, 1, 3, 4

B 2, 4, 1, 3

C 3, 1, 2, 4

D 3, 2, 4, 1

Question 4

The following image shows the fluctuation in levels of four hormones involved in the ovulation and menstrual cycle: follicle-stimulating hormone (FSH), luteinising hormone (LH), oestrogen and progesterone.

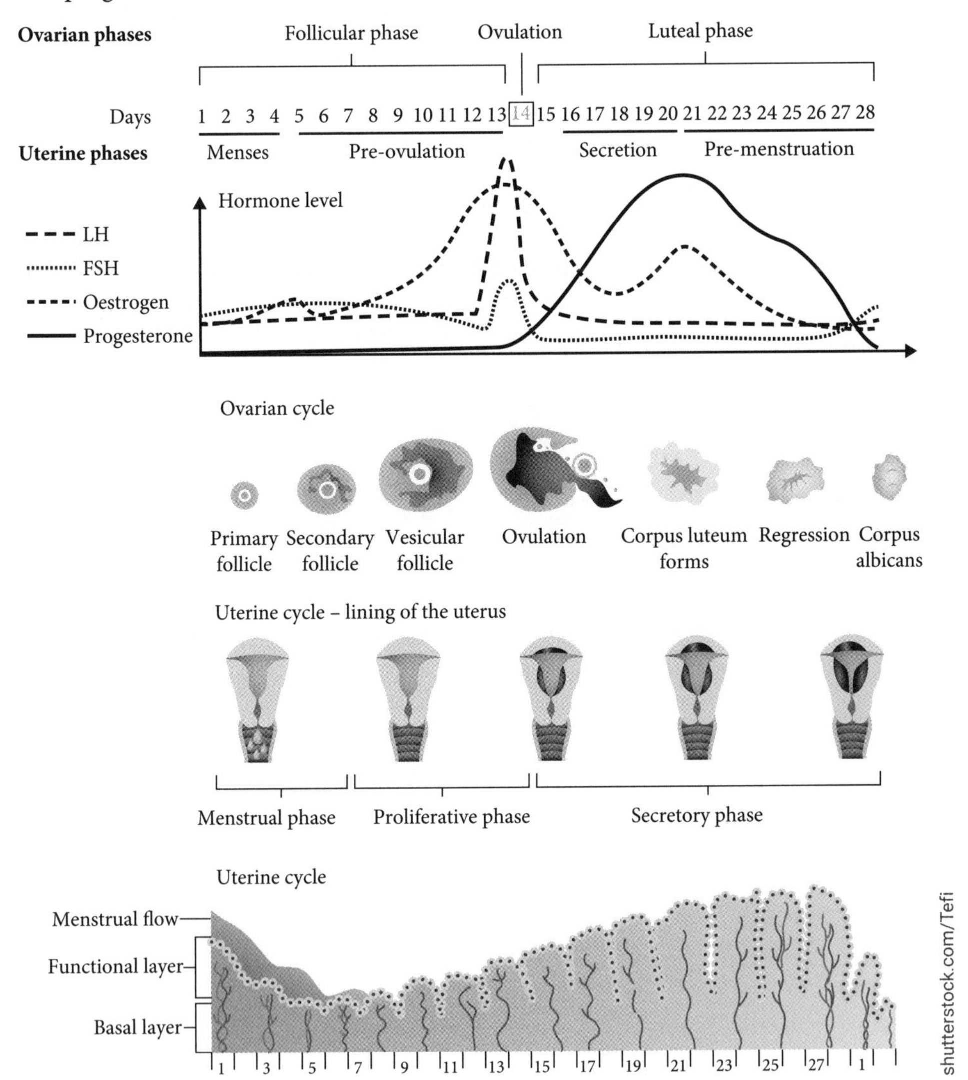

Based on the image, what is the most likely role of FSH and progesterone?

	FSH	Progesterone
A	Initiates release of egg from ovary	Initiates thickening of uterine wall
B	Inhibits the release of progesterone	Inhibits the release of oestradiol
C	Inhibits the release of LH	Initiates thickening of uterine wall
D	Initiates thickening of uterine wall	Initiates release of egg from ovary

Cell replication

Question 5

The diagram on the right represents a segment of DNA.

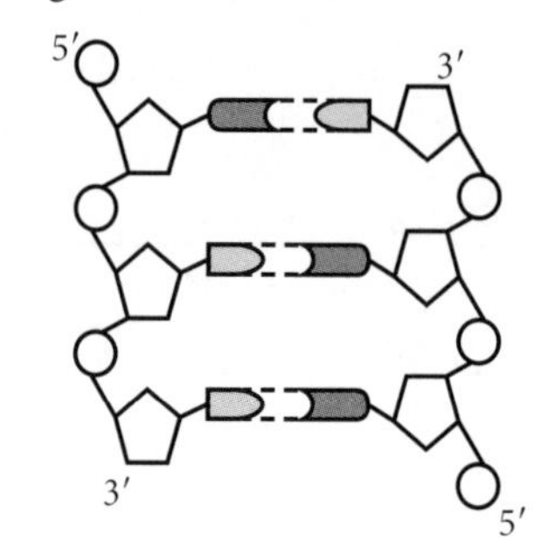

How many nucleotides are present in the segment of DNA?

A 2

B 3

C 6

D 12

Question 6

Analysis of a segment of DNA determined that guanine made up 22% of the total amount of nitrogenous bases in DNA. What percentage (%) of the nitrogenous bases in this segment are thymine?

A 22%

B 28%

C 44%

D 78%

DNA and polypeptide synthesis

Question 7

Which of the following is a distinguishing feature of DNA in prokaryotic cells compared to eukaryotic cells?

A Prokaryotic cells have a single linear chromosome.

B Prokaryotic cells have a single circular chromosome.

C DNA in prokaryotic cells is tightly wrapped around histone proteins.

D Prokaryotic cells lack chromosomes. The DNA is found only in their plasmids.

Question 8 ©NESA 2019 SI Q14 (ADAPTED)

The following DNA template strand codes for a sequence of four amino acids.

GAT ATC GAT CTA

Which of the following correctly represents the anticodons on the transfer RNA during synthesis of the polypeptide chain?

A CUA UAG CUA GAU

B GAU AUC GAU CUA

C GAT ATC GAT CTA

D CTA TAG CTA GAT

Use the following mRNA table to answer Questions 9 and 10.

Second base

First base	U	C	A	G	Third base
U	UUU, UUC: Phe UUA, UUG: Leu	UCU, UCC, UCA, UCG: Ser	UAU, UAC: Tyr UAA Stop UAG Stop	UGU, UGC: Cys UGA Stop UGG Trp	U C A G
C	CUU, CUC, CUA, CUG: Leu	CCU, CCC, CCA, CCG: Pro	CAU, CAC: His CAA, CAG: Gln	CGU, CGC, CGA, CGG: Arg	U C A G
A	AUU, AUC, AUA: Ile AUG Met/Start	ACU, ACC, ACA, ACG: Thr	AAU, AAC: Asn AAA, AAG: Lys	AGU, AGC: Ser AGA, AGG: Arg	U C A G
G	GUU, GUC, GUA, GUG: Val	GCU, GCC, GCA, GCG: Ala	GAU, GAC: Asp GAA, GAG: Glu	GGU, GGC, GGA, GGG: Gly	U C A G

Question 9

The base sequence in a section of mRNA is given below.

UAC UAC AAG CAU

Identify the amino acid coded for by the second codon.

A Methionine

B Threonine

C Tyrosine

D Stop

Question 10

The base sequence in a section of template DNA is given below.

GAT ACG ATG CAT

If a point mutation occurred where thymine (T) replaced guanine (G) at the third nitrogen base of the 2nd base triplet, what would be the consequence for the amino acid sequence?

A The sequence would stop.

B Threonine would replace cysteine.

C Tryptophan would replace cysteine.

D No effect, as it would produce the same amino acid.

Genetic variation

Question 11 ©NESA 2020 SI Q20

This chart illustrates three correlation patterns indicating the influence of genes and environment on different traits in individuals.

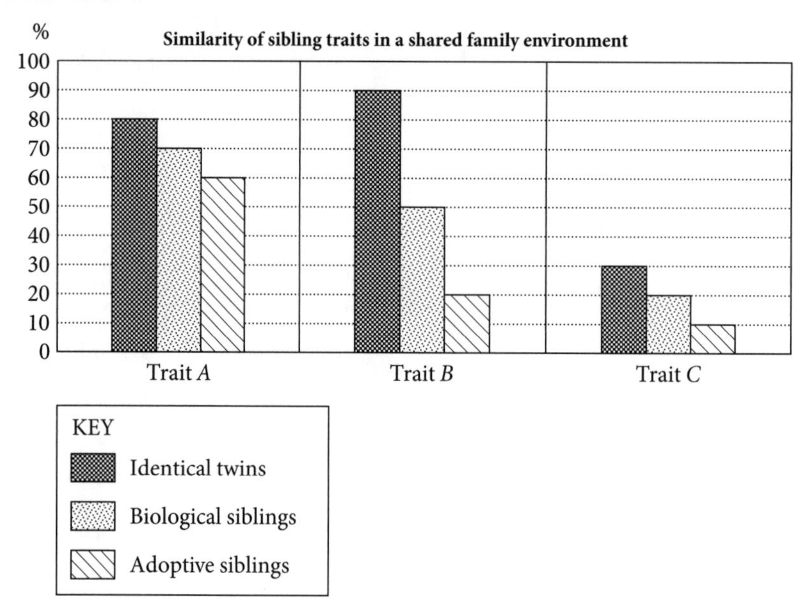

What does the data show about how genes and family environment affect the three traits?

	Trait A		Trait B		Trait C	
	Genes	**Family environment**	**Genes**	**Family environment**	**Genes**	**Family environment**
A	Low	High	High	Low	Low	Low
B	Low	High	High	Low	High	High
C	High	Low	Low	High	Low	Low
D	High	Low	Low	High	High	High

Question 12

The diagram at right shows two pairs of homologous chromosomes and the arrangement of the alleles of *four* genes on them.

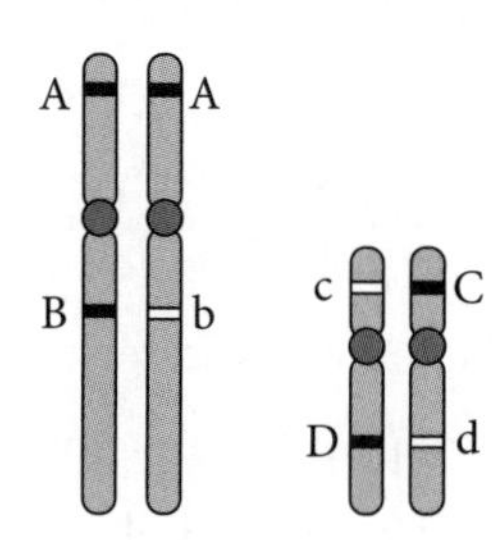

If crossing over does not occur, what are the possible genotypes of the resulting gametes?

A ABAb, cDCd

B ABcD, ABCd, AbcD, AbCd

C ABcd, ABCD, Abcd, AbCD

D AAcC, AADd, BbcC, BbDd

Question 13

The diagram shows the early stages of meiosis.

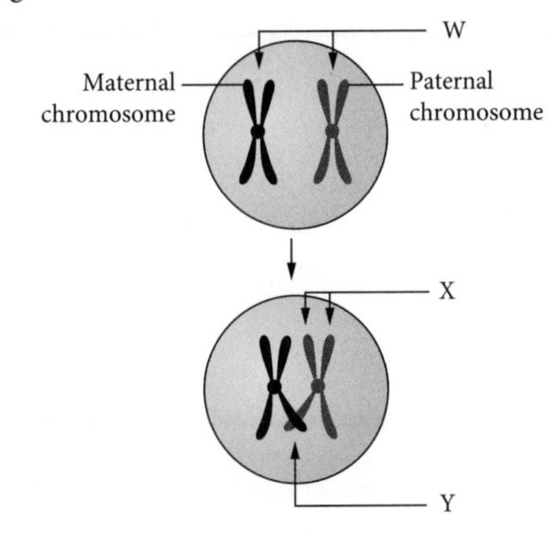

Which of the following options correctly identifies W, X and Y?

	W	X	Y
A	Homologous chromosomes	Sister chromatids	Independent assortment
B	Sister chromatids	Homologous chromosomes	Crossing over
C	Homologous chromosomes	Sister chromatids	Crossing over
D	Sister chromatids	Homologous chromosomes	Independent assortment

Question 14

Which of the following options introduces new alleles into a species?

A Mutation

B Fertilisation

C Crossing over

D Polypeptide synthesis

Question 15

The pedigree shows the inheritance of a disorder.

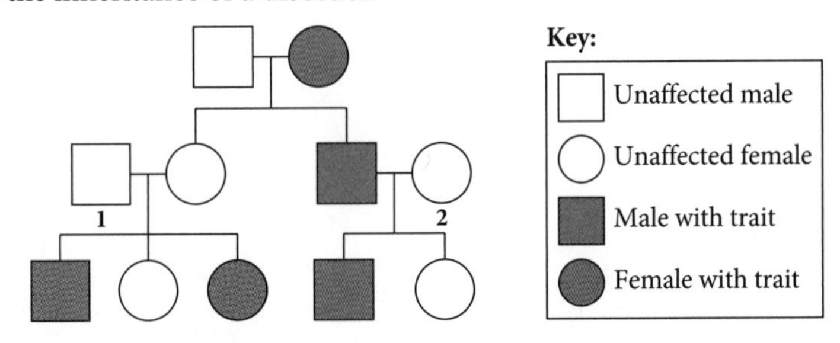

Which row of the table shows the genotypes of individuals 1 and 2?

	Individual 1	Individual 2
A	Tt	TT or Tt
B	Tt	TT
C	Tt	Tt
D	X^TY	X^TX^t

Question 16

In humans, if an individual who was homozygous for blood group A (I^AI^A) had a child with someone who was homozygous for blood group B (I^BI^B), the child would express both gene products on the surface of their red blood cells.

An inheritance study involving 1000 couples was performed. All couples consisted of one homozygous group A and a heterozygote.

Which graph below best represents the distribution of phenotypes you would expect from this cross?

A

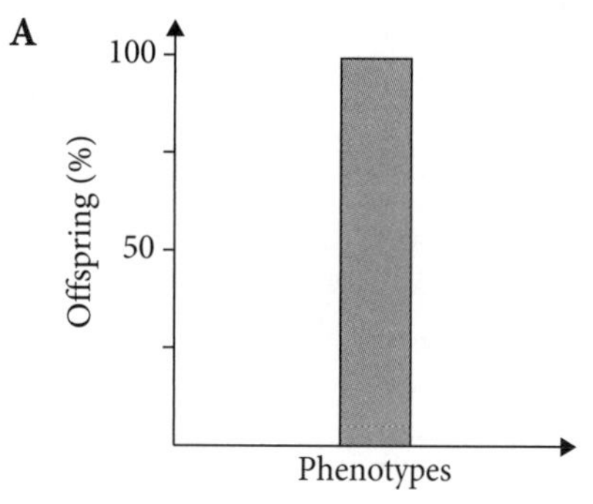

B

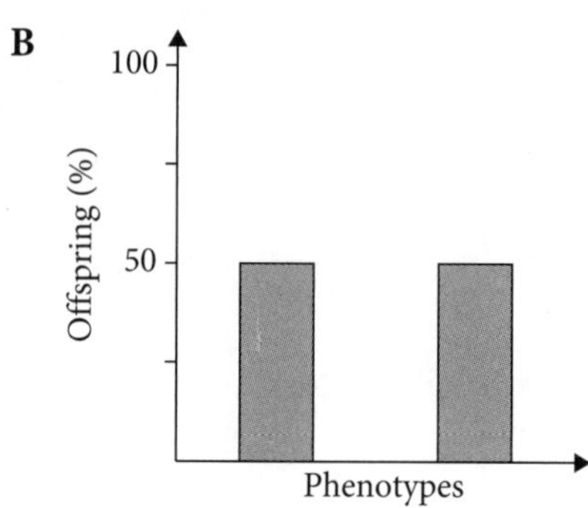

C

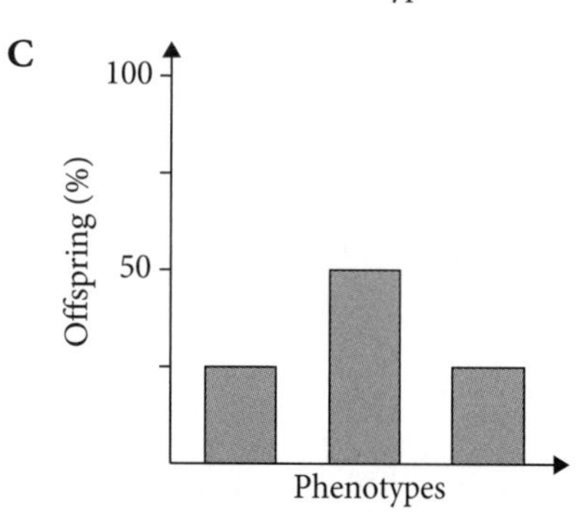

D

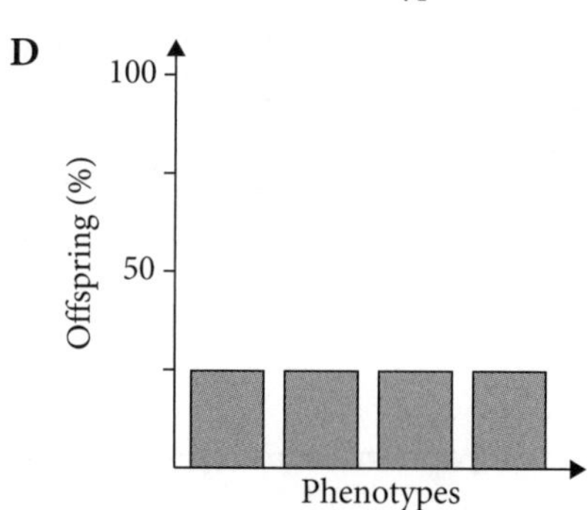

Question 17

In a population of 200 individuals, it is known that 36% are homozygous dominant (TT) and 44% are heterozygous (Tt). What are the respective allele frequencies for T and t?

	T (%)	t (%)
A	36	64
B	64	36
C	80	20
D	58	42

Question 18 ©NESA 2021 SI Q10

Cystic fibrosis is an autosomal recessive disorder caused by mutations in the *CFTR* gene. Many different recessive alleles cause cystic fibrosis.

The four most common alleles of the *CFTR* gene and their frequencies in the Australian population are shown in the table.

Allele	Frequency of allele
A	98.33
a1	1.13
a2	0.08
a3	0.07

What will be the most common genotype of cystic fibrosis patients in Australia?

A a1/a1

B a1/a2

C A/a1

D A/A

Inheritance patterns in a population

Question 19

A genome-wide association study explored the link between a single nucleotide polymorphism (SNP) and a disease. The original nucleotide base was guanine (G), which was replaced by thymine (T) when the mutation originally occurred. The SNP frequency was compared between 1000 individuals with the disease (study group) or without the disease (control group).

The following data was presented.

Comparison of SNP between control group and study group

Percentage of individuals (%)
100
75
50
25
49%
51%
27%
73%
G T
Control group
G T
Study group

Which conclusion is most appropriate?

A All individuals who have the SNP will develop the disease.

B Having the SNP increases the likelihood of developing the disease.

C If an individual does not have the SNP, they will not develop the disease.

D Having the SNP does not increase the likelihood of developing the disease.

Question 20 

Cystic fibrosis is an autosomal recessive disease caused by a three-nucleotide deletion.

A PCR and gel electrophoresis were performed on the relevant gene sequence of two parents who do not have the disease, to determine the likelihood of their child inheriting the disease. The image shows the DNA profile of the two prospective parents.

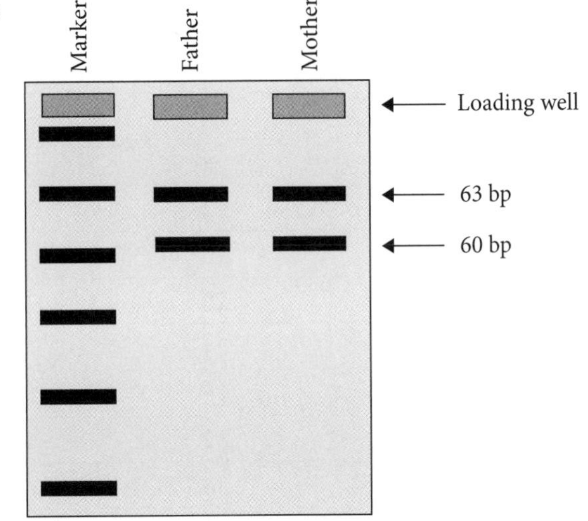

What is the likelihood of one of their children having cystic fibrosis?

A 0% **B** 25%

C 50% **D** 100%

Question 21

The diagram shows the DNA profiles from a crime scene where the victim was assaulted by one of four people.

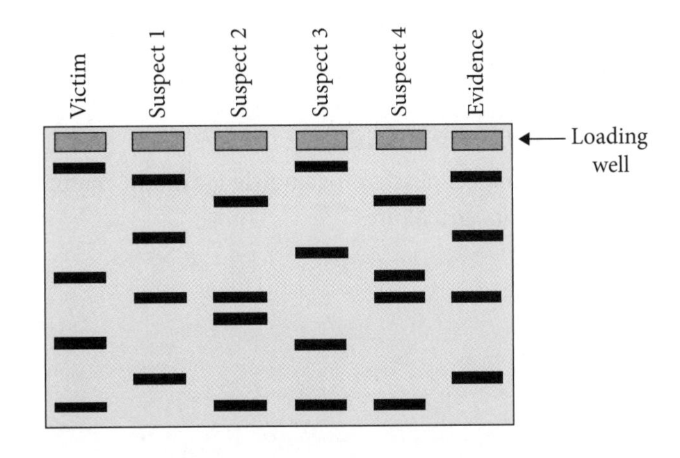

Identify which suspect is most likely to have committed the crime.

A Suspect 1

B Suspect 2

C Suspect 3

D Suspect 4

Short-answer questions

Solutions start on page 247.

Reproduction

Question 22 (5 marks)

Bacteria reproduce by binary fission.

a What type of reproduction is binary fission? 1 mark

b Outline a procedure that could be used to test the effect of temperature on the reproduction of bacteria. 4 marks

Question 23 (9 marks) ©NESA 2019 ADDITIONAL SAMPLE EXAM MOD 5 Q11 (ADAPTED)

a Complete the table below, indicating the location where each of the listed processes occur within the female reproductive tract. 3 marks

Process	Location
Site of egg development	
Site of fertilisation	
Site of implantation of blastocyst	

b The following data shows the average amount of human chorionic gonadotropin (hCG) produced by pregnant women.

Weeks of pregnancy	hCG (ng/mL)
0	0
4	85
8	185
12	185
16	80
20	65
24	60
28	65
32	75
36	65
40	35

Use the information provided to graph the levels of hCG in a normal pregnancy. 3 marks

c Describe the role and changes in levels of one hormone involved in controlling pregnancy. 3 marks

Question 24 (6 marks)

Evaluate the impact of scientific knowledge on the manipulation of plant and animal reproduction in agriculture.

Cell replication

Question 25 (11 marks) ©NESA 2012 SII Q29 (ADAPTED)

a Provide two features that distinguish mitosis from meiosis. 2 marks

b Assess the effect of mitosis and meiosis on the continuity of species. 4 marks

c Complete the following diagram to show the process by which gametes are formed. 3 marks

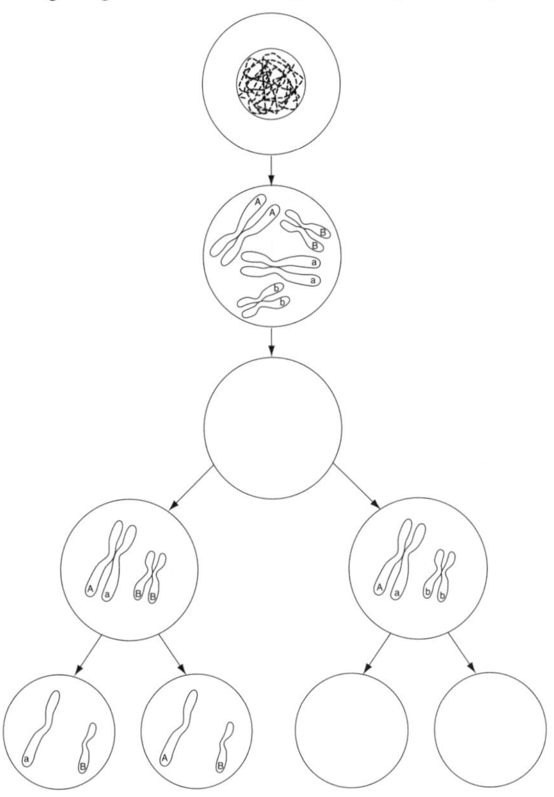

d How does the segregation of chromosomes during meiosis lead to a wide variety of gametes being produced? 2 marks

Question 26 (4 marks)

Draw a labelled flow chart to outline the steps involved in DNA replication.

DNA and polypeptide synthesis

Question 27 (3 marks)

Construct a table that compares DNA in prokaryotic and eukaryotic cells.

Question 28 (10 marks)

Insulin is a small protein hormone released from the beta cell of the pancreas in response to rising blood glucose levels. It is coded for by the *INS* gene located on chromosome 11. Insulin works by binding to receptors on muscle and fat cells to increase their ability to absorb glucose from the blood.

a Describe how the *INS* gene controls the production of insulin. 6 marks

b With reference to the *INS* gene, assess the importance of correct gene expression. 4 marks

Question 29 (4 marks)

With reference to a named example, assess how genes and environment affect phenotypic expression.

Genetic variation

Question 30 (4 marks)

The diagram shows chromosomes from a diploid cell ($2n = 6$) with the genotype for five genes.

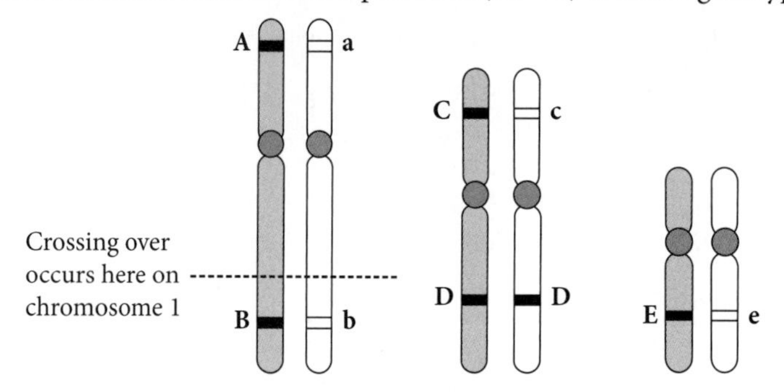

a How many chromosomes will be present in daughter cells after meiosis? 1 mark

b What is the genotype of the parent cell? 1 mark

c Write four possible genotypes of daughter cells if crossing over occurs at the indicated location on chromosome 1. 2 marks

Question 31 (7 marks) 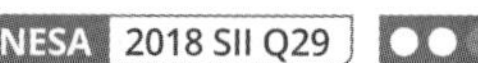

The diagram models the process of meiosis.

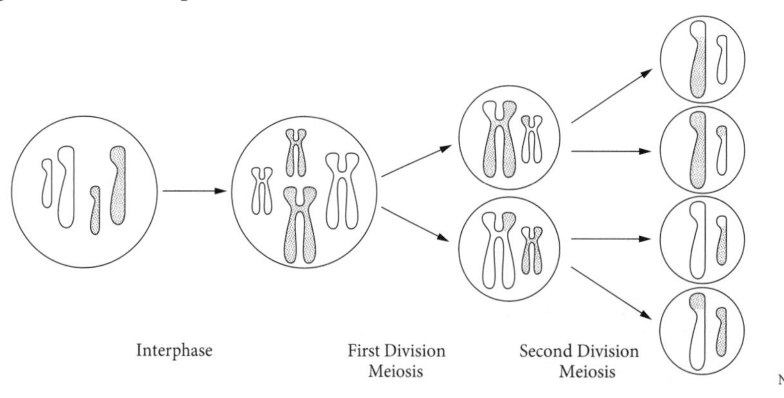

a Describe the process that accounts for the changes shown in the model during interphase. 2 marks

b Explain the structure and behaviour of chromosomes in the first division of meiosis. Include detailed reference to the model. 5 marks

Question 32 (6 marks)

Andrew's wife Angela has a history of red–green colour blindness in her family. Kai, their two-year-old son, may be red–green colour blind. Two of Angela's brothers, Vincent and Milton, are colour blind but her other brother, Paul, is not. Angela's mother, Mary, is a carrier of red–green colour blindness. Her father, Joe, is unaffected.

a Construct a family pedigree to show the inheritance of this sex-linked genetic disorder. 4 marks

b Predict whether Kai will be colour blind. Justify your answer. 2 marks

Question 33 (5 marks) ©NESA 2019 SII Q30

Experiments were conducted to obtain data on the traits 'seed shape' in plants and 'feather colour' in chickens. In each case, the original parents were pure breeding and produced the first generation (F_1). The frequency data diagrams below relate to the second generation offspring (F_2), produced when the F_1 generations were bred together.

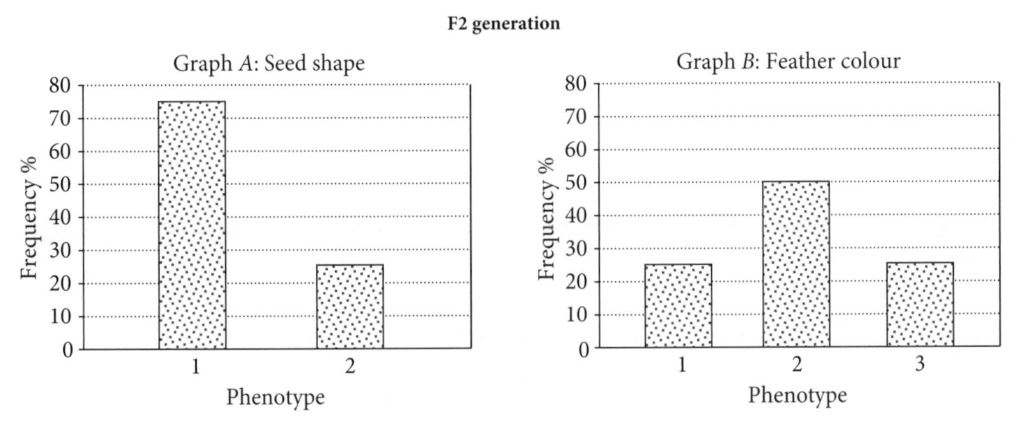

Explain the phenotypic ratios of the F_2 generation in both the plant and chicken breeding experiments. Include Punnett squares and a key to support your answer.

Inheritance patterns in a population

Question 34 (2 marks)

What is a single nucleotide polymorphism?

Question 35 (9 marks)

a Distinguish between genetic sequencing and DNA profiling. 2 marks

b What is the name given to DNA sequences located in the introns and known to contain repeating nucleotide sequences, typically 2–5 base pairs (bp) long? 1 mark

c Explain how technologies can be used to determine inheritance pattens within a population. 6 marks

Question 36 (7 marks)

With reference to a large-scale collaborative population genetics project, explain how the use of data analysis has led to the identification of trends, patterns and relationships in a specific field.

CHAPTER 2
MODULE 6: GENETIC CHANGE

2

Chapter 2
Module 6: Genetic change

Module summary

Module 6 covered types of genetic mutations, their causes and their effects on an organism. The influence of mutations, meiosis and fertilisation on the gene pool of populations was also examined. Through observation, trial and error and, more recently, sophisticated genetic technologies, we have harnessed biological processes to improve human health and food production, and to find solutions to environmental issues. This is known as biotechnology. This module also addressed the uses and implications of biotechnology for medicine, agriculture, industry and the environment, and how different viewpoints in our society influence the development and use of a range of technologies.

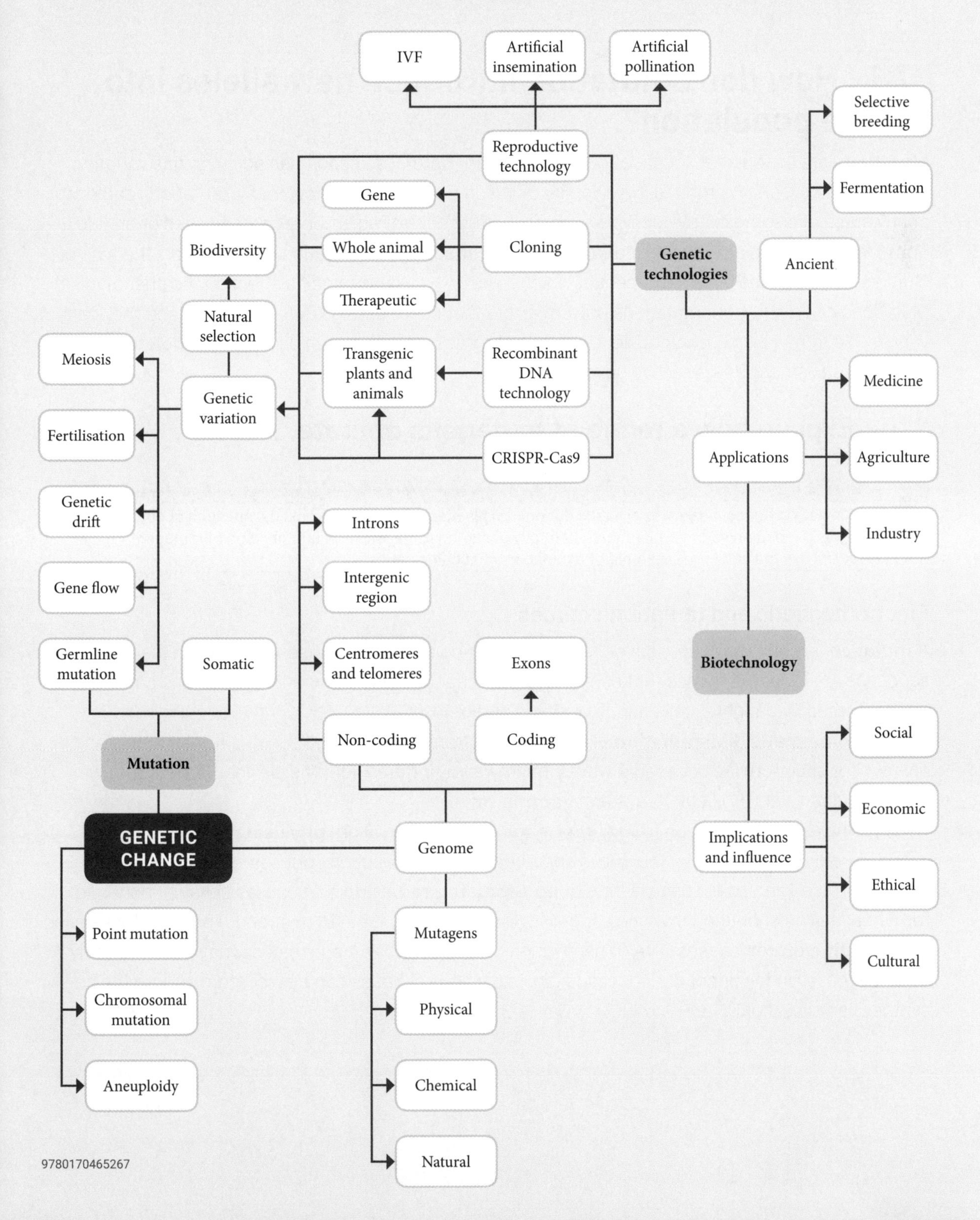

Outcomes

On completing this module, you should be able to:

- explain natural genetic change and the use of genetic technologies to induce genetic change.

Key science skills

The key science skills demonstrated in this module require you to:

- solve scientific problems using primary and secondary data, critical thinking skills and scientific processes
- communicate scientific understanding using suitable language and terminology for a specific audience or purpose.

2.1 How does mutation introduce new alleles into a population?

Mutations are the original source of genetic variation. Natural selection has acted on this variation over billions of years to produce the diversity of life on Earth that we know today. Mutations can occur at any stage of life and in any cell type, such as the first cells of an embryo or the tissues of an adult. Some mutations can also be passed from parent to offspring. The location of a mutation, the type of change, and the stage of development at which it occurs will determine its effect on the phenotype. Mutations can result from mutagens or from errors that occur during **DNA** replication, and they can vary in size from a single **nucleotide** to a whole chromosome. They can be harmless, **deleterious** or provide organisms with a selective advantage.

2.1.1 Explain how a range of mutagens operate

Note

This is an 'including but not limited to' syllabus dot point. This means you may be asked a question about electromagnetic radiation sources, chemicals, naturally occurring mutagens and another type of mutagen. *Particle radiation* is the additional example of a mutagen you could use.

Electromagnetic and radiation sources

A **mutagen** is a physical, chemical or naturally occurring agent that causes a change in the DNA sequence. Cells have enzymes that can repair these changes, but uncorrected mutations can kill cells or be inherited by daughter cells during **meiosis** and mitosis.

Electromagnetic (EM) radiation exists as a spectrum of wavelengths with different amounts of energy. The spectrum includes radio waves, microwaves, infrared light, visible light (the colours we can see), ultraviolet (UV) light, X-rays and **gamma rays**.

Three types of EM radiation (UV light, X-rays and gamma rays) are **physical mutagens**. Most of the EM radiation emitted by the Sun is absorbed by the atmosphere, but some UV light reaches Earth's surface. Ultraviolet light is a type of **non-ionising radiation** that causes covalent bonds to form between neighbouring thymines and/or cytosines on the same DNA strand. This structure, called a **pyrimidine dimer**, causes DNA to bend or kink. As a result, DNA polymerase may not accurately 'read' the bases in the dimer during replication, and incorrect bases can be incorporated into the newly synthesised strand of DNA.

Note

Most HSC questions are pitched at the 'explain' level. To prepare for the exam, practise using conjunctions. For example, 'Electromagnetic radiation knocks electrons out of atoms in DNA. *As a result*, this breaks the chemical bonds between bases', or 'Mutagens change the DNA sequence. *This leads* to genetic diseases.'

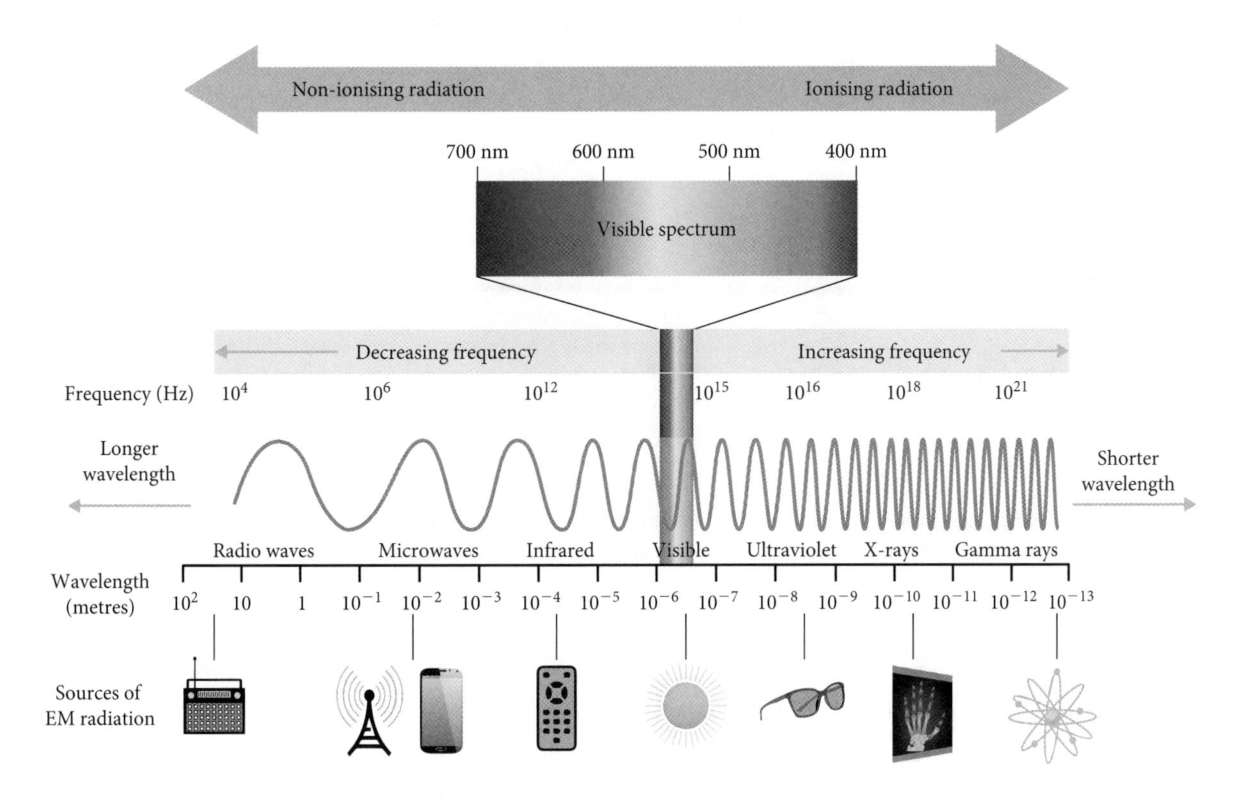

FIGURE 2.1 The electromagnetic radiation spectrum

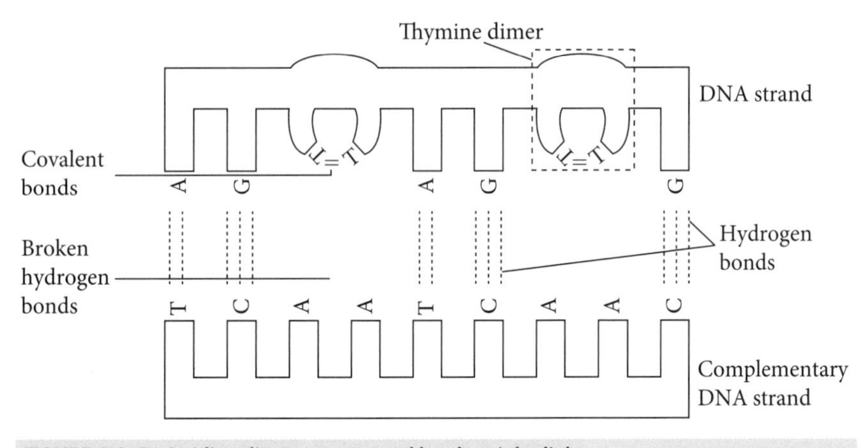

FIGURE 2.2 Pyrimidine dimers are caused by ultraviolet light.

X-rays and gamma rays are **ionising radiation**. When these waves enter a cell and hit DNA, their high level of energy is absorbed, which knocks electrons out of atoms and breaks the chemical bonds within DNA. Exposure to X-rays is generally restricted to medical procedures. The limited exposure and low doses used means the likelihood of mutation is low.

Gamma rays are produced from the radioactive decay of certain radioisotopes used in some medical tests and as fuel in nuclear power plants. It is only on rare occasions, such as a nuclear reactor accident, that humans are exposed to high doses of gamma rays. **Particle radiation** (e.g. alpha and beta particles) is another type of physical mutagen that is produced with gamma rays in the process of radioactive decay.

Ionising and particle radiation can break one or both strands of DNA, which can result in the loss of large regions of a chromosome if the breaks are not repaired by the cell.

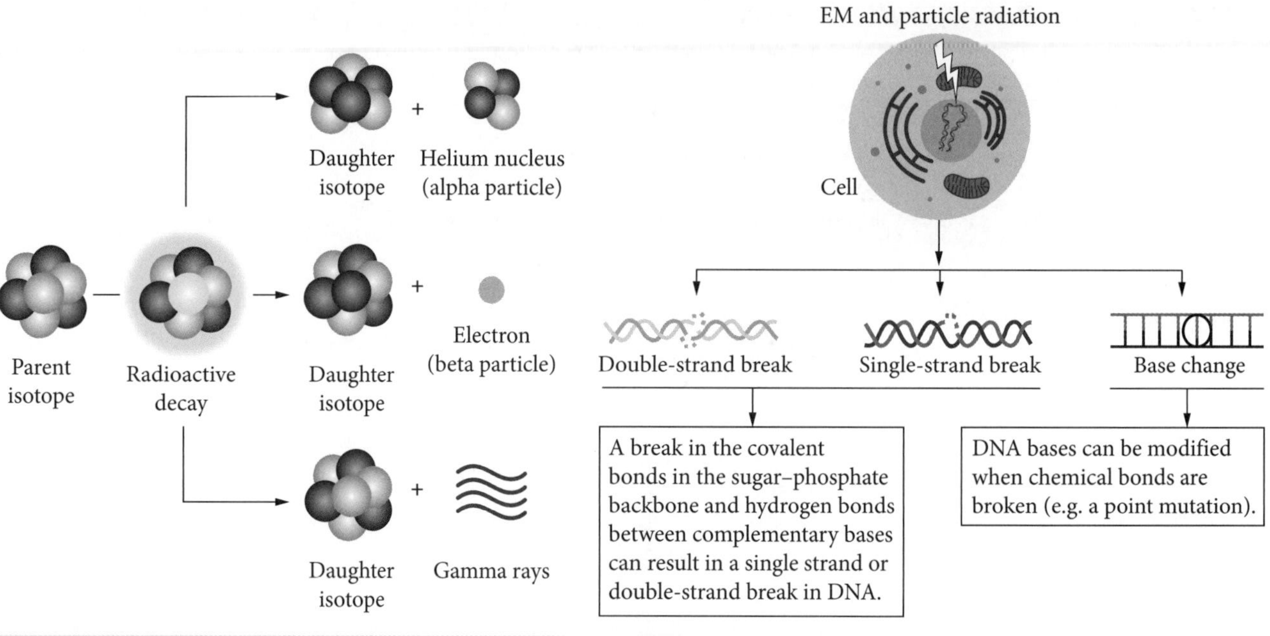

FIGURE 2.3 Gamma rays and particle radiation are physical mutagens emitted during radioactive decay.

FIGURE 2.4 Types of DNA damage caused by electromagnetic and particle radiation

Chemicals

Chemical mutagens are compounds that cause mutations by altering the chemical structure of DNA bases. They are classified based on how they interact with DNA to cause mutations. The three main types of chemical mutagens are modifiers, **base analogues** and **intercalating agents**.

Base modifiers change the chemical structure of a base (e.g. alcohol, nitrates in processed meats such as hot dogs, and nitric oxide). **DNA polymerase** then incorporates a base that is complementary to the modified base during mitosis or meiosis.

A base analogue is a molecule that resembles or mimics a base that is incorporated into DNA (e.g. some chemotherapy drugs and 5-bromouracil). In the next cell division, DNA polymerase incorrectly matches a base with the analogue.

An intercalating agent is a molecule that inserts between adjacent bases on the same DNA strand (e.g. some chemicals in cigarettes and acridine). This causes the DNA to bend or kink. DNA polymerase then deletes or adds an extra base during mitosis or meiosis.

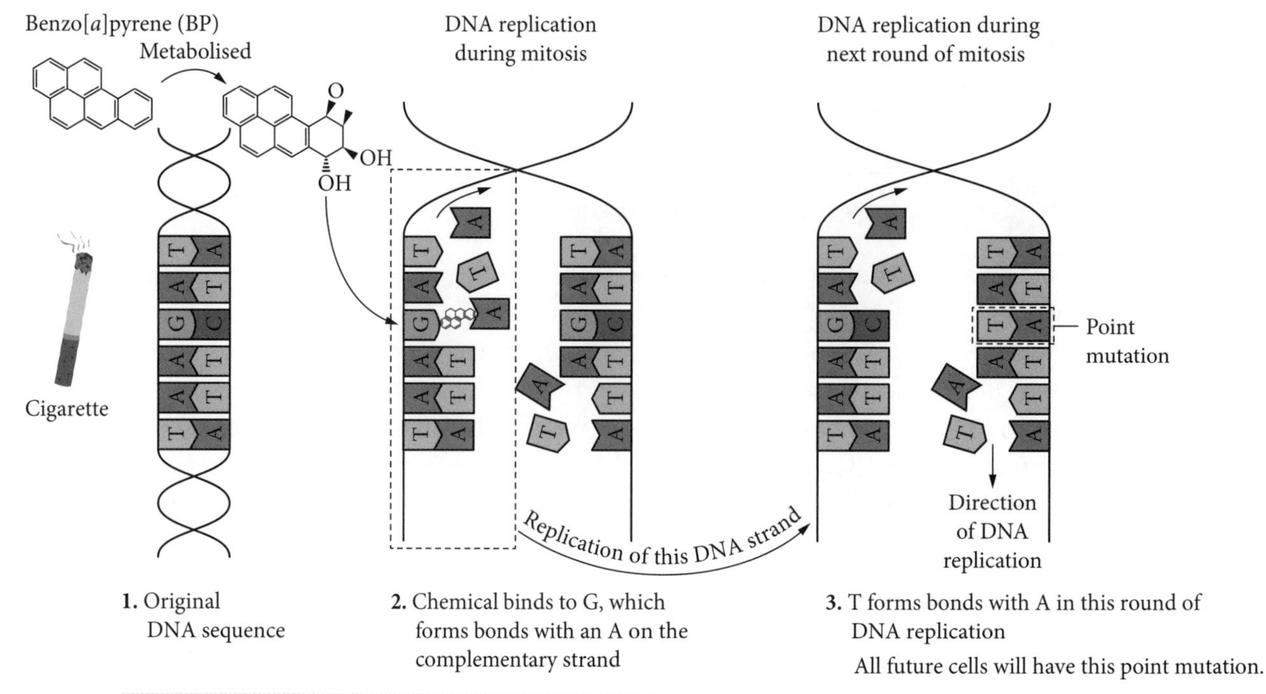

FIGURE 2.5 An example of a chemical mutagen in cigarette smoke causing a point mutation

Naturally occurring mutagens

Naturally occurring mutagens include **transposons**, **viruses**, chemical compounds produced by other organisms, and **reactive oxygen species**. Each of these operate as mutagens in different ways.

Transposons are DNA sequences that are able to move and insert themselves at a new location in the genome, which can disrupt genes. For example, the variegated colouring of maize cobs is caused by transposable elements that insert themselves near a gene that encodes for kernel colour.

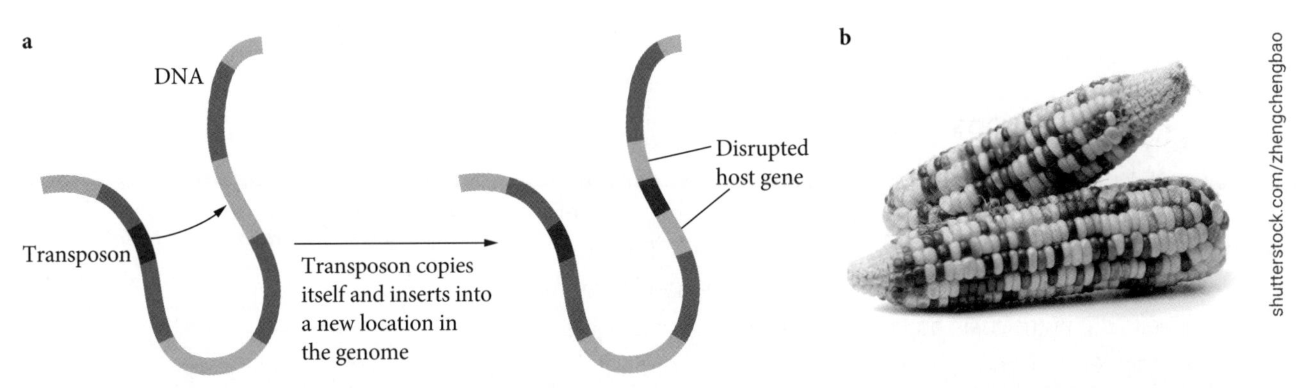

FIGURE 2.6 a Transposable elements cause mutations by changing their location in the genome. b Different-coloured kernels in maize caused by transposons

Retroviruses (RNA viruses) can directly cause mutations in the cell of an infected host when they insert into DNA. After entering a cell, these viruses convert their RNA genome into DNA, in a process called **reverse transcription**. This copied DNA inserts into the host's genome, changes the DNA sequence and can inactivate a gene. Other viruses such as the human papillomavirus (HPV), which causes cervical cancer, can cause indirect mutations. Proteins expressed by tumour suppressor genes (e.g. p53) in our cells normally help to repair mutations. They can be inactivated when proteins produced by HPV bind to them, resulting in mutations at sites across the genome.

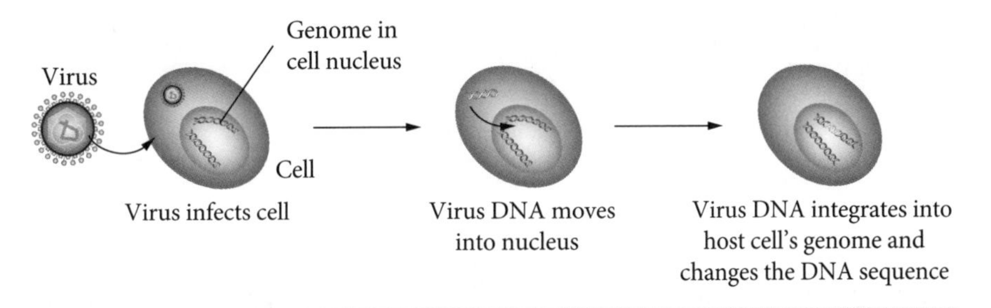

FIGURE 2.7 How viruses can cause mutations

Mycotoxins are chemicals produced by some fungi, such as *Aspergillus* and *Penicillium*, which can contaminate food. When they are ingested, mycotoxins enter cells, where they are metabolised (broken down), and the chemical products bind to DNA to cause point mutations. For example, the breakdown product of aflatoxin B1, a type of mycotoxin, causes guanine to mutate to thymine.

Finally, reactive oxygen species are naturally occurring chemicals that are produced during cellular respiration. These molecules contain oxygen and are highly unstable (e.g. superoxide and hydroxyl), which means they react easily with other molecules, including DNA. They can break the chemical bonds between bases and cause point mutations.

2.1.2 Comparing the causes, processes and effects of different types of mutation, including but not limited to point and chromosomal mutations

Note

To 'compare', you need to identify the similarities and differences between the different types of mutations. While the different types of point and chromosomal mutations are not explicitly mentioned, a knowledge of them will address the 'including but not limited to' component of the syllabus dot point.

Point mutations

Cells have a variety of safety checkpoints to ensure that genetic mutations are detected and repaired before cell division. Occasionally mutations are not repaired and these errors are copied into the genome of new daughter cells during meiosis and mitosis. **Point mutations** are changes in a single nucleotide and can be broadly classified as **transitions** and **transversions**.

All point mutations are similar, as only a single nucleotide is changed. However, they differ in the way they affect the amino acid sequence of the resulting polypeptide.

Point mutations that involve the substitution of a single nucleotide are classified as **silent**, **missense** or **nonsense mutations**.

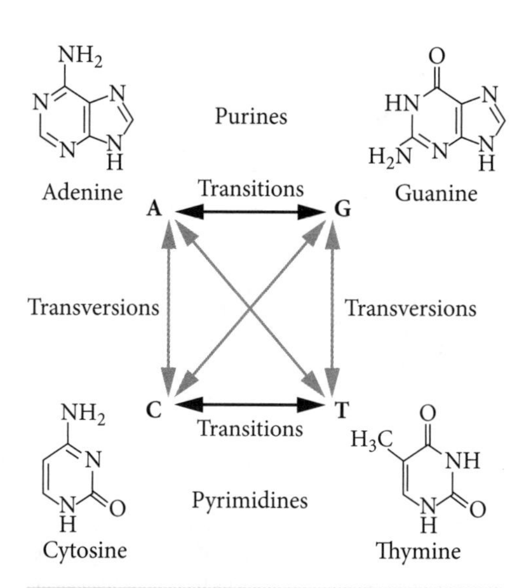

FIGURE 2.8 Transitions and transversions

Single nucleotides can also be inserted into or deleted from DNA. When this occurs in coding DNA, it results in a **frameshift mutation**, which leads to a complete change in the translated polypeptide sequence from the site of the mutation onwards.

TABLE 2.1 Causes, processes and effects of different types of point mutations

Type of point mutation	Causes	Processes	Effects
Silent	• Physical, chemical or naturally occurring causes • DNA polymerase inserting the incorrect base into a new DNA strand during DNA replication	• Single nucleotide substitution that does not change the encoded amino acid	• No change in the polypeptide sequence • No functional effects and evolutionarily neutral
Missense		• Single nucleotide substitution that results in a change in a single amino acid	The resulting polypeptide may: • have reduced or increased activity • not form the correct tertiary or quaternary structure • be non-functional
Nonsense		• Single nucleotide substitution that changes the amino acid to a stop codon	• Translation is terminated prematurely, producing a shortened polypeptide • The polypeptide does not function or is degraded by the cell
Frameshift		• Insertion or deletion of a nucleotide, or a number of nucleotides not divisible by three, into the sequence of an **exon**	• The entire sequence following the mutation will not be read correctly • Stop codons are often introduced at a point in the sequence after the mutation site

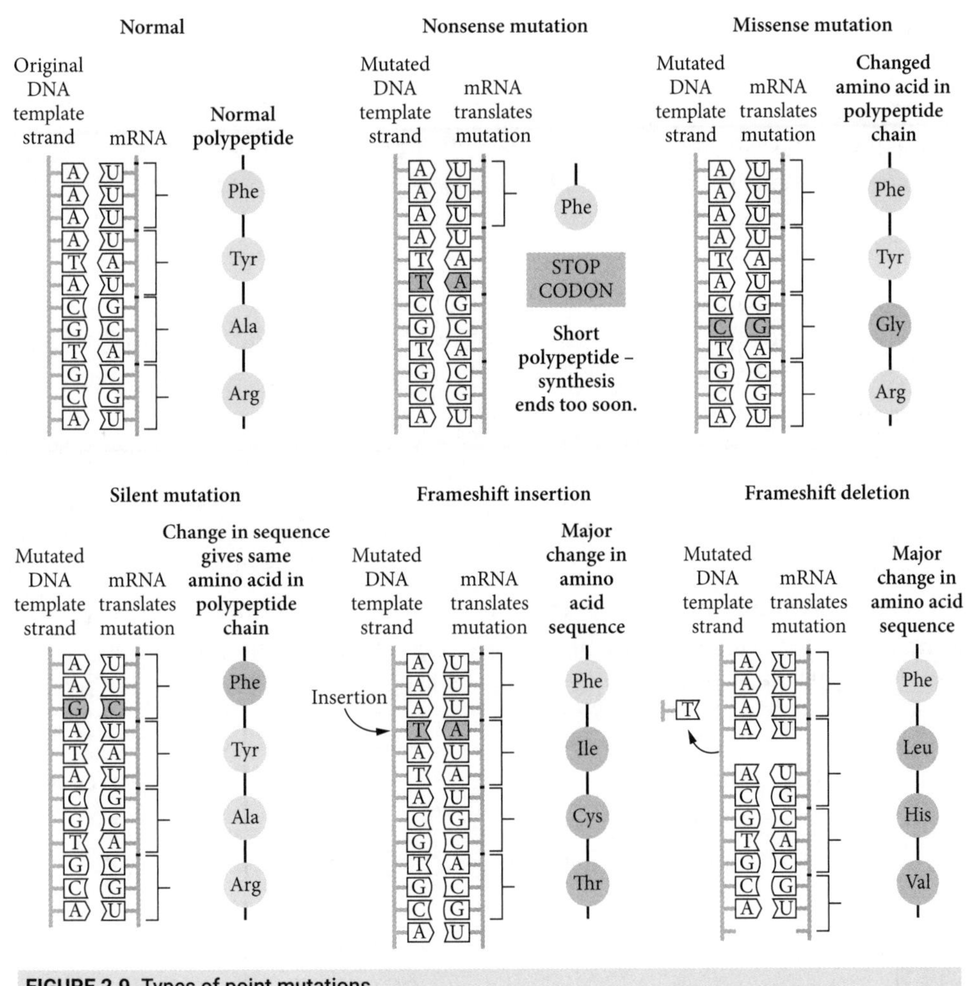

FIGURE 2.9 Types of point mutations

Chromosomal mutations

Chromosomal mutations affect the structure or inheritance of a chromosome and usually involve more than one gene. This is why they are sometimes called 'block' mutations. The main types of chromosomal mutation are translocations, inversions, duplications, insertions and deletions. They are similar in that they all affect the level of gene expression. They differ in the way they affect chromosomal structure.

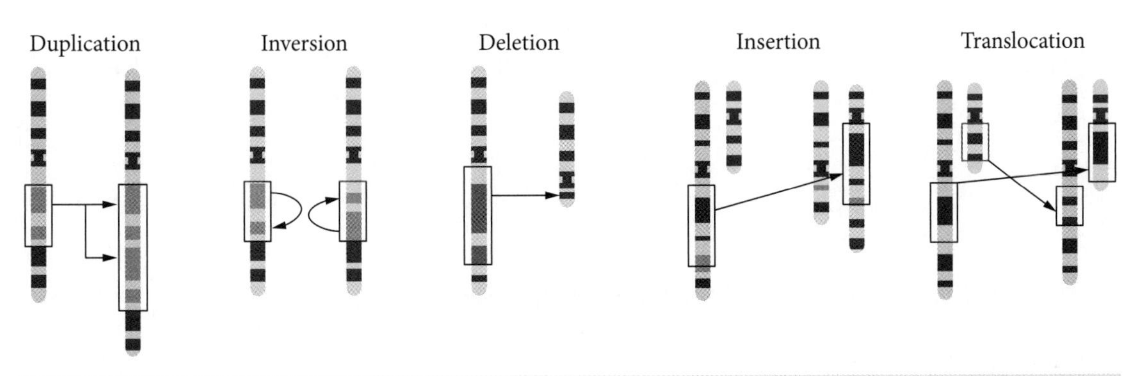

FIGURE 2.10 Types of chromosomal mutations

Translocation means to move something from one place to another. In genetics, a translocation is when one or more chromosomes break, and the chromosomal fragments re-join with a different chromosome or to a new site on the same chromosome. Translocations can disrupt gene expression by splitting genes, separating **regulatory elements** from a gene or fusing genes together. The Philadelphia chromosome is a famous example of a translocation that creates a new hybrid **oncogene**, *BCR-ABL*, which causes a type of blood cancer called chronic myeloid leukaemia.

An **inversion** refers to a region of a chromosome that breaks at two ends, rotates 180° and reinserts in the reverse orientation on the same chromosome. Inversions can involve a few nucleotides, parts of a gene, a whole gene or multiple genes. Break points within a gene reverse the normal order of the exons, which will result in a non-functional protein. For example, the sex-linked recessive disorder, haemophilia A, is commonly caused by the inversion of **introns** 1 to 22 of the *F8* gene on the X chromosome.

Chromosomal mutations also include **duplications, insertions** and **deletions**. A duplication occurs when one or more genes, or a chromosomal region, are copied into the genome. An insertion occurs when a large section of one chromosome is moved and incorporated into a different chromosome or into a different region of the same chromosome. Finally, a deletion, also called a **partial monosomy**, occurs when a large section of a chromosome is lost. Each of these mutations can result in the gain or loss of large numbers of genes, which can disrupt the delicate balance of gene products inside a cell.

Aneuploidy is different from the other chromosomal mutations, because the sequence doesn't change but the chromosome number does. The gain or loss of a whole chromosome results from the process of **chromosome non-disjunction** in which sister **chromatids** or **homologous chromosomes** fail to move to opposite poles of the daughter cells during cell division. This can occur during mitosis, meiosis I or meiosis II, so aneuploidy can be observed in somatic cells and gametes.

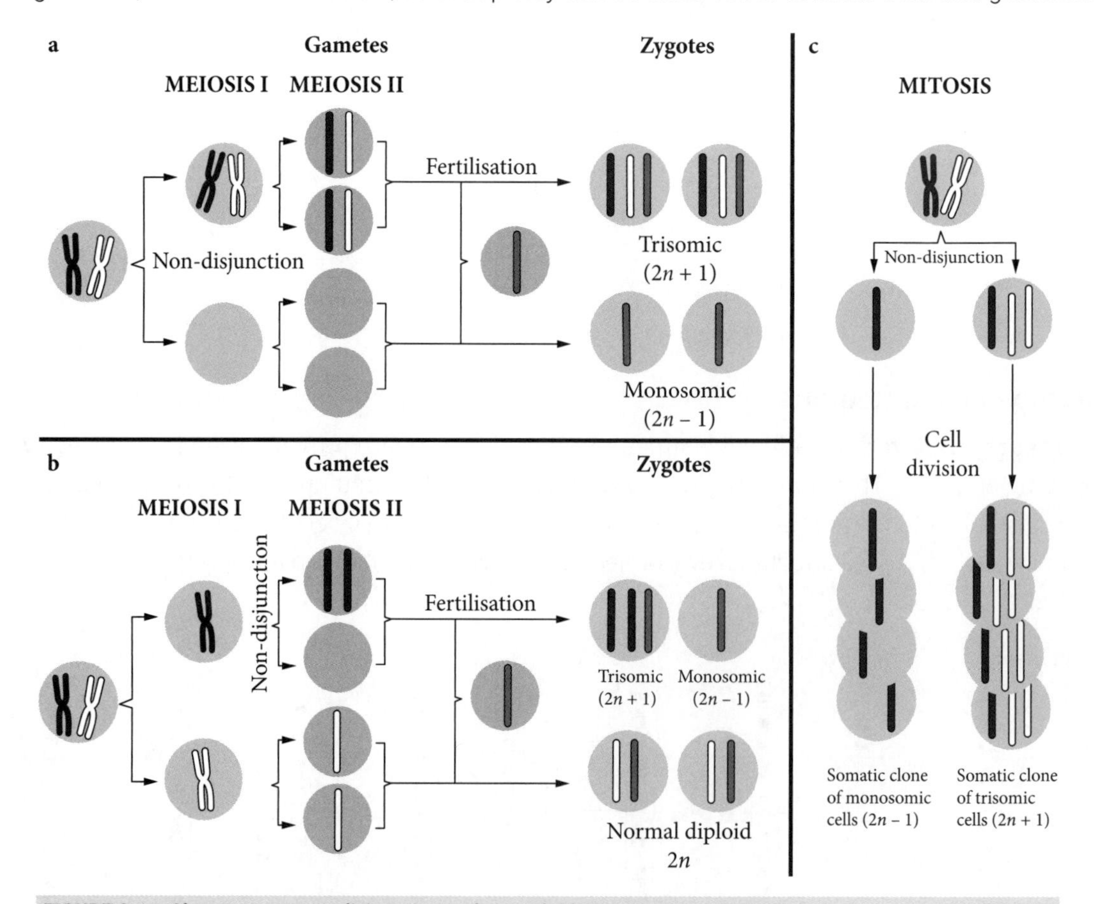

FIGURE 2.11 Chromosome non-disjunction and aneuploidy in meiosis and mitosis

Aneuploidy can occur in both autosomal and sex chromosomes. If inherited as a **germ-line mutation**, it is usually lethal or results in intellectual and physical disabilities. If aneuploidy occurs during mitosis, it causes somatic mosaicism (see section 2.1.3). The most common form of aneuploidy in humans is Down syndrome, which involves the gain of an extra chromosome 21 (trisomy 21). Klinefelter syndrome is the most common sex-linked aneuploidy and involves the gain of an extra X chromosome in males. Monosomies (loss of a chromosome) are rare and typically fatal, except for Turner syndrome.

TABLE 2.2 Causes, processes and effects of aneuploidy

Condition	Causes	Processes	Effects
Down syndrome	Usually caused by segregation of the incorrect number of chromosomes in gametes. Can also occur in somatic cells.	Additional copy of chromosome 21 (trisomy 21) segregates in a daughter cell	A $2n = 47$ genotype. Cognitive impairment and a range of other disabilities
Klinefelter syndrome		Additional copy of the X chromosome segregates in a daughter cell	A $2n = 47$ genotype (XXY) in males. Reduced sexual development and fertility but a normal life span; can be treated with hormone therapy
Turner syndrome		X chromosome fails to segregate into a daughter cell	A $2n = 45$ genotype (X) in females. Short stature, infertility, no menstruation, congenital heart defects, hearing and spatial awareness problems; can be treated with hormone replacement therapy

2.1.3 Distinguish between somatic mutations and germ-line mutations and their effect on an organism

Note
To 'distinguish', you need to identify the differences between somatic and germ-line mutations.

Somatic mutations

Sexually reproducing organisms have two main types of cells: somatic cells and gametes (sperm or egg). A **somatic cell** is any cell in the body other than the **gametes**. The somatic cells of animals are **diploid**, which means they have two copies of each **autosome** and two sex chromosomes. Mutations in somatic cells only affect the individual in which they occur, and are not passed on to the next generation.

In theory, every somatic cell in an organism has an identical genotype. However, if a mutation occurs during mitosis it results in **somatic mosaicism** (two populations of cells with distinct genotypes).

Note
Think of mosaicism as being like a mosaic tile artwork. The tiles are the cells, and the different colours are the genotypes.

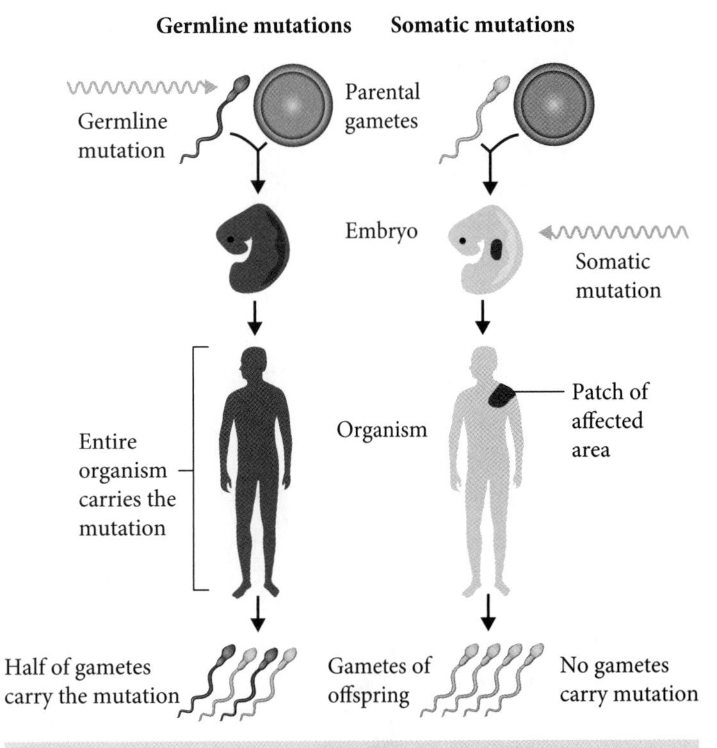

FIGURE 2.12 The origin of germ-line and somatic mutations

The effect of the mutation depends on how early in development it occurs. If a mutation occurs early in embryonic development, the mutation will be present in a higher proportion of cells than if the mutation occurs in adulthood, when it may only affect a few cells. About 1% of Down syndrome cases occur in a somatic mosaic form. They are caused by non-disjunction of chromosome 21 in a somatic cell after **fertilisation**. This produces a mixture of cells in the body, with some having two copies of chromosome 21 ($2n = 46$) and some having three copies of chromosome 21 ($2n = 47$).

9780170465267

Cancer is another example of somatic mosaicism. Most cancers are caused by mutations in somatic cells as people get older. The mutation results in two populations of cells in the body: healthy cells with a normal genotype, and cancer cells with the mutation. The effects of cancer on an organism are covered in more detail in Module 8 (section 4.2.1).

Germ-line mutations

Gametes are the reproductive cells of the body (sperm in males and eggs in females). They have a **haploid** (*n*) set of chromosomes. If a mutation occurs in a gamete, it is called a germ-line mutation. Germ-line mutations can be passed on to the next generation and will be present in the offspring. The effect of a germ-line mutation on the body depends on the gene affected and the pattern of inheritance.

> **Note**
> If an offspring inherits a germ-line mutation, the mutation will be present in every cell of its body.

> **Note**
> Refer to Module 5 for examples of different patterns of inheritance. Module 8 provides examples of germ-line mutations and their effects on the phenotype.

2.1.4 Assessing the significance of 'coding' and 'non-coding' DNA segments in the process of mutation

> **Note**
> To 'assess', you need to use your knowledge of coding and non-coding DNA to make a value judgement about the effect a mutation will have on the function of a cell or the health of an organism.

The human genome is three billion base pairs long and consists of about 25 000 genes. The basic structure of a gene consists of a **promoter**, where **transcription factors** and DNA polymerase bind to start transcription, followed by alternating exons and introns and a transcription termination site. **Coding DNA** only makes up about 1% of the entire human genome. Mutations in coding DNA have the most significant impact on the phenotype of an organism as they directly affect protein structure (e.g. enzymes and hormones), which is essential to their function.

> **Note**
> Coding DNA refers to the exons in genes. Any section of DNA that is not translated into a polypeptide is classified as non-coding.

So, what is the rest of the genome? The rest of the genome is **non-coding DNA** as it is not translated into polypeptides. It includes introns, intergenic regions, telomeres and centromeres. Mutations in non-coding DNA can also affect gene expression and cellular processes. Their significance is **assessed** below.

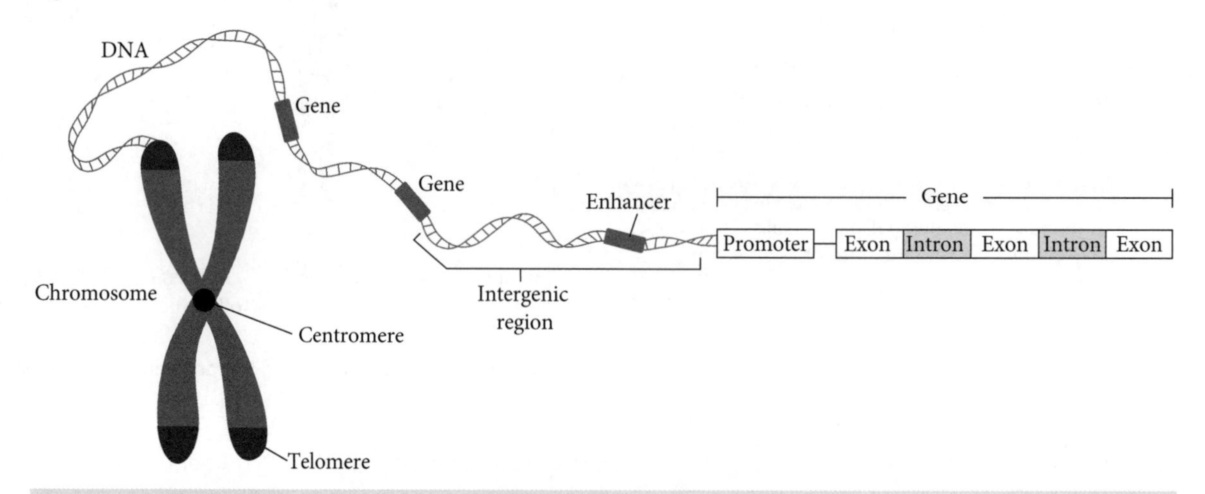

FIGURE 2.13 Coding and non-coding regions of the genome

Introns

Introns are located inside the genes of eukaryotic organisms. They alternate with exons in an exon–intron–exon–intron (etc.) pattern in the gene. Introns are transcribed into pre-mRNA but are then spliced out when mRNA is formed prior to translation. A mutation within an intron will have no impact on the final polypeptide, unless it occurs in a **splice site**. Splice sites are the sequences at the junctions between introns and exons. They are recognised by the spliceosome, a complex of enzymes that cuts introns out of pre-mRNA and re-joins the exons. A mutation may cause the spliceosome to skip the splice site and remove an exon, which will result in an incorrect mRNA sequence and polypeptide.

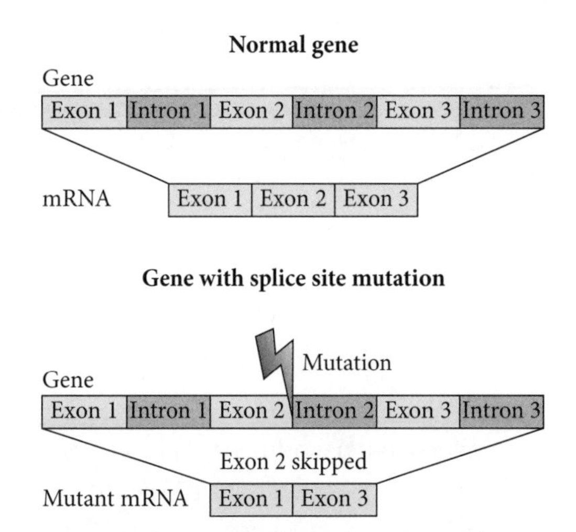

FIGURE 2.14 The effect of a splice site mutation on an mRNA sequence

Intergenic regions

Intergenic regions, or 'gene deserts', are the long regions of DNA in between genes. They make up most of the genome. Before whole-genome sequencing existed, intergenic DNA was thought to have no function and was referred to as 'junk' DNA. However, we now know that intergenic regions (as well as introns) contain regulatory elements such as **enhancers** that 'talk' to genes and tell them when to be switched on and how much gene product to make. Mutations in enhancers affect the binding of transcription factors, which can increase or decrease the amount of gene product. For example, a point mutation that changed a C to a T in a non-coding enhancer of the *LCT* gene resulted in lactase persistence and the ability to digest lactose in dairy products in adult humans (covered in Module 5, Section 1.4.3).

Telomeres and centromeres

Telomeres are located on the tips of chromosomes in eukaryotic organisms. They consist of repeated stretches of short pieces of non-coding DNA. Human telomeres range in size from 2 to 50 kilobases and consist of approximately 300–8000 repeats of the sequence CCCTAA. They act as protective 'caps' that prevent the deletion of genes on the chromosomes. Over the human lifetime, these repeats are gradually deleted, which is one of the main reasons we age.

Centromeres are the specialised regions of a repetitive non-coding DNA that separate the short and long arm of a chromosome. They join sister chromatids together when a chromosome is copied during DNA replication. During cell division, they also act as an anchor point for the spindle fibres to pull sister chromatids apart into the daughter cells. It is essential that mutations do not occur in centromeres, otherwise chromosomes may break during cell division.

2.1.5 Investigating the causes of genetic variation relating to the processes of fertilisation, meiosis and mutation

Earlier in this module you compared types of mutation, and in Module 5 you looked at the processes in meiosis and the different mechanisms of sexual reproduction. This section investigates how these processes (mutation, meiosis and fertilisation) cause genetic variation. Genetic variation is the different combination of alleles between individuals in a population or species as a whole.

Mutations

Note
Genetic variation is only concerned with genetic information that is inherited from one generation to the next. Somatic mutations are not counted as genetic variation.

Point and chromosomal mutations in germ-line cells create new alleles that may be passed on to offspring. Mutations occur independently in individuals within a population, and over time this creates a mixture of new alleles. Germ-line mutations are the original source of genetic variation in populations, and natural selection acting on these mutations has created the diversity of life on Earth.

Meiosis

Three key processes cause genetic variation in meiosis.

- Genetic recombination (crossing over) – new combinations of alleles are created when genetic material is exchanged between non-sister chromatids of homologous chromosomes.
- Random segregation of chromosomes – a parent randomly gives one allele for each gene to its offspring.
- Independent assortment of chromosomes – the genotype of each gamete has an equal probability of occurring, because the allele a gamete receives for one gene has no influence on the alleles it receives for other genes.

The combined effect of these three processes is that every gamete has a unique combination of alleles.

Fertilisation

Finally, fertilisation brings together male and female gametes that have undergone the three processes discussed above. The mixing of haploid sets of chromosomes from two parents is an additional event that creates new allele combinations.

2.1.6 Evaluating the effect of mutation, gene flow and genetic drift on the gene pool of populations

The **gene pool** is the total amount of genetic variation in a population, measured by allele and genotype frequencies. The gene pool can be changed by genetic mutation, gene flow, genetic drift and natural selection. Evolution occurs when changes in the gene pool happen over a long period of time.

Note
To 'evaluate', you need to make a value judgement based on 'for and against' criteria. Think of advantages, disadvantages and point of view.

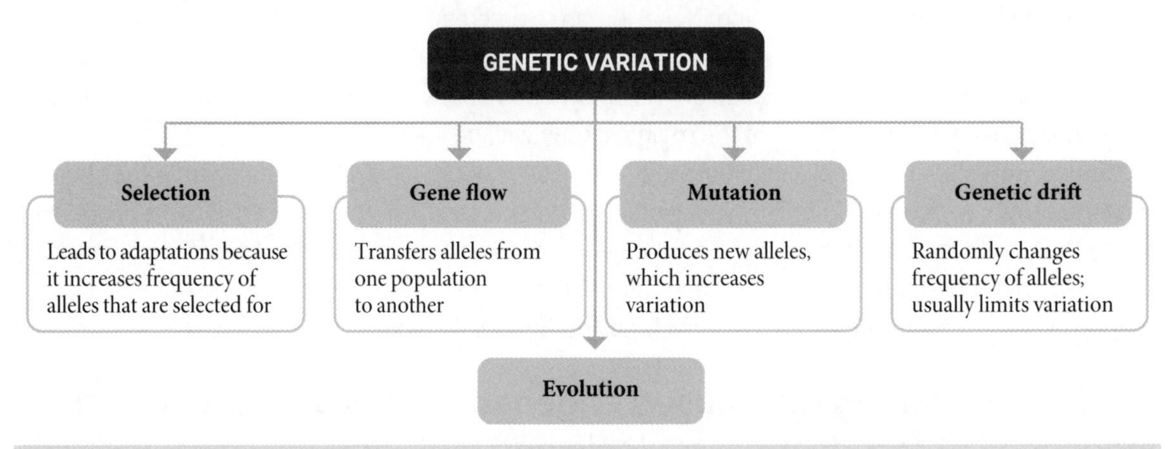

FIGURE 2.15 The effect of different processes on genetic variation

Note
While you don't need to remember the following equations, they could be a stimulus in an exam question about genotype frequencies.

Before evaluating these factors, let's examine how geneticists calculate genotype and allele frequencies in a population. The sum of the allele frequencies for two alleles at a given locus must be 1 (assuming there are only two alleles).

$$p + q = 1$$

where p is the frequency of one allele (T) and q is the frequency of the recessive allele (t).

From p and q, the genotype and allele frequencies can be determined using the Hardy–Weinberg equation:

$$p^2 + 2pq + q^2 = 1$$

where p^2 is the frequency of one homozygous dominant genotype (TT), q^2 is the frequency of the homozygous recessive genotype (tt), and $2pq$ is the frequency of the heterozygous genotype (Tt).

The **Hardy–Weinberg equilibrium** principle states that the allele and genotype frequencies in a population will remain constant from one generation to the next if mutation, gene flow, genetic drift and natural selection are not influencing the gene pool. You can use the Hardy–Weinberg equation to measure genotype and allele frequencies between generations. If there is a change over generations, this indicates that one or more of these factors is affecting the gene pool, which may be a sign of evolution in action.

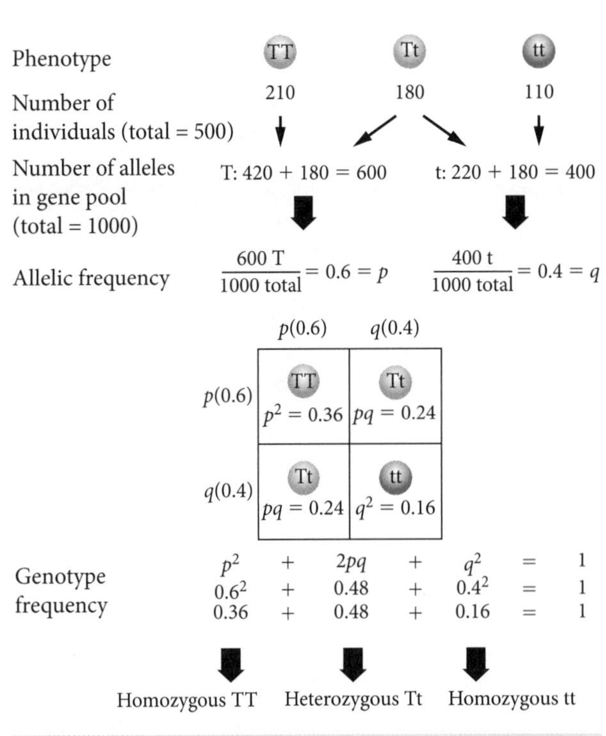

FIGURE 2.16 Calculating allele and genotype frequencies using the Hardy–Weinberg equation

Mutations

The chance of a new mutation occurring in a gamete and being passed on to offspring is very low, and new mutations do not have an immediate or significant effect on allele frequency in a population. For example, in humans the mutation rate at a locus per generation is estimated to be one change per 10 000–1 000 000 gametes. Over long periods of time, however, the frequency of a new mutation can increase or decrease in a population through the process of natural selection. This is illustrated by the deer mice case study.

FUR COLOUR IN DEER MICE

CASE STUDY

The *agouti* gene is responsible for variation in fur colour in many mammals. The deer mouse (*Peromyscus maniculatus*) lives in central USA. Ancestral populations of these mice had dark fur, which provided camouflage against the dark soils in their habitat. Approximately 10 000 years ago, a glacier deposited light-coloured sand in the region, which created two adjacent habitats – one with dark soil and one with light, sandy soil. Today, mice in the sandy habitat have lighter-coloured fur, the result of a germ-line mutation in the *agouti* gene. The light-coloured phenotype is extremely rare in mice living on dark soil and vice versa. It is believed that natural selection resulted in this distribution, with the respective fur colour providing camouflage and protection from predators such as birds.

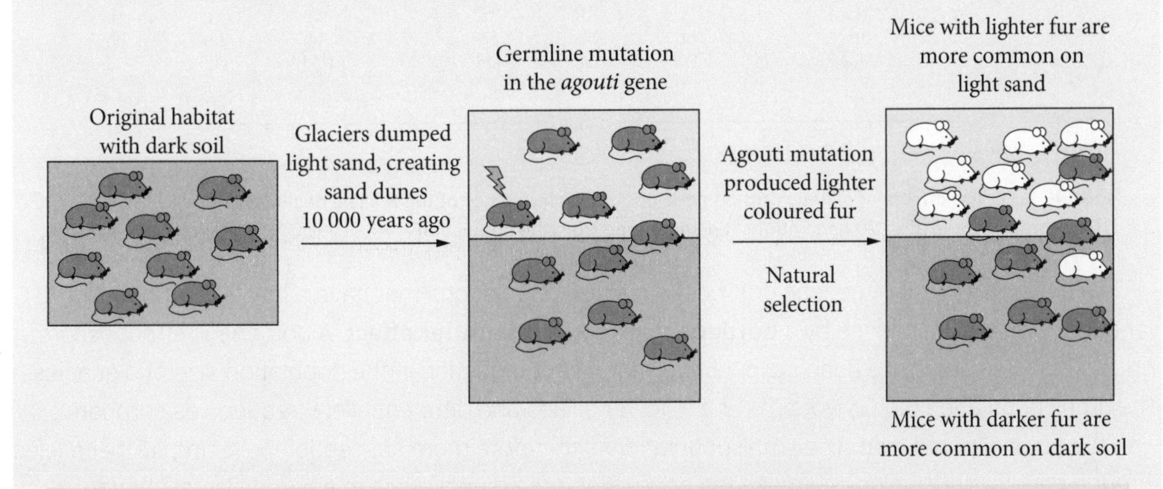

FIGURE 2.17 Natural selection in the deer mouse after a mutation in the *agouti* gene

Gene flow

Gene flow is the movement of genes into and out of a population when organisms migrate between areas. In the case of plants (which can't move), this involves the movement of pollen, seeds or spores by wind or pollinators. The mixing of individuals within a species may introduce new alleles to the population; or the proportion of an existing allele may change if multiple individuals move and reproduce. This is a common way for allele frequency to change, and it plays an important role in changing variation in the gene pool. Many endangered species, such as the cheetah, exist in fragmented populations, which prevents gene flow, reduces genetic variation and places them at greater risk of extinction.

RR Bb DD
RR Bb DD
Rr Bb DD
rr BB DD
RR Bb DD
Rr BB DD
Rr bb DD
Rr Bb DD
RR bb DD
rr bb DD
RR Bb DD
Population 1 before immigration
Rr Bb dd
RR Bb DD
RR Bb DD
Rr Bb DD
Rr Bb dd
rr BB DD
RR Bb DD
Rr BB DD
Rr bb DD
Rr Bb DD
RR bb DD
rr bb DD
RR Bb DD
Population 1 after immigration of new narrow allele
Alleles: R = round, r = square, B = grey, b = white, D = broad, d = narrow

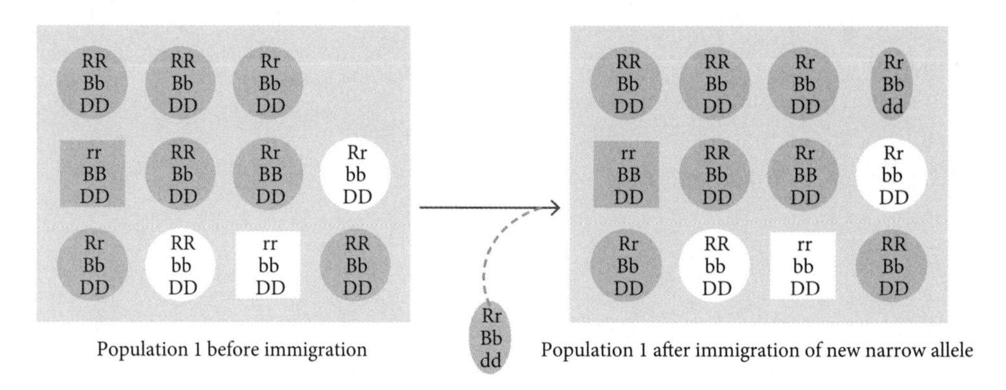

FIGURE 2.18 A migrating individual introduced the d allele into the population.

Genetic drift

Genetic drift is the random fluctuation in allele frequencies in a population over time. It is the result of chance events that influence the survival and reproduction of an organism. These fluctuations occur in all populations, but the effects of genetic drift are more pronounced in smaller populations. Over many generations, an allele may be lost from a population. When only one allele remains at a locus, geneticists say the allele has become 'fixed' – that is, it has a frequency of 1.

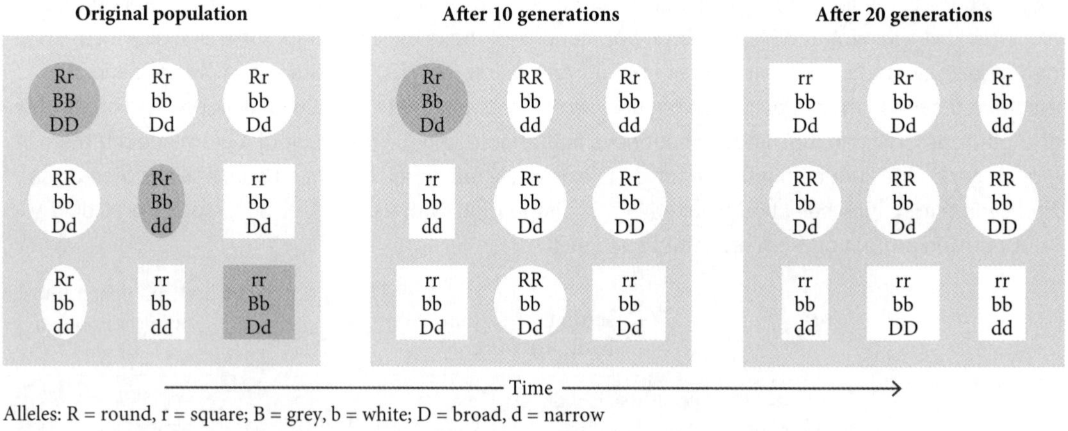

Original population
After 10 generations
After 20 generations
Rr BB DD
Rr bb Dd
Rr bb Dd
RR bb Dd
Rr Bb dd
rr bb Dd
Rr bb dd
rr bb dd
rr Bb Dd
Rr Bb Dd
RR bb dd
Rr bb dd
rr bb dd
Rr bb Dd
Rr bb DD
rr bb Dd
RR bb Dd
rr bb Dd
rr bb Dd
Rr bb Dd
Rr bb dd
RR bb Dd
RR bb Dd
RR bb DD
rr bb dd
rr bb DD
rr bb dd
Time
Alleles: R = round, r = square; B = grey, b = white; D = broad, d = narrow

FIGURE 2.19 The effects of genetic drift on gene pools. The frequency of the B allele declined after 10 generations, and after 20 generations it disappeared from the population.

The extreme effects of genetic drift occur in small populations that have a low level of genetic variation. This can be caused by a **bottleneck effect** or a **founder effect**. A bottleneck effect occurs when a disaster, such as a bushfire, causes a significant reduction in the population size of a species. The small population of survivors may have fewer alleles and different allele frequencies compared with the original population. The small population is therefore more susceptible to losing further alleles through genetic drift, which can make the population more susceptible to environmental change.

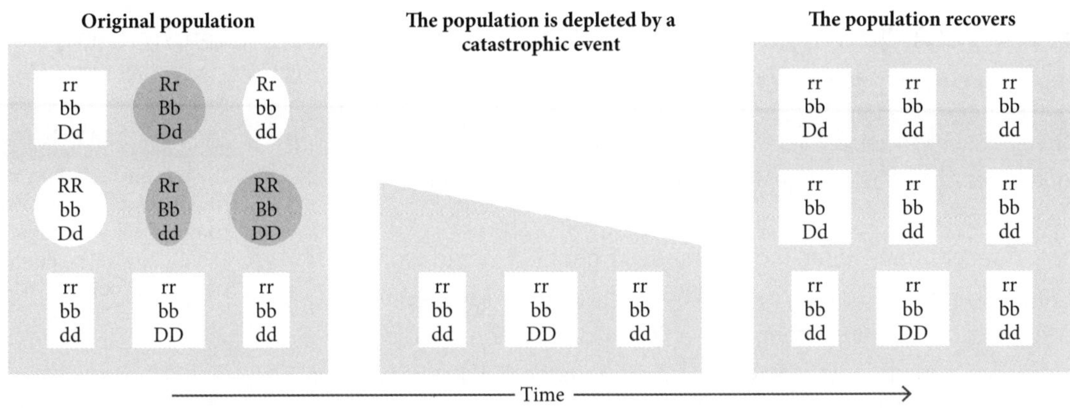

FIGURE 2.20 The bottleneck effect. The R and B alleles were lost from the population after the catastrophic event.

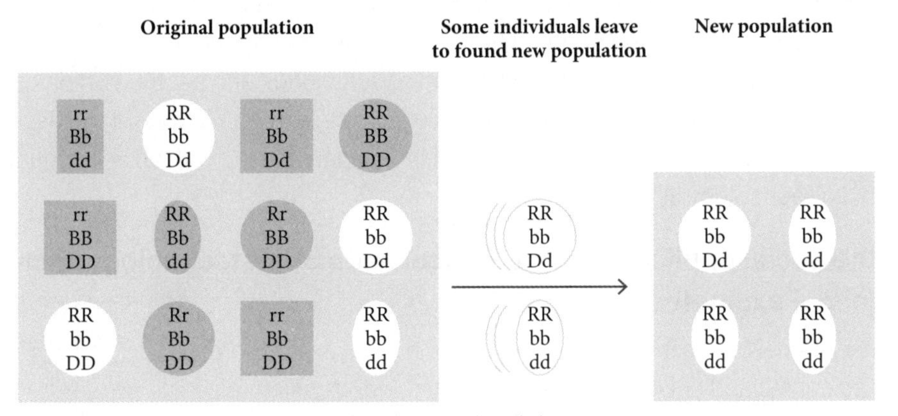

FIGURE 2.21 The founder effect. The r and B alleles are absent in the new population.

A founder effect occurs when a small group of individuals separates from a larger population to establish a new colony. An example is when a small number of individuals fly or swim to an island and become isolated from the original population on the mainland (e.g. birds flying to a newly formed volcanic island). The small size of the new colony means it will experience strong genetic drift.

In summary, genetic drift can have a strong effect on a gene pool if the population is small, but it has less effect in a large, randomly mating population.

2.2 How do genetic techniques affect Earth's biodiversity?

Biotechnology combines our understanding of biology with technology to create new and useful products and processes that aim to improve our quality of life and the environment. The term 'biotechnology' was coined in the 1960s when there were major advances in genetic engineering. While we normally associate biotechnology with state-of-the-art genetic techniques used in the laboratory, processes such as fermentation and selective breeding date back to ancient civilisations, well before the discovery of DNA. However, it is modern biotechnology, especially techniques that manipulate the genomes of living organisms, that raises social, cultural and ethical concerns, which shape its use around the world.

2.2.1 Investigating the uses and applications of biotechnology (past, present and future)

Before discussing past, present and future biotechnologies, it is important to define these periods of time. Biotechnology from the past can be divided into ancient (pre-1800) and classical (1800–1945) periods, while modern (present-day) biotechnology ranges from post-World War II (1945) until today. These periods are defined by the practices, discoveries and knowledge of biological systems at the time. Ancient biotechnology involved common observations of nature, when people didn't necessarily understand the molecular basis of their practices.

> **Note**
> This is an 'including' dot point, which means you may be asked a question about any of the content in this section.

Examples that have their origins in the ancient past include selective breeding, which began between 10 000 and 5000 years ago, and fermentation used to make beer, bread, vinegar and dairy products. These practices allowed humans to maintain reliable, high-energy sources of food and fibre near settlements. Our understanding of heredity, natural selection and the mechanisms of infection improved in the classical period, and discoveries such as antibiotics and pasteurisation continued to improve our quality of life. The discovery of the double helix structure of DNA and the development of recombinant DNA technology revolutionised modern biological research, which will continue to shape the direction of biotechnology in the future.

Analysing the social implications and ethical uses of biotechnology, including plant and animal examples

The development of new biotechnology often results in ethical and moral dilemmas for scientists and society, so there needs to be discussion to determine whether the benefits outweigh any detriments and costs of the new product. Table 2.3 provides an analysis of social and ethical implications associated with the use of biotechnology in microorganisms, plants, animals and humans.

> **Note**
> An 'implication' is a positive or negative outcome that is likely to happen.

TABLE 2.3 Social and ethical implications of using biotechnology

Organism	Examples	Social implications	Ethical issues
Microorganisms	• Bacteria are used to produce recombinant proteins like insulin to treat diabetes • Yeast is used to naturally make food, alcoholic drinks and antibiotics	• Threat of bioterrorism or epidemics • Violence and disease from alcohol abuse • Higher standard of health	• Most people have no objection to using unmodified microorganisms because they are considered a lower form of life • Humans may be interfering with the process of evolution by creating GMOs
Plants	• Artificial pollination and recombinant DNA technology are used to express desirable traits • Examples include domesticated crops or GMOs like Bt cotton and Golden Rice	• Food labelling for consumer choice • Patents to protect the intellectual property of companies that make a new GMO • Testing to ensure the safety of GMOs for consumption and impacts on ecosystems • Reliable supplies of food and fibre • Improvements in nutrition and health	• Modification of plants is considered more ethical than animals, because plants don't have thoughts and feelings • Humans may be interfering with the process of evolution by creating GMOs
Animals	• Artificial insemination and recombinant DNA technology are used to express favourable traits, e.g. transgenic salmon and biopharming		• Animal rights need to be considered • Humans may be interfering with the process of evolution by creating GMOs

››

TABLE 2.3 cont.

Organism	Examples	Social implications	Ethical issues
Humans	• Products are developed to prevent, diagnose and treat diseases, e.g. genetic testing, genetic engineering and cloning	• Consumer input needed to guide research priorities to benefit users • Better education needed so citizens can make informed decisions • Equity issues need to be addressed to make products accessible to all people • Privacy laws are required to prevent free access to a person's genetic information • Anxiety caused by knowledge of harmful mutations	• Animal rights need to be considered in experiments on animals that test the effect of genetic mutations or new medicines • Therapeutic cloning requires robust debate about when life begins and whether the benefits of this research outweigh the cost of terminating a potential life

Researching future directions in the use of biotechnology

The future directions of biotechnology will be influenced by new scientific discoveries, economics, the changing environment, ethical issues and the needs of society. This section investigates examples of biotechnology that are likely to be important in medicine, agriculture, the environment and industry in coming decades.

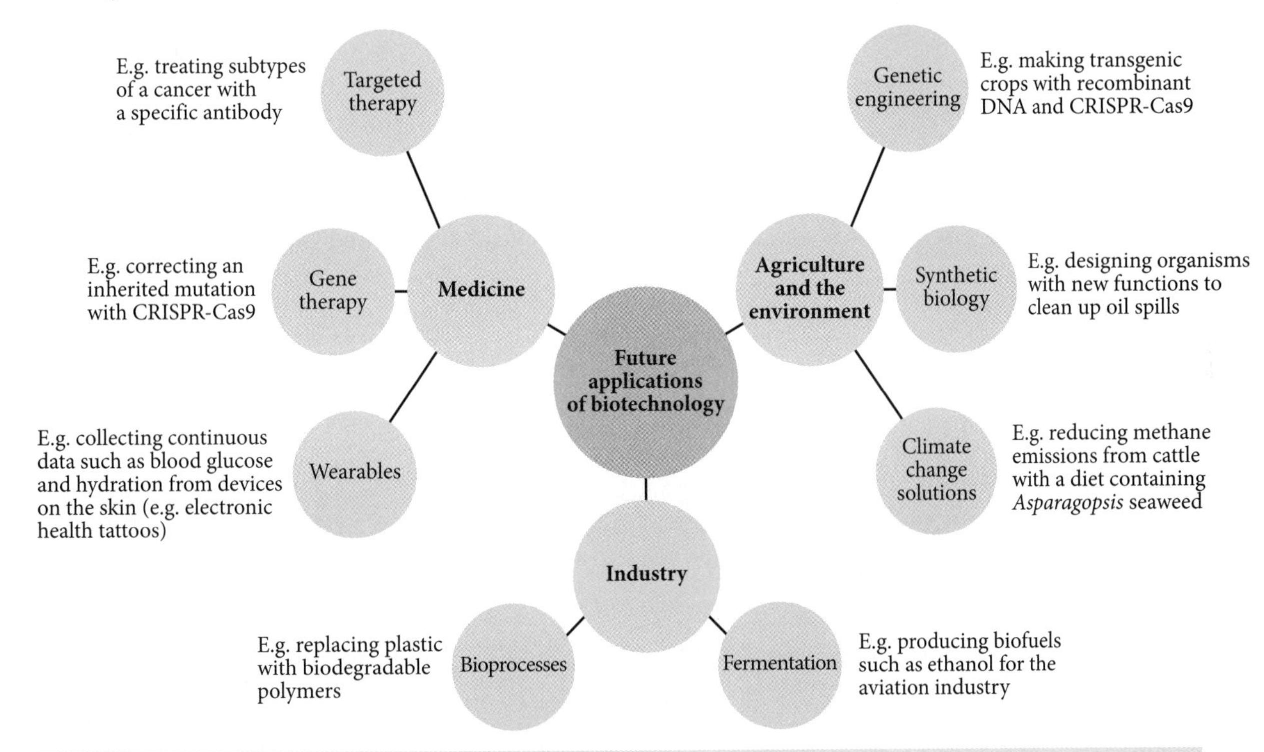

FIGURE 2.22 Some possible future directions of biotechnology

Medicine

The affordability and speed of whole-genome sequencing will have a major impact on human health in the future. **Personalised medicine** will use information about a person's own genes to tailor treatments to achieve the best outcome for them. An example is specific treatments for a subtype of cancer. This is different from the traditional 'one size fits all' approach to medical treatment, which may benefit some people more than others; for example, people with HER2 and breast cancer (see Figure 2.23).

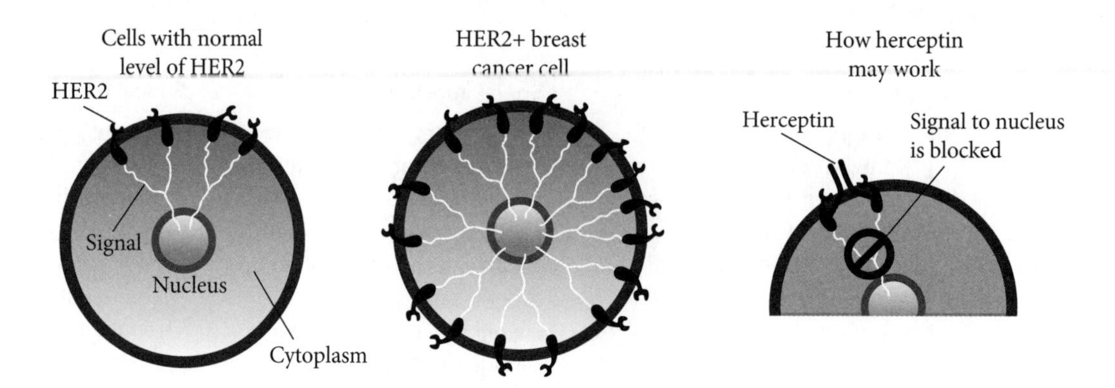

FIGURE 2.23 Personalised treatment of HER2-positive breast cancer

> **Note**
> An understanding of transcription, translation and the immune system helped to develop mRNA vaccines.

mRNA **vaccines** are a new type of vaccine that can prevent illness from viral infections. With traditional vaccines, a person is injected with viral proteins or an inactivated version of a virus that triggers an immune response specific to that pathogen. The design, testing and production of these vaccines can take a long time. mRNA vaccines, however, take advantage of normal processes within a cell to generate an immune response. The COVID-19 pandemic saw the first widespread use of mRNA vaccines, such as the Pfizer and Moderna vaccines. mRNA vaccines are likely to become more common because they can be designed more quickly. This is important, as scientists predict that humans will be exposed to more new viruses through climate change, urbanisation, global travel and increased human contact with animals.

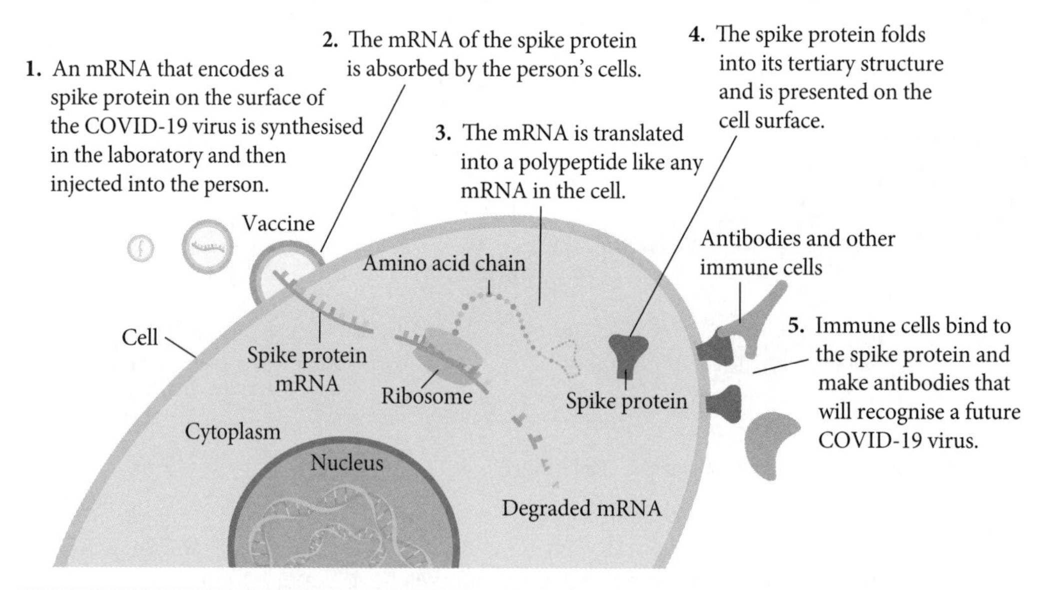

FIGURE 2.24 How an mRNA vaccine works

Agriculture and the environment

Genetic engineering is used to introduce or edit a single gene to create a desirable trait in an organism. Genetic engineering will work alongside **synthetic biology**, which uses online databases to combine multiple genes or gene parts to redesign an organism to produce a substance for human use, or to create an organism with a new ability. This might include the design of new crop varieties that are resistant to a range of abiotic (drought, heat, frost and salinity) and biotic (pests and disease) factors, or bacteria to clean up human pollution (**bioremediation**).

The evidence from climate science indicates that, to prevent the worst effects of climate change, we need to limit global warming to 1.5°C relative to pre-industrial levels. To achieve this, greenhouse gas emissions from human activities and farming practices must be reduced significantly. Biotechnology will produce some future solutions; for example, methane emissions from cattle can be reduced by growing 'clean meat' in the laboratory or by feeding cattle an *Asparagopsis* species of seaweed, which prevents the formation of methane by inhibiting a specific enzyme in cows' guts.

Industry

Industry uses chemicals during polymer production, food and fibre processing, mining and metal refining and the production of fuels. A shift from chemical processes to **bioprocesses** would achieve a more sustainable future. For example, biotechnology may help to replace plastics with biopolymers, such as polylactic acid, which is recyclable, biodegradable and compostable. Alternative sources of heat, electricity and fuel that do not emit greenhouse gases will also be needed in the next decade to minimise the effects of climate change. Biofuels such as ethanol, produced by fermentation, will also be used more widely as a fuel. Airline companies such as Boeing aim to have all commercial craft capable of operating on biofuels by 2030.

Evaluating the potential benefits for society of research using genetic technologies

Table 2.4 summarises the potential advantages and disadvantages for society of research using genetic technologies in medicine, agriculture and the environment.

Note

This is another 'evaluate' dot point. A good way to address this would be to construct a table that lists the advantages and disadvantages of different genetic technologies and use this to make a value judgement on their respective benefits to society.

TABLE 2.4 Advantages and disadvantages of using genetic technology

Area	Genetic technology	Advantages	Disadvantages
Medicine	Vaccinations, genetic testing, genetic engineering	• Vaccinations prevent infectious diseases • Early diagnosis of diseases (e.g. genetic testing) results in a better health outcome • New treatments for diseases (e.g. **gene therapy** and recombinant proteins, like insulin for diabetes) prevent illness or death	• Rare allergic reaction to vaccinations • Ethical concerns about access to personal genetic information • Genetic technologies are expensive
Agriculture	Transgenic plants and animals	• Better food security • More food variety, longer shelf life and better nutritional value • Healthy food is accessible to more people	• Hybridisation with species in the wild, and possible decline in genetic variation, may have an impact on ecosystem services
Environment	Bioremediation, production of biofuels	• Improve the quality of ecosystem services, e.g. cleaning up pollution in soil and the ocean • Reducing CO_2 emissions to prevent climate change	• Public concern about the environmental risk of uncontrolled GMOs

Evaluating the changes to Earth's biodiversity due to genetic techniques

There are three levels of **biodiversity**: genetic, species and ecosystem. These interact to create the complexity of life on Earth. Genetic diversity is the variety of genes within a species, which differs between individuals and populations. Species diversity is the variety of species in a habitat. Some habitats, such as tropical rainforests, have high species diversity, while a polluted river may have low species diversity. Ecosystem diversity is the variety of ecosystems (the interaction between organisms and their environment) in an area.

The following examples illustrate how genetic techniques affect Earth's biodiversity in a positive and negative way.

Genetically modified plants and animals

Genetic techniques introduce one or more genes into a plant or animal to change the phenotype of the organism. One of the main fears about genetic techniques is 'genetic contamination' of the wild population with DNA from a **genetically modified organism (GMO)**. If a GMO 'escaped' into the wild it could have an impact on biodiversity, if it hybridised with a closely related species and outcompeted other species in the ecosystem. This possibility is more likely in plants because pollen can be distributed by wind, water and animals. Buffer zones (areas with no crops) are sometimes placed next to conservation areas to minimise the risk of cross-breeding with native species in habitats that are adjacent to the farmland.

The risk of hybridisation also exists for GM animals. Genetically modified salmon, called AquAdvantage salmon, were developed by combining genes from the Chinook salmon and the ocean pout. This results in the production of growth hormone all year round and the fish grow 25% faster than natural salmon (see section 2.3.4 for recombinant DNA techniques). The modified salmon could have a selective advantage if they escaped into the wild, because they grow to adult size three times faster than wild salmon.

Scientists have minimised the effect on genetic diversity by only fertilising eggs from unmodified salmon with the modified sperm, and on species diversity by creating sterile adults (by 'shocking' the eggs after fertilisation), farming the fish in tanks on land, and by adding protective sieves on plumbing to prevent the escape of eggs into waterways.

Conservation genetics

Conservation genetics is the use of genetic techniques to manage endangered species and to reduce the risk of extinction. DNA sequencing, SNPs and STR markers are used to measure genetic diversity. If a species has low genetic diversity, or if populations are fragmented, scientists may intervene and relocate animals or build habitat corridors between populations to increase gene flow. For example, the Koala Kiss project aims to build uninterrupted forest habitat for koalas along the eastern coast of Australia.

In summary, genetic techniques can affect biodiversity in different ways. Most evidence indicates that GMOs are no better adapted for survival than their unmodified relatives, but there are legitimate concerns about their possible effects on the different levels of biodiversity. While risk management can reduce the likelihood of a GMO escaping into the wild and affecting the gene pool of the wild population and affecting ecosystem biodiversity, it cannot guarantee that this will not happen.

2.3 Does artificial manipulation of DNA have the potential to change populations forever?

The rapid development of technologies that can quickly and cheaply sequence the genome of any organism has significantly improved our understanding of the structure and function of DNA. This, combined with advances in techniques such as cloning, artificial insemination and pollination, transgenics and DNA editing, has resulted in a wide range of new applications in science, medicine, agriculture and industry. But where do we stop? Advances in technology need to be balanced by cultural and societal values and concerns about unforeseen impacts on the environment and human health.

2.3.1 Investigating the uses and advantages of current genetic technologies that induce genetic change

Recombinant DNA technology and CRISPR-Cas9 are two current technologies used to create GMOs by inducing genetic change in a genome.

Note
Genetic technologies are sometimes called 'genetic engineering'. They are techniques used by humans to intentionally change the DNA sequence of an organism.

Recombinant DNA technology

Recombinant DNA technology has existed for approximately 40 years. This technology uses **restriction enzymes** and **vectors** to 'cut and paste' DNA molecules from two different species together. A piece of foreign DNA can be inserted into a host genome at a random location or an exact location, or it can be used to inactivate a gene. Some of the advantages of this technique include cloning genes to produce medically important proteins (e.g. insulin to treat diabetes), creating a **transgenic organism** (e.g. Golden Rice to prevent vitamin A deficiency), knocking-in or knocking-out a gene to study its biological function, or the preparation of a vaccine to prevent an infectious disease (e.g. measles).

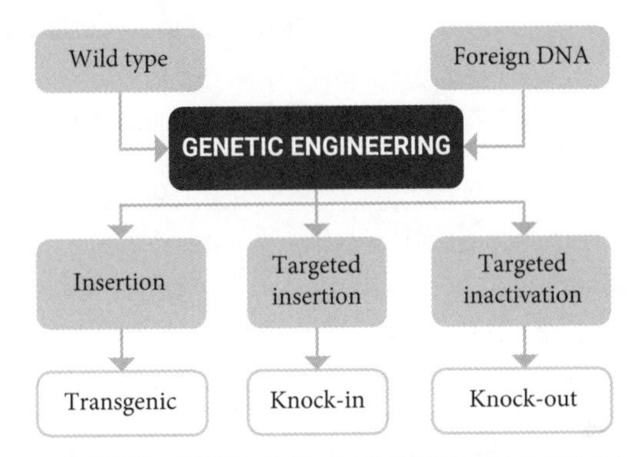

FIGURE 2.25 A summary of the applications of genetic technologies

Recombinant DNA technology (see sections 2.3.3 and 2.4.4) can be time-consuming and expensive, and some applications are being replaced by a new genetic technology, called CRISPR-Cas9.

CRISPR-Cas9

Before investigating the uses and advantages of CRISPR, it is helpful to understand its natural function in bacteria. CRISPR-Cas9 is a natural defence mechanism used by bacteria to attack viruses that infect them (**bacteriophages**). The genome of a bacterium contains a CRISPR locus, which consists of sequences called **C**lustered **R**egularly **I**nterspaced **S**hort **P**alindromic **R**epeats. A gene that encodes the Cas9 enzyme is also in a bacterium's genome.

When a bacterium is infected by a virus, the Cas9 enzyme locates the viral DNA in the cell, cuts it up into pieces and inserts it between the repeats in the CRISPR locus. This is the bacterium's 'memory' of the viral attack. If the bacterium is infected by the virus again, it transcribes the viral DNA to make CRISPR RNA (crRNA). The crRNA, which is complementary to the viral DNA, combines with Cas9 and guides it to the virus and cuts it up.

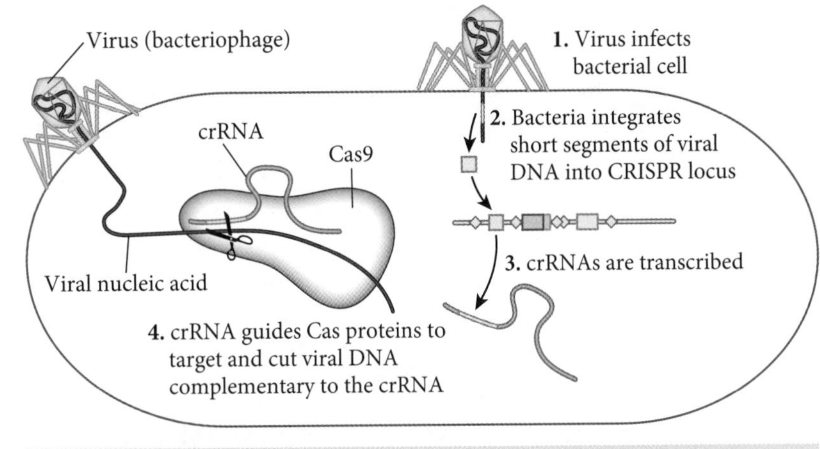

FIGURE 2.26 CRISPR-Cas9 is a bacterial defence mechanism against viruses.

Scientists have engineered CRISPR-Cas9 to precisely 'edit' specific locations in a human, animal or plant genome. Guide RNA can be synthesised to target a specific gene to create a knock-out (inactivate a gene), a knock-in (correct a mutation) or a transgenic organism (insert a gene).

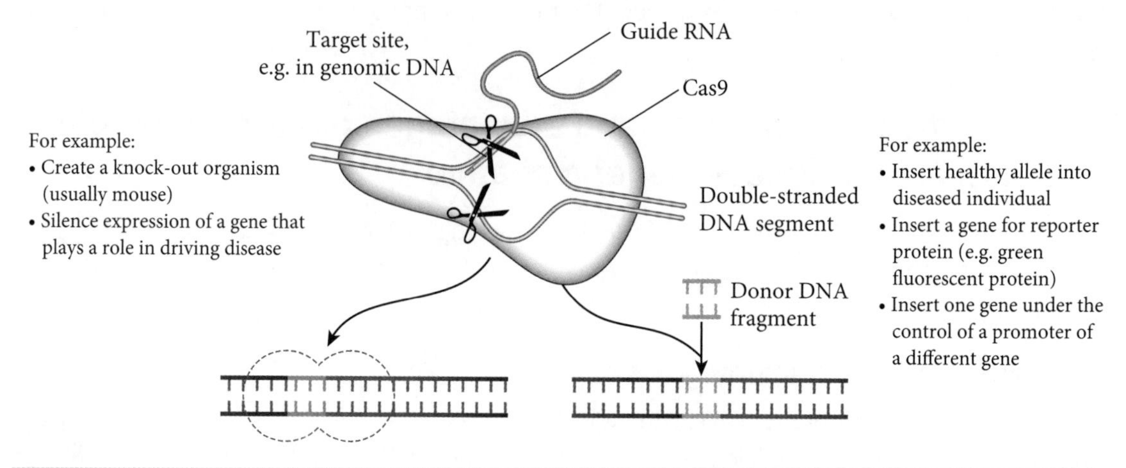

FIGURE 2.27 Application of CRISPR-Cas9 to 'edit' the genomes of other organisms

The advantage of CRISPR-Cas9 is that it is relatively quick, easy and inexpensive to design for a gene of interest. CRISPR-Cas9 has already been used to create model organisms for research and to alter crops, and has been trialled in humans to cure genetic diseases (e.g. correction of a mutation in the gene that causes sickle-cell disease).

> **Note**
> CRISPR-Cas9 could also be used as an example of a future direction of biotechnology.

2.3.2 Comparing the processes and outcomes of reproductive technologies

> **Note**
> This is an 'including but not limited to' dot point. This means you may be asked a question about artificial insemination, artificial pollination and another type of reproductive technology. *In vitro* fertilisation is the additional example you could use.

Artificial insemination and pollination

Artificial insemination is a type of assisted reproductive technology that uses any method other than sexual intercourse to introduce semen into the reproductive tract of a female. Assisted reproduction can help people who might not otherwise be able to have children. It is also widely used in agriculture to increase the likelihood of passing on desirable traits in livestock, and the technique is also beneficial in helping to increase the population size of endangered species.

Artificial pollination is the intentional human transfer of pollen from the male to the female reproductive organ of a plant. It is also used to select for desirable traits in horticulture (e.g. a particular flower colour) and agriculture (e.g. sweetness of fruit), to increase crop yields when natural pollinators are rare, and to save endangered plant species.

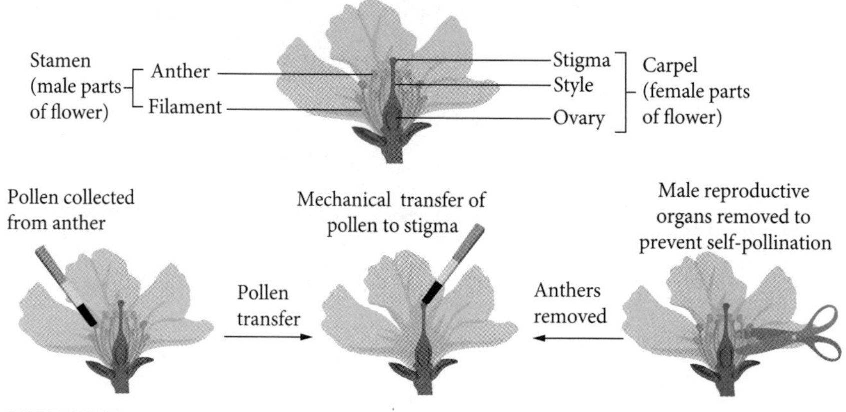

FIGURE 2.28 The process of artificial pollination

TABLE 2.5 A comparison of artificial insemination (animals) and artificial pollination (plants)

	Processes	Outcomes
Similarities	• Human intervention. Sperm (animals) and pollen (plants) are collected from a male and transferred to a female • Fusion of genetically unique male and female gametes • Sperm and pollen need to be cryopreserved for use in future	• Offspring with desirable traits (e.g. faster growth rate, higher seed production or bigger fruit) • Increases the chance of fertilisation when animals don't mate or if there is a lack of pollinators (e.g. insects, birds, mammals) • Reduction in genetic variation, because one male may be used to sire many offspring • Used to save endangered plants (e.g. Australian orchids without insect pollinators) and animals in captive breeding programs (e.g. in 2021 twin Giant Pandas were born at the Smithsonian National Zoo) • Better control of passing on a particular trait to offspring compared to random mating
Differences	• Sperm is collected with an artificial vagina or by electroejaculation. Pollen is collected with a brush • In flowering plants, pollen is delivered from anther (male) to stigma (female). In animals, sperm is introduced into the vagina • The male reproductive organs of a plant may be removed to prevent self-pollination. Animal reproductive organs are not affected	• Much higher yields in plants than animals because pollen can be quickly transferred • Artificial insemination is usually restricted to fewer individuals (e.g. to pass on traits of a prize-winning racehorse). Pollen can be transferred between many plants in a large crop

In vitro fertilisation

***In vitro* fertilisation (IVF)** is a more invasive type of assisted reproductive technology used when a female is unable to conceive naturally. The outcome of this technology is a fertilised embryo in parents who are unable to conceive naturally. However, the success rate of IVF declines with age and also depends on underlying fertility issues. IVF is also used to help save endangered animals from extinction (e.g. in 2020 the first IVF cheetah cubs were born in Columbus Zoo in the USA). The success rate of IVF in wild animals is limited by our knowledge of the reproductive system and cycle of each species.

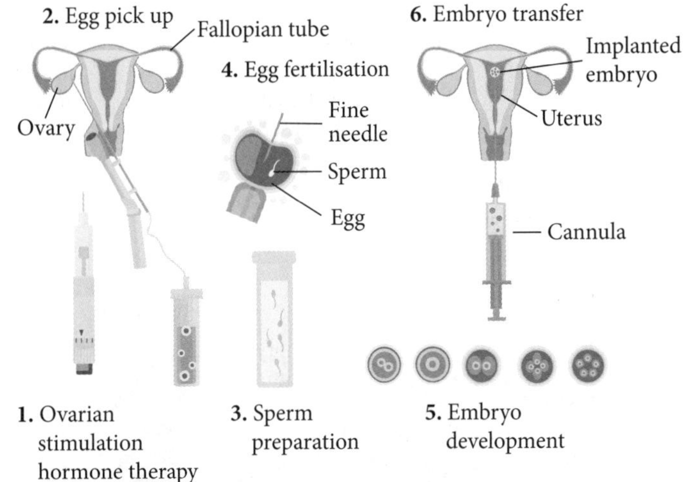

FIGURE 2.29 The IVF fertilisation cycle is a form of assisted reproduction.

2.3.3 Investigating and assessing the effectiveness of cloning

Note

This is an 'including but not limited to' dot point. You may be asked an 'assess' (don't forget the value judgement!) question about whole-organism cloning and gene cloning, or another type of cloning. Therapeutic cloning is another example of cloning you can use.

Cloning is the process of making a genetically identical copy of a gene, cell, tissue or whole organism. This section investigates and assesses the effectiveness of three types of cloning: **whole-organism cloning** (also known as reproductive cloning), gene cloning and therapeutic cloning.

Whole-organism cloning

Whole animals can be cloned using two techniques: **somatic cell nuclear transfer (SCNT)** and **embryo splitting**. In SCNT, the genome from a somatic cell is transferred to an egg that has had its nucleus removed. The embryo is then implanted into a **surrogate mother**, where it develops until birth. The offspring are genetically identical to the parent from which the somatic cell nucleus was taken.

Cloning by embryo splitting begins by creating an embryo using IVF. At the six-cell stage, the embryo is split into two. When the two embryos have developed to the **blastocyst** stage, they are implanted into the uterus of a surrogate mother. This procedure mimics the natural process that gives rise to identical twins. The two offspring are identical because they both have the same genetic material. They share 50% of their DNA with each parent used in the original IVF.

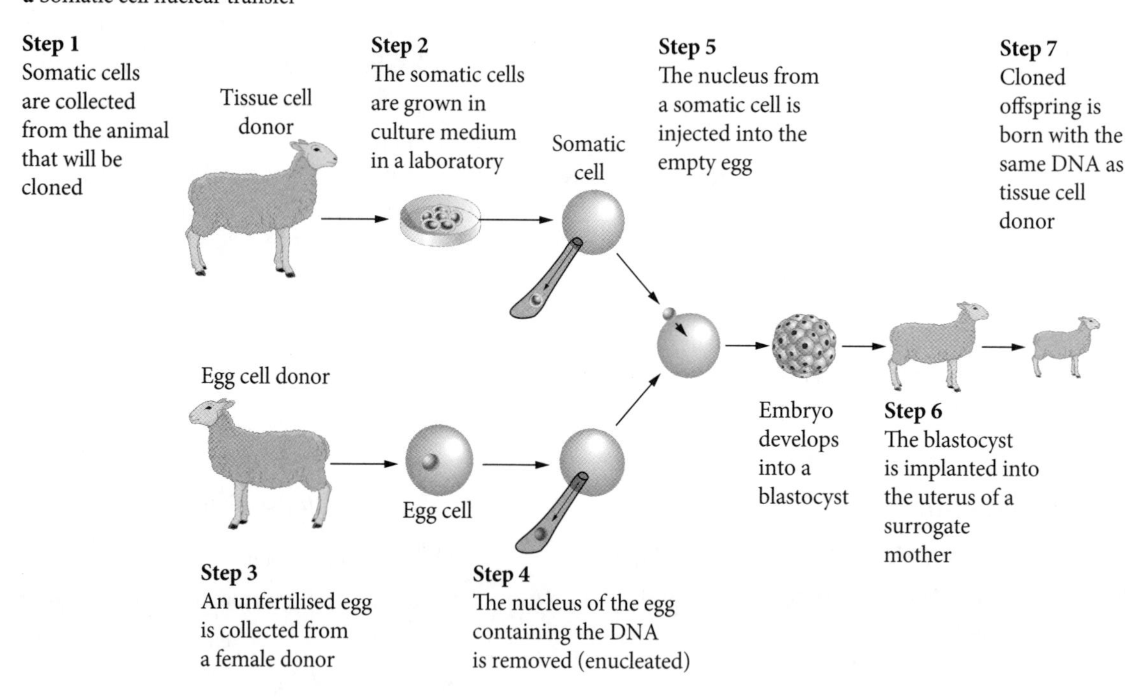

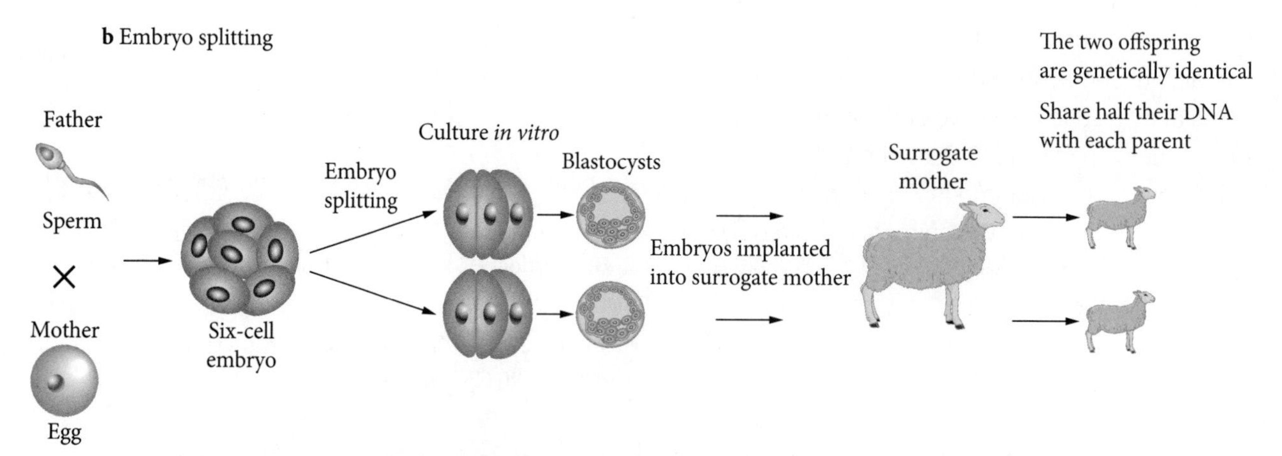

FIGURE 2.30 Two types of whole-animal cloning: a SCNT and b embryo splitting

Whole-animal cloning is very inefficient, as the clones often have developmental abnormalities or die soon after birth. Dolly the sheep was the first animal to be cloned from an adult somatic cell in 1996. The adult cell came from the udder of a six-year-old sheep. The research team made 276 attempts before they succeeded. This low efficiency is because the biology of early stages of embryonic development is not completely understood. Other animals have been cloned in the past two decades, mainly for agricultural and laboratory purposes, but the success rate for cloning is still only about 1%.

Whole plants can be cloned by taking a cutting or by producing **explants** in tissue culture. Cuttings are widely used in horticulture. Tissue culture is used to conserve rare species, such as the Wollemi pine. Whole-plant cloning is very efficient and is used to produce plants quickly and cheaply on a large scale.

In summary, when assessing whole-organism cloning you could conclude that animal cloning is expensive, time consuming and inefficient. Plant cloning, on the other hand, is efficient, and therefore less expensive and time consuming.

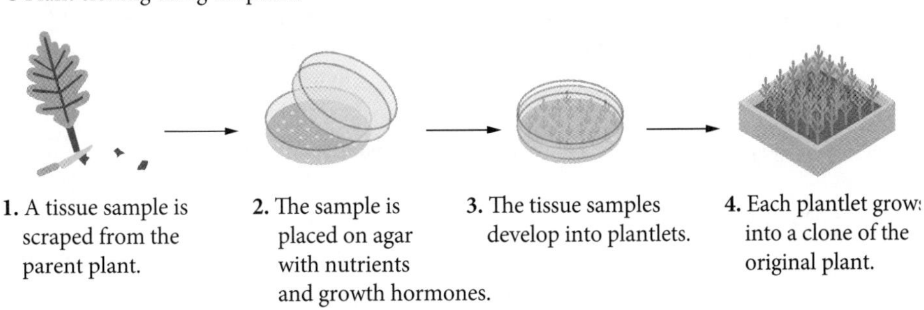

FIGURE 2.31 Cloning a plant: a taking a cutting, b growing exoplants in tissue culture

Gene cloning

Gene cloning is the process of making identical copies of a segment of DNA that codes for a polypeptide. The process involves making a copy of a gene from one organism and inserting it into the genetic material of a vector, usually a circular bacterial plasmid. The recombinant plasmid is then inserted into bacterial cells, which reproduce via binary fission and act as 'mini factories' to make many copies of the cloned gene. Gene cloning is a common technique that has a wide range of applications, including biopharmaceuticals, gene therapy and gene analysis. It is a very efficient process and produces consistent results, because scientists use commercially available enzymes, vectors and kits to do their experiments. The technique of gene cloning is described in more detail in section 2.3.4.

Therapeutic cloning

Therapeutic cloning is the creation of a cloned embryo for the purpose of producing **embryonic stem cells**. Embryonic stem cells (ES cells) are one of the earliest types of cells that form in a developing embryo. They are **pluripotent**, which means they can differentiate into almost any type of cell in the body. Therapeutic cloning therefore has potential medical applications such as the repair and replacement of damaged tissue caused by injury or disease.

The initial stages of therapeutic cloning are identical to whole-animal cloning. The nucleus of a somatic cell from a human is combined with an enucleated donor egg. Then, when the embryo has developed to the blastocyst stage, instead of being implanted into a surrogate mother, the ES cells are harvested, which destroys the embryo in the process. The ES cells are then grown in culture medium with special growth factors that cause them to differentiate into a desired tissue. Concerns about the potential for tumours to form from these cells, and ethical issues raised by the destruction of embryos, have inhibited progress in this field of cloning.

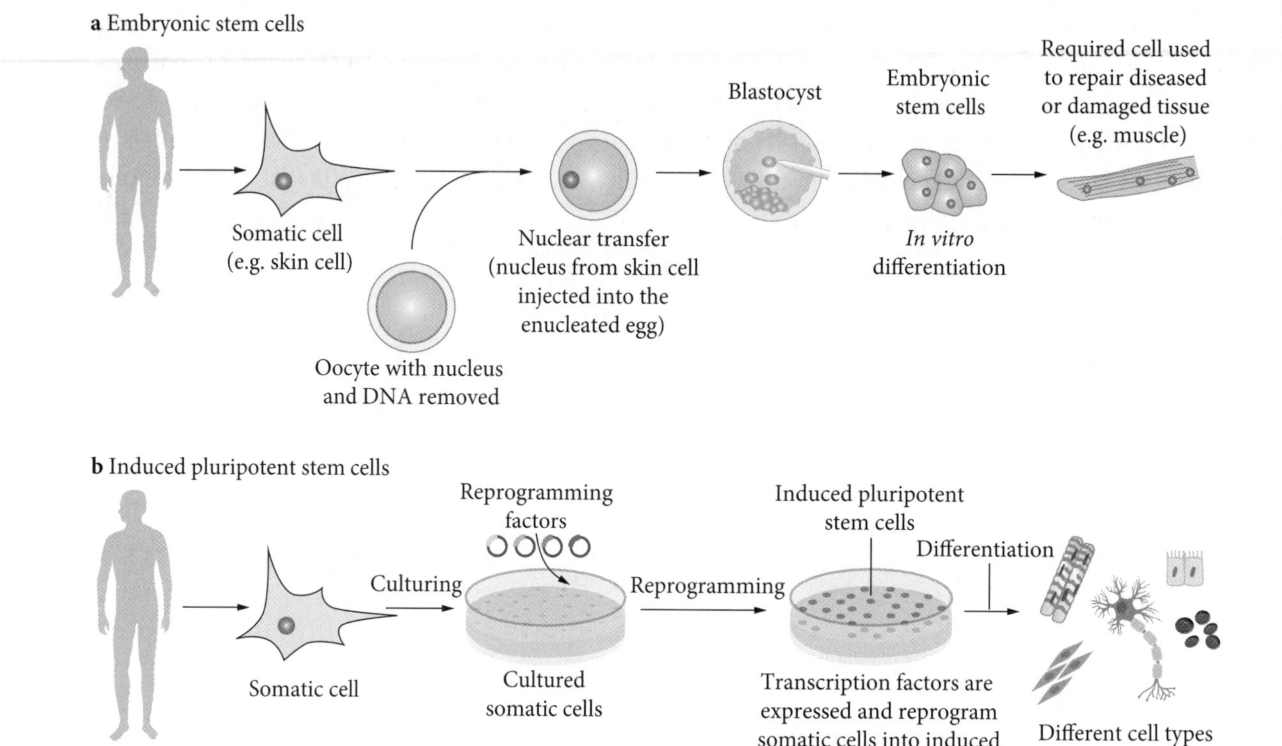

FIGURE 2.32 Therapeutic cloning: a embryonic stem cells, b induced pluripotent stem cells

Scientists have recently overcome the ethical issues by reprogramming somatic cells directly into ES cells using a special 'cocktail' of transcription factors. These reprogrammed cells are called **induced pluripotent stem cells** (iPSC). The added advantage of iPSCs is that they are derived from the person receiving the treatment, so they are less likely to cause an immune response when introduced back into the body. The generation of cloned ES cells is relatively efficient, but it is challenging to get them to differentiate into the desired cell types. This is still an early field of research but neurons, blood, liver cells, and egg, sperm and bone precursor cells have already been produced from iPSCs.

2.3.4 Describing techniques and applications used in recombinant DNA technology; for example, the development of transgenic organisms in agricultural and medical applications

Before describing some examples of recombinant DNA technology, it is important that you understand the steps involved in gene cloning, which is always the first step in the creation of a transgenic organism.

Note

You should aim to be able to describe the techniques and applications of recombinant DNA technology for one example of a transgenic animal, a transgenic plant and a medical application.

The process of gene cloning

For a gene to be cloned, it must be inserted into a vector, which is usually a circular piece of DNA called a **plasmid**. Plasmids are found naturally in the cytoplasm of bacteria.

Plasmids replicate themselves independently, express genes they carry, and can be exchanged between bacteria (recall the process of conjugation from Module 5). Scientists have taken advantage of these three features to artificially introduce plasmids containing a desired gene into bacteria

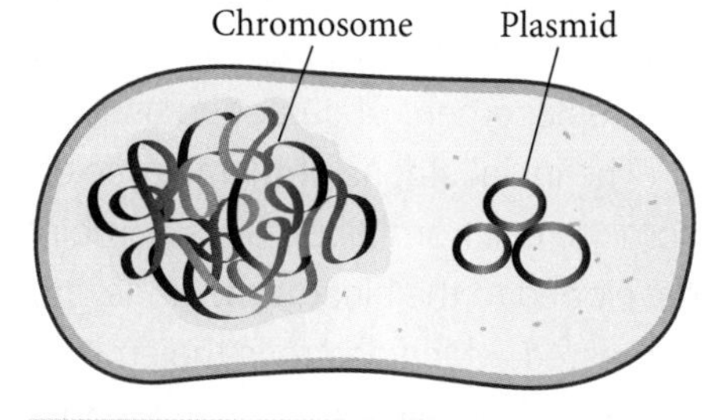

FIGURE 2.33 Plasmids are circular pieces of double-stranded DNA in bacteria.

(**transformation**) and to use them to produce many copies of a gene (e.g. to later introduce into an animal or plant to make a transgenic organism) or to express a cloned gene (e.g. to make a medically important protein such as insulin). Plasmids that are genetically engineered to transcribe and translate a gene in a specific organism are called **expression vectors**.

Cloning begins by preparing the gene to insert into the plasmid. This can be done by using a **restriction endonuclease** to cut a gene out of a genome, or by making a copy of the gene from mRNA using reverse transcription. Restriction endonucleases are named after the bacteria from which they were isolated, and they create 'sticky' or 'blunt' ends after they cut through both strands of a specific DNA sequence. Sticky ends are usually used for gene cloning because they make it easier to join two pieces of DNA together.

TABLE 2.6 Restriction endonucleases: source, site of action and type of cut ends they produce

Name of enzyme	Source	Recognition site and cleavage site	Nature of cut ends
*Eco*R1	*E. coli* RY13	5′ - G|AATTC - 3′ 3′ - CTTAA|G - 5′	Sticky
*Hind*III	*Haemophilus influenzae* Rd	5′ - A|AGCT - 3′ 3′ - TTCGA|A - 5′	Sticky
*Bam*HI	*Bacillus amyloliquefaciens* H	5′ - G|GATCC - 3′ 3′ - CCTAG|G - 5′	Sticky
*Bal*I	*Brevbacterium albidum*	5′ - TGG|CCA - 3′ 3′ - ACC|GGT - 5′	Blunt
*Hae*III	*Haemophilus aegyptius*	5′ - GG|CC - 3′ 3′ - CC|GG - 5′	Blunt

The plasmid is cut with the same restriction endonuclease that was used to prepare the ends of the gene insert. An enzyme called **DNA ligase** then catalyses the formation of bonds between overhanging nucleotides. In other words, a restriction endonuclease and DNA ligase 'cut and paste' two pieces of DNA together (**ligation**).

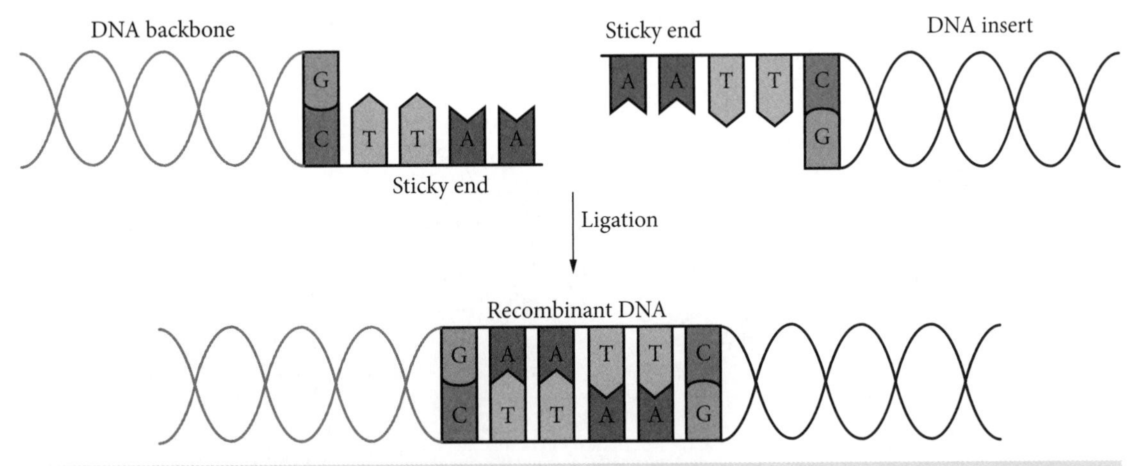

FIGURE 2.34 Ligation: joining two pieces of DNA together with DNA ligase

The recombinant plasmid is then introduced into bacteria, which are grown on agar containing an antibiotic until small colonies can be seen. Antibiotics are used because the plasmid is also engineered to contain an antibiotic resistance gene. Only bacteria that have taken up a plasmid will grow. Each colony contains thousands of individual bacteria containing a plasmid with the foreign gene.

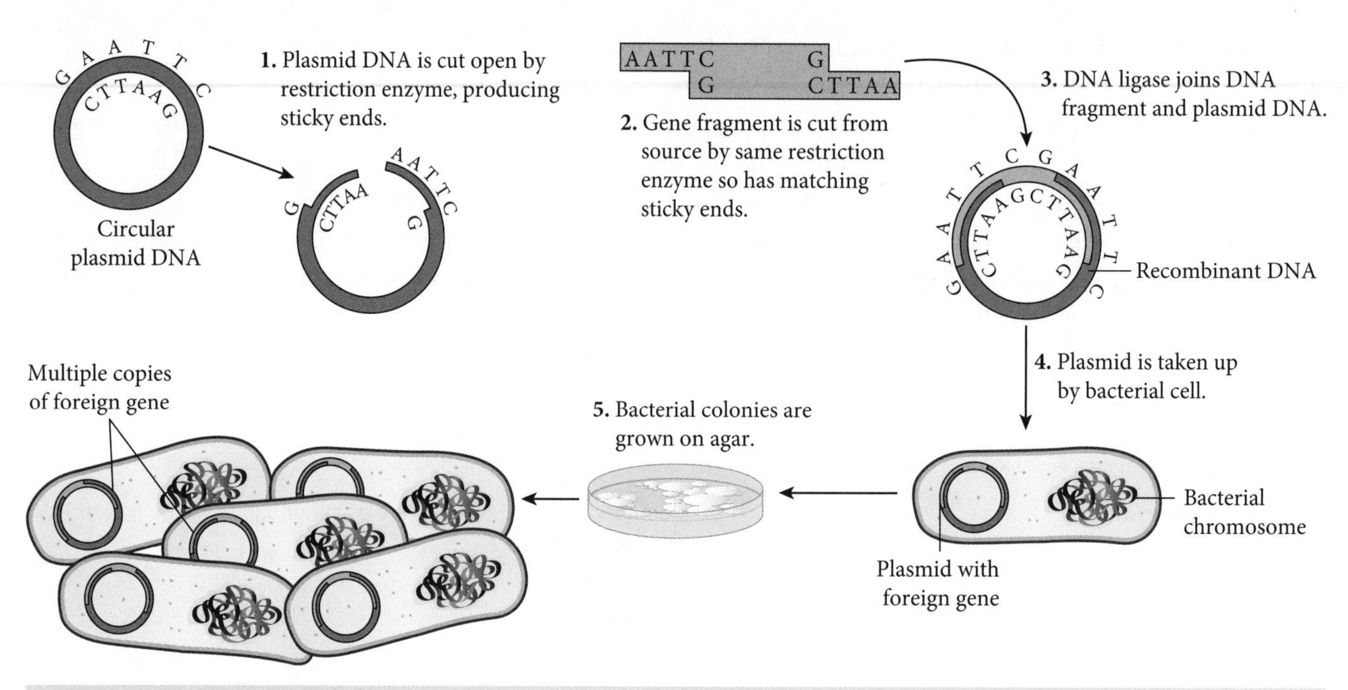

FIGURE 2.35 The process of gene cloning

Medical applications of genetically modified bacteria

The production of insulin is an example of how a GMO has been applied effectively in medicine. Type I diabetes is a non-infectious autoimmune disease in a person who cannot produce insulin to regulate their blood glucose level. It is investigated in more detail in Module 8 (section 4.2.1). The injection of insulin is the only available treatment for the disease.

The human insulin gene was cloned into a plasmid that expresses the gene in bacteria or yeast. The microorganisms transcribe and translate large amounts of the insulin polypeptide in fermentation tanks. Scientists then harvest and purify the insulin and use the hormone to treat diabetes.

Before the development of recombinant DNA technology, insulin was collected from the pancreases of livestock. Insulin derived from animals saved many lives, but it had erratic effects on glucose levels in the blood and caused allergic reactions, because the human immune system made antibodies against the foreign protein. Modern technology has overcome these issues, and has also eliminated the ethical concerns related to using animals for human benefit.

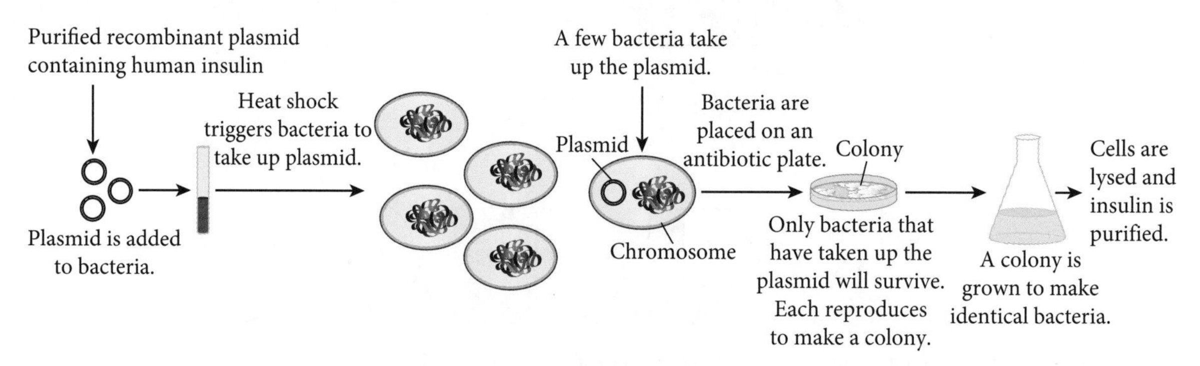

FIGURE 2.36 Production of human insulin using recombinant DNA technology

Study of gene function in biomedical research

To understand the function of a gene, scientists need to know in which cells, tissues or organs it is expressed. Recombinant DNA technology can be used to fuse the promoter of a gene of interest to the coding region of a gene that can be easily visualised. This is called a **reporter gene**. Green fluorescent protein (GFP), which glows green when it is stimulated with UV light, is an example of a reporter gene. This technique provides insight into the causes of disease and assists in the development of treatments.

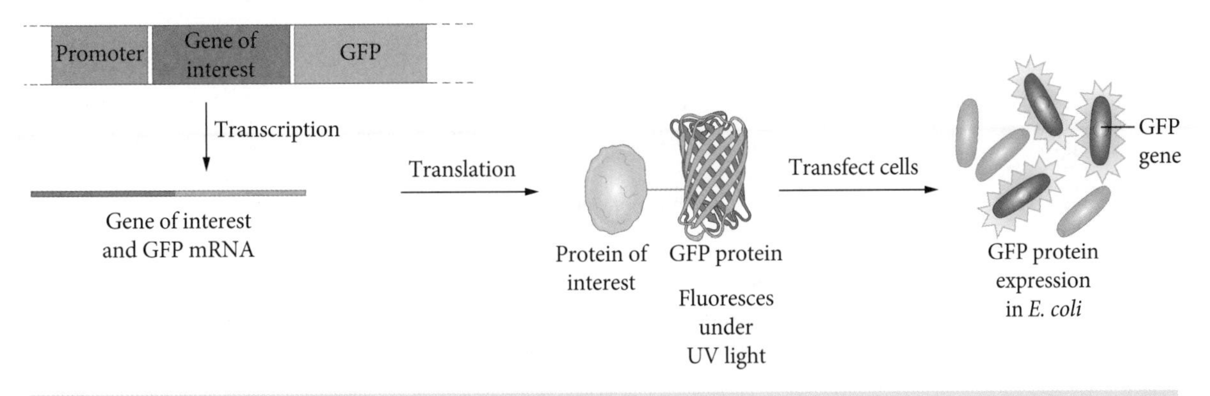

FIGURE 2.37 The use of reporter genes to study gene function

Use of transgenic animals in medicine

Transgenic animals have been developed for medical applications mainly to act as convenient producers of recombinant proteins. This has been nicknamed 'bio-pharming'. To create a transgenic animal, a cloned gene is inserted into the DNA of the host organism. Transgenic animals are created by injecting a vector, either a plasmid or a virus that contains the transgene, into a fertilised egg. The cloned gene inserts into one of the host organism's chromosomes at a random location. The egg is allowed to develop until the blastocyst stage, at which point it is implanted into the uterus of a surrogate mother. A small number of offspring will express the transgene when they are born.

For example, Tracy the transgenic sheep produced human protease inhibitor α1-antitrypsin in her milk, which is used to treat people with emphysema and can potentially alleviate some of the symptoms of cystic fibrosis. Mammalian milk is convenient to collect without sacrificing the animal.

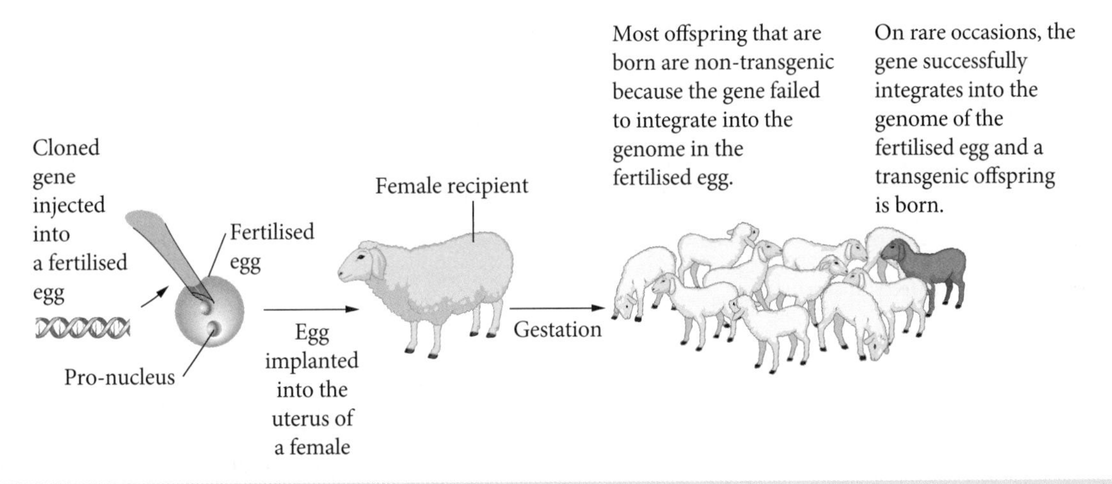

FIGURE 2.38 The use of recombinant DNA technology to make a transgenic animal

Use of transgenic animals in agriculture

It has been difficult to make transgenic mammalian livestock with a desirable trait (e.g. more muscle in beef cattle), as growth and development are complex traits controlled by many genes. However, there have been successful applications in aquaculture, such as AquAdvantage salmon (described in section 2.2.1). The transgenic fish was developed by combining a growth hormone gene from the Chinook salmon with a gene promotor from the ocean pout.

Note
Increased growth rate is also a benefit of a genetic technology.

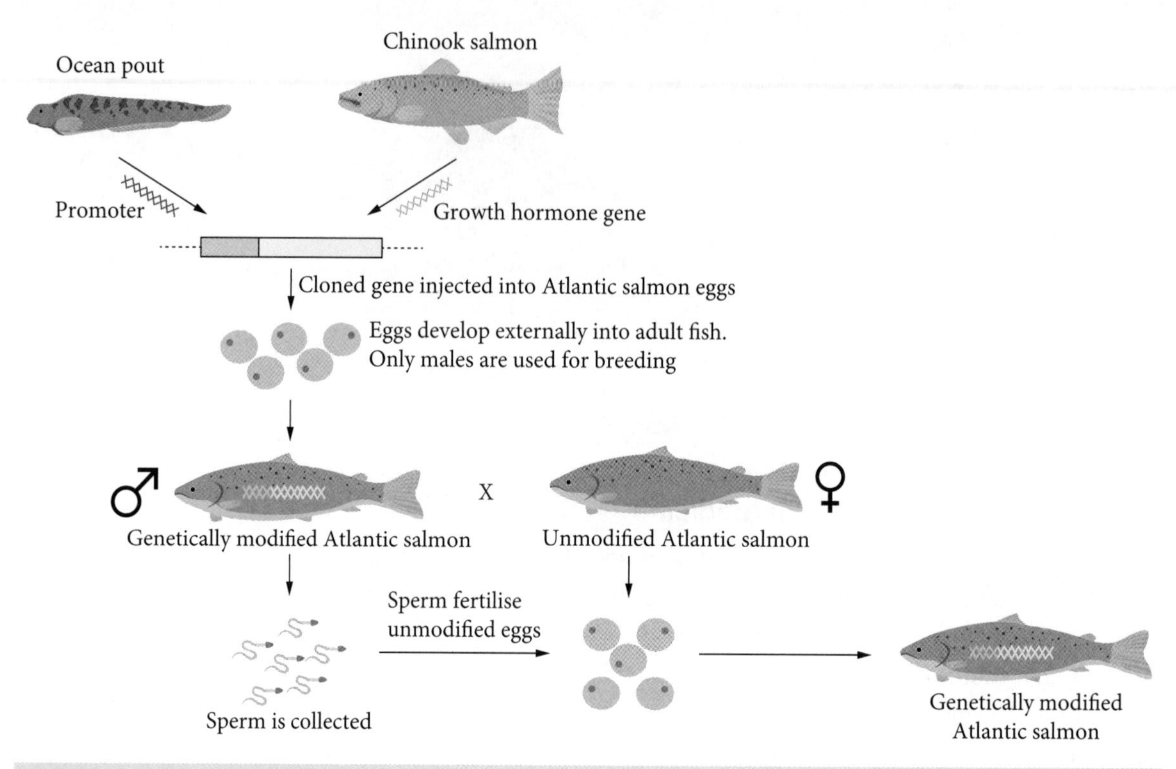

FIGURE 2.39 AquAdvantage salmon is a transgenic animal.

Use of transgenic plants in agriculture

Transgenic plants are produced to be more resistant to insect pests or herbicides, grow faster, or be better adapted to abiotic factors (e.g. frost and salinity). A desired gene can be introduced into a plant by two methods. In the first method, the gene is cloned into a plasmid, called a tumour-inducing (Ti) plasmid, that naturally exists in a plant pathogen called *Agrobacterium tumefaciens*, and plant cells are infected with this. In the second method, called the 'gene gun method', metal particles coated with the gene are fired at plant tissue. Both methods incorporate the gene into a chromosome at a random location.

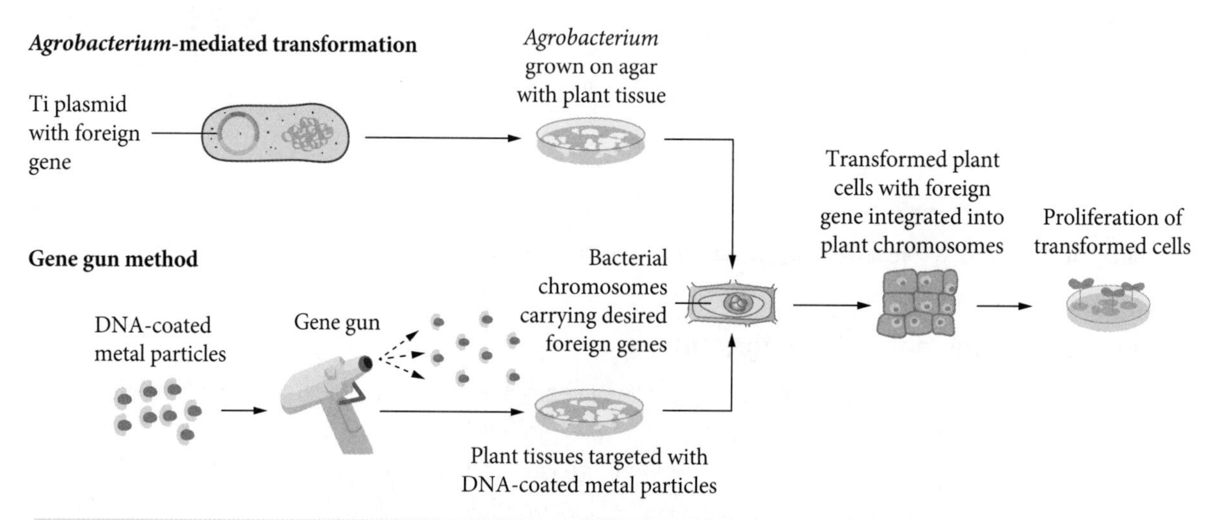

FIGURE 2.40 Two methods used to make transgenic plants: the Ti plasmid and *Agrobacterium tumefaciens*, and the 'gene gun' method

Bt cotton and Golden Rice are examples of transgenic plants with very different applications: resistance to insect pests and vitamin production for human consumption, respectively. Cotton crops can be eaten by insect pests, such as caterpillars of certain moth species (e.g. the cotton bollworm,

Helicoverpa armigera). Bollgard III, a recently developed variety of Bt cotton, contains a 'cocktail' of three insecticidal transgenes from the common soil bacterium *B. thuringiensis* (Cry1Ac, Cry2Ab and Vip3A). The toxic proteins accumulate in the leaves and kill feeding caterpillars, and therefore reduce the need for pesticides.

Note

Pest resistance is another benefit of a genetic technology, because it increases crop yields and reduces the use of pesticides that kill non-target species.

Many people living in developing countries may not have access to nutritious food and may suffer from vitamin deficiencies that cause disease. Vitamin A deficiency causes blindness and exacerbates other conditions such as diarrhoea and measles, because vitamin A is important in the immune response. Golden Rice is a genetically modified variety of normal white rice (*Oryza sativa*) that contains a gene from maize called phytoene synthase (*psy*) and a gene from the bacterium *Erwinia uredovora* called carotene desaturase (*crtI*). The proteins from these two genes create a biochemical pathway in the rice cells that produces beta-carotene (the precursor of vitamin A) in rice grains. Vitamin A is therefore made accessible to more people in this cheap and readily available food source.

Note

Vitamin production is yet another benefit of a genetic technology, because it prevents malnutrition and disease.

2.3.5 Evaluating the benefits of using genetic technologies in agricultural, medical and industrial applications

Note

In the last dot point you described the techniques and applications of genetic technologies. For this dot point, you need to 'evaluate' such examples. This means you must determine the advantages/disadvantages or benefits/limitations of a technology and then form your own opinion based on the evidence.

We have already addressed the potential benefits of biotechnology for society. There are also benefits to producers, such as farmers and industries that use biotechnology, and organisations that provide health services, such as hospitals. The advantages and disadvantages of examples are shown in Table 2.7.

Note

It is up to you to conclude whether each technology should be used, based on the evidence collected. Each person will have a different opinion, based on their own values.

TABLE 2.7 Advantages and disadvantages of biotechnology in agriculture, medicine and industry

Application	Example	Advantages	Disadvantages
Agriculture (crops)	Transgenic crops *Examples*: • Bollgard III cotton (insecticide resistance) • 'Siren' canola (herbicide resistance) • Golden Rice (vitamin A)	• Reduced use of chemicals • Less likely to kill non-target pests • Reduced soil compaction because machinery is on the land for less time • Increased crop yield because of reduced competition from pests and weeds • Less tillage and soil degradation • Decrease in labour and fuel usage	• Killing weeds reduces food supply for wild animals • Negative impact on food chains • Potential cross-breeding and contamination of wild species • Development of resistance in target species • Non-target invertebrate species may also be killed • Possible decline in crop biodiversity • High cost of seeds

››

TABLE 2.7 cont.

Application	Example	Advantages	Disadvantages
Agriculture (animals)	Aquaculture *Example*: AquAdvantage salmon	• Faster growth rates • Increased revenue for farmers	• Research is time consuming and expensive • Difficult to create because traits are often controlled by multiple genes • Potential side effects such as developmental abnormalities • Transgenic organism may escape into the wild and introduce transgenes into the environment
Medical	Genetic testing, vaccines and personalised medicine	• More efficient testing and treatment procedures • Reduced cost to the healthcare system	• Some treatments are too expensive for most people
Industrial	• Production of chemicals, polymers and biofuels using bioprocessing • Food and fibre bioprocessing • Extraction of metals	• Reduction of waste such as chemical solvents and heavy metals • Waste by-product less likely to require treatment • Environmentally sustainable products (e.g. biofuels and recyclable polymers) • Less water used in production	• The use of biological processes may not be as efficient as chemical processes • Developing new techniques is expensive in the short term

2.3.6 Evaluating the effect on biodiversity of using biotechnology in agriculture

Note

This is another 'evaluate' syllabus dot point. This means you need to compare biotechnology (e.g. a GM crop) to the traditional variety or conventional farming practice it will replace. Don't forget to make a value judgement!

TABLE 2.8 The effects of conventional agriculture and agricultural biotechnology on biodiversity

	Conventional agriculture	Agricultural biotechnology
Crops	• Biodiversity is reduced because habitat is destroyed to access fertile land • Thousands of traditional crops varieties used to be grown • Biodiversity is reduced because herbicides and pesticides kill non-target species and pollute waterways • Aquatic biodiversity is reduced because large-scale irrigation drains rivers and waterways • Pollination depends on wind or animals, which maintains genetic diversity	• GM crops may limit loss of biodiversity by growing on marginal land • GM crops are grown as **monocultures**. Over 1000 varieties face extinction • GM crops require less herbicides and pesticides, which reduces negative effects on food chains • Drought-resistant GM crops protect biodiversity because crops need less water • Artificial pollination reduces genetic diversity because pollen from only one plant is used
Livestock	• Animals mate randomly with each other, which maintains genetic diversity	• Artificial insemination reduces genetic diversity because sperm from one male fertilises the eggs of many females
Aquaculture	• Overfishing of natural fish stocks reduces genetic and species diversity • Genetic diversity is high because of gene flow between individuals in the wild	• Genetic, species and ecosystem diversity is reduced because of diseases and nutrient pollution • Species diversity may decline if escaped GM fish out-compete wild species

2.3.7 Interpreting a range of secondary sources to assess the influence of social, economic and cultural contexts on a range of biotechnologies

Before addressing this dot point, it is important to understand three key terms: **society**, **culture** and **economy**. A society is a large group of people who have consistent interactions in a particular place. It might be a town, city or a country. It can consist of many cultures, which are the characteristics and knowledge of a particular group of people. Finally, the economy is the consumption of goods and services involving the exchange of money.

It is important to distinguish between the *influence* of society, the economy and culture on biotechnology, and the *implications* and *benefits* of biotechnology, which were covered in section 2.2.1. Table 2.9 uses examples in a range of contexts to interpret how society, the economy and culture may *influence* a range of biotechnologies.

TABLE 2.9 The influence of society, economy and culture on biotechnologies

Context	Influence
Society	**Education** Individuals have different levels of education. This can influence a person's ability to make decisions about the risks and benefits of a biotechnology; for example, vaccination hesitancy can be caused by limited understanding of how vaccines work. **Social media** Information on social media is often personal opinion rather than evidence based. Targeted digital advertising also reinforces a person's current belief system. For example, a person may have a strong view about a transgenic organism without understanding the thorough safety tests involved in producing it. **Geography** Countries have different priorities based on their environment. Australia has one of the highest rates of skin cancer, so most people would support skin cancer research.
Culture	**Religion** Many religions view fertilisation as the beginning of life, and therefore regard cloning and the use of embryos in biotechnology as destroying a life. The dominant religious beliefs in a society may therefore limit the progress of research into therapeutic cloning. Some religions prohibit the consumption of certain animal products and this may extend to their use in biotechnology. For example, cattle are sacred in Hinduism and pork consumption is forbidden in Judaism and Islam. Jehovah's Witnesses refuse blood transfusions based on biblical teaching. It is therefore important for doctors to determine individual preferences when treating individuals of some religious groups.
Economic	**Personalised cancer therapy** Personalised cancer therapy using antibodies costs $100 000–$200 000 per year for each patient. Most people cannot afford this, and it is often too expensive to be subsidised by the public healthcare system. This may limit advances in this technology if costs do not come down. **Public health and prevention** Governments invest in screening and vaccination programs to reduce the long-term high costs of treating patients with a disease. For example, the National HPV vaccination program in Australia provides free vaccination to females and males aged up to 19 years, to prevent the development and spread of cervical cancer.

Glossary

aneuploidy A condition in which an individual organism has one or a few chromosomes above or below the normal chromosome number

artificial insemination A type of assisted reproductive technology that uses any method other than sexual intercourse to introduce semen into the reproductive tract of a female

artificial pollination The mechanical transfer of pollen by humans from the stamen to the stigma of a flower when natural pollinators are absent or to select for favourable traits

assess Make a judgement of value, quality, outcomes, results or size

autosome Any chromosome in a eukaryotic cell that is not a sex chromosome

bacteriophage A type of virus that infects bacteria

base analogue A chemical mutagen with a similar chemical structure to a DNA base that incorporates into DNA during DNA replication

biodiversity The interaction between genetics, species and ecosystems that creates the variety of life on Earth

bioprocess A specific process that uses living cells or their components (e.g. bacteria and enzymes) to obtain desired products

bioremediation The use of microorganisms to clean up chemical pollution in the environment

biotechnology The field of science that combines our understanding of biology with technology to create new and useful products and processes that aim to improve our quality of life and the environment

blastocyst A cluster of cells formed at an early stage in development of a mammal and consists of an inner cell mass and an outer layer of trophoblast cells

bottleneck effect A rapid reduction in the size of the population of a species

centromere A specialised region of DNA that separates the short and long arm of a chromosome

chemical mutagen A molecule that undergoes a chemical reaction with DNA, resulting in a base change in the DNA sequence

chromatid One of two halves of a replicated chromosome

chromosomal mutation A genetic change affecting a whole chromosome or a large section of a chromosome containing many genes

https://get.ga/aplus-hsc-bio-u34

A+ DIGITAL FLASHCARDS

Revise this topic's key terms and concepts by scanning the QR code or typing the URL into your browser.

chromosome non-disjunction When chromosomes fail to separate correctly during meiosis or mitosis, resulting in aneuploidy

cloning A technique used to make exact genetic copies of genes, cells, tissues or whole organisms

coding DNA A sequence of DNA that encodes a polypeptide

culture The characteristics and knowledge of a particular group of people

deleterious Describes a mutation results in an abnormal, or lack of, protein that has a negative impact on the phenotype

deletion A genetic change that involves the loss of one or more bases from a DNA sequence

diploid Describes a cell or organism that has two copies of each chromosome

DNA A molecule that consists of two polynucleotide chains that wind around each other to form a double helix structure; carries the genetic instructions in all living things

DNA ligase An enzyme that catalyses the formation of bonds between nucleotides of two sections of single-stranded DNA

DNA polymerase An enzyme responsible for the addition of nucleotides during the replication of DNA

duplication A genetic change that involves the gain of one or more copies of a gene or region of a chromosome

economy The system of trade and industry by which the wealth of a region or country is made and used

electromagnetic radiation A form of energy that consists of a pair of electric and magnetic fields that propagate together at the speed of light

embryonic stem cell A type of cell derived from the inner cell mass of the blastocyst

embryo splitting The process of dividing an embryo at the cleavage of blastocyst stage to produce two or more genetically identical organisms

enhancer A short sequence of DNA that is bound by transcription factors and physically interacts with a promoter to regulate gene expression

exon A portion of a gene that codes for amino acids that form part of a polypeptide sequence

explant A small piece of tissue removed from a plant and grown in sterile growth medium

expression vector A plasmid or virus that has been engineered to transcribe a cloned gene

fertilisation The fusion of an egg cell and a sperm cell to produce a zygote

founder effect The loss of genetic variation that occurs when a small number of individuals from a large population form a new colony

frameshift mutation The insertion or deletion of a nucleotide, which changes the open reading frame of a protein-coding gene

gamete A haploid reproductive cell of an organism

gamma rays A form of high-energy ionising radiation

gene cloning A molecular biology technique in which a copy of a gene is made by inserting it into a vector such as a circular plasmid

gene flow The transfer of alleles from one population to another population of the same species through the movement of individuals

gene pool The total genetic diversity within a population or a species

genetic drift Random fluctuations in the numbers of gene variants (alleles) in a population

genetically modified organism (GMO) An organism whose genome has been altered using genetic engineering techniques in the laboratory

germ-line mutation A heritable change in the DNA sequence that originates in gametes (egg or sperm) during meiosis

genetic engineering The use of recombinant DNA technology to alter the DNA of an organism

gene therapy A technique that modifies a person's genes to treat a disease

haploid Describes a cell that contains a single set of chromosomes

Hardy–Weinberg equilibrium A principle that states that the genetic variation in a population will remain unchanged unless there is influence from external factors

homologous chromosomes A pair of chromosomes, inherited from separate parents, with the same sequence of genes

induced pluripotent stem cell A somatic cell that has been reprogrammed back into a pluripotent stem cell

intercalating agent A chemical mutagen that can insert between bases in the sugar phosphate backbone of DNA

insertion A type of mutation that involves the addition of DNA to a chromosome

intergenic region A non-coding region of the genome located in between genes

intron A non-coding section of an mRNA transcript, or the DNA encoding it, that is spliced out before the mRNA is translated into a protein

inversion A chromosomal mutation in which a region of DNA breaks at two points and the segment bounded by the two breakpoints inserts back into the chromosome in the reverse orientation

***in vitro* fertilisation (IVF)** A type of reproductive technology in which fertilisation occurs outside the body and the early-stage embryo is introduced into the uterus to complete development

ionising radiation Electromagnetic radiation or subatomic particles that can detach electrons from atoms

ligation The process of joining together two pieces of DNA

meiosis A type of cell division that reduces the number of chromosomes in the parent cell by half and produces four gamete cells

missense mutation A single nucleotide substitution in an exon that changes the encoded amino acid

monoculture The cultivation of a single crop species

mutation A change in a DNA sequence that can result from mistakes during DNA replication or exposure to a mutagen

mutagen A chemical, physical or biological agent that has the ability to change a DNA sequence

mycotoxin A toxic compound produced by some species of fungi

non-coding DNA A region of DNA that does not code for a polypeptide

non-ionising radiation Long-wavelength electromagnetic radiation that does not have the energy to detach electrons from atoms

nonsense mutation A single nucleotide substitution in an exon that results in a stop codon

nucleotide A building block of DNA and RNA consisting of a base, a sugar molecule and one phosphoric acid

oncogene A mutated gene that contributes to cancer by increasing the rate of mitosis

partial monosomy The presence of only one copy of a gene that results from a deletion

particle radiation A form of radiation that consists of fast-moving subatomic particles that have both energy and mass

personalised medicine The use of a person's genetics to predict disease development and to tailor treatments

physical mutagen Ionising or particle radiation that changes the DNA sequence by removing electrons and breaking chemical bonds

plasmid A small, circular, double-stranded DNA molecule that is distinct from chromosomal DNA

pluripotent Describes a cell that has the potential to differentiate into any type of cell in the body

point mutation A change in a single nucleotide at the same position in the DNA sequence

promoter A region of DNA near the beginning of a gene that is bound by transcription factors to initiate transcription

pyrimidine dimer A covalent bond that forms between thymine and/or cytosine in the same strand of DNA

reactive oxygen species An unstable molecule containing oxygen that is highly reactive with other molecules in a cell

recombinant DNA technology A method of producing DNA by combining DNA from two different organisms

regulatory element A section of non-coding DNA that controls the level of transcription from a gene

reporter gene A gene that is attached to the regulatory region of another and allows visualisation of its expression

restriction endonuclease (enzyme) An enzyme that recognises a specific DNA sequence and cuts through both strands

reverse transcription The process by which the enzyme reverse transcriptase converts mRNA to complementary DNA

silent mutation A single nucleotide substitution in an exon that does not change the encoded amino acid

society A large group of people who have consistent interactions in a particular place

somatic cell Any cell in the body other than a gamete

somatic cell nuclear transfer The technique of transferring the nucleus of a somatic cell to an enucleated oocyte to create a cloned embryo

somatic mosaicisim When two or more populations of somatic cells with different genotypes exist in a body

somatic mutation A non-heritable change in the DNA sequence in any cell of the body that occurs during mitosis

splice site The genetic sequence at the boundary between an exon and an intron

surrogate mother A woman who carries a baby through pregnancy on behalf of another woman

synthetic biology A field of science in which genetic engineering is used to redesign organisms for useful purposes

telomere The repetitive, non-coding DNA associated with specialised proteins at the tips of linear chromosomes in eukaryotes

therapeutic cloning The creation of a cloned embryo for the purpose of harvesting embryonic stem cells

transcription factor A protein that binds to DNA to initiate and control the rate of gene transcription

transformation The horizontal transfer of genetic material between bacteria

transgenic organism A living thing that has had its genome altered through the introduction of a gene or genes from a different species

translocation A mutation that results from one segment of chromosome transferring to a non-homologous chromosome or to a different site on the same chromosome

transition A point mutation in which a pyrimidine base (cytosine or thymine) is replaced by a different pyrimidine base, or a purine base (adenine and guanine) is replaced by a different purine base

transposon A sequence of DNA that is able to copy and insert itself at a new position in the genome

transversion A point mutation in which a pyrimidine base is replaced with a purine base, or vice versa

whole-organism cloning The process of making a genetically identical copy of an individual organism

vaccine A substance used to stimulate the production of antibodies to prevent illness from a viral infection in the future

vector A plasmid or virus that carries and transfers genetic material between organisms

virus A small, infectious agent that can only replicate inside the cells of a living organism

xenotransplantation The transfer of cells, tissues or organs from another species into a human body

Exam practice

Multiple-choice questions

Solutions start on page 253.

Mutations

Question 1

Which of the following is a type of ionising electromagnetic radiation?

A X-rays

B Microwaves

C Radiowaves

D Infrared light

Question 2

Gamma rays, X-rays and particle radiation cause genetic mutations through which mechanism?

A Addition of atoms to nucleotides

B Insertion of hydrogen bonds into the DNA backbone

C Addition of chemical modifications to nitrogenous bases

D Ionisation, which breaks chemical bonds in the DNA double helix

Question 3

Base analogues are a type of chemical mutagen. Which statement most accurately describes how they cause genetic mutations?

A They cause double strand breaks in DNA with high-energy radiation.

B They cause a chemical modification to bases, which are then converted into new bases.

C They have a chemical structure like natural bases. DNA polymerase matches an incorrect base with them during DNA replication.

D They insert into the DNA backbone and distort the shape of the DNA. Additional bases are added by DNA polymerase at the site of insertion.

Question 4 ©NESA 2019 SI Q15

A germ-line mutation is known to have occurred. How is it possible that there has been no noticeable change in the phenotype of the offspring?

A The mutation occurred in a stretch of RNA.

B The mutation occurred in a protein-coding region.

C The mutation occurred in a stretch of non-coding DNA.

D The mutation did not affect the DNA sequence of any gametes.

Question 5

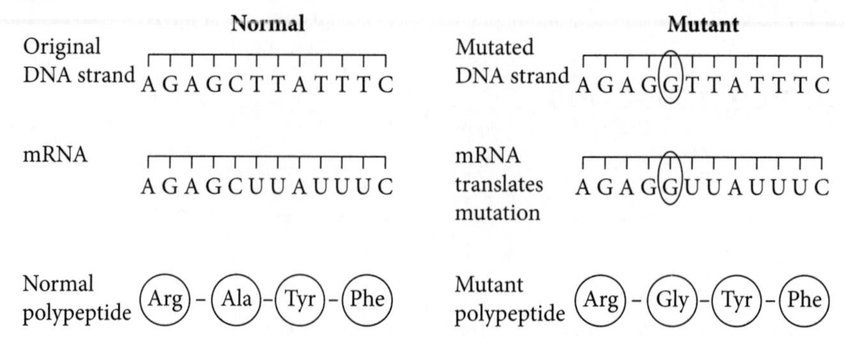

The left half of the diagram shows the normal DNA sequence of an exon, the transcribed mRNA sequence, and the translated polypeptide sequence. The right half of the diagram shows a mutation that has generated a new polypeptide sequence. What is this type of mutation?

A Nonsense mutation

B Frameshift mutation

C Insertion mutation

D Missense mutation

Question 6

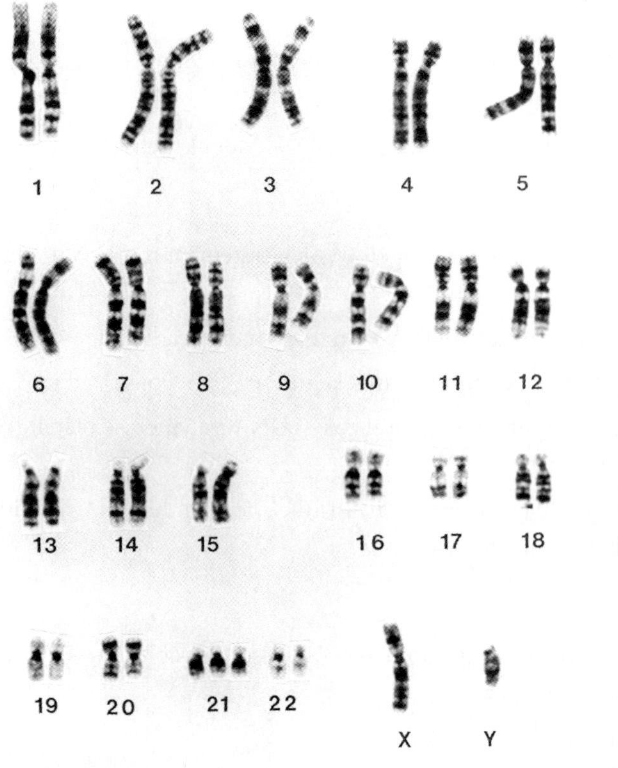

The image above is a karyotype of human chromosomes from a somatic cell. What condition is shown in the karyotype?

A Polyploidy

B Aneuploidy

C Duplication

D Translocation

Question 7

Which of the following sentences correctly describes the features of somatic mutations?

A They occur in somatic cells and cause gamete mosaicism.

B They can affect the individual and can be passed on to offspring.

C They can occur in any cell that isn't a gamete and only affect the individual.

D They can be present in every cell of the body and can be passed on to offspring.

Question 8

Genetic recombination is a mechanism that increases genetic variation and can occur during meiosis. Which of the following statements about genetic recombination is correct?

A Genetic material is exchanged between non-homologous chromosomes.

B Genetic material is exchanged between sister chromatids of homologous chromosomes.

C Genetic material is exchanged between sister chromatids of non-homologous chromosomes.

D Genetic material is exchanged between non-sister chromatids of homologous chromosomes.

Question 9

A population of koalas consisted of 100 individuals, and the frequency of two alleles (P and p) at a gene locus was P = 0.6 and p = 0.4. A bushfire drastically reduced the population to 10 individuals, which consisted of 2 heterozygotes (Pp), 7 homozygotes (PP) and 1 homozygote (pp). What are the new frequencies of the P and p allele in the population?

A P = 0.6, p = 0.4

B P = 0.2, p = 0.8

C P = 0.8, p = 0.2

D P = 0.4, p = 0.6

Biotechnology

Question 10

Which of the following can be classified as an ancient form of biotechnology?

A Fermentation to clone bacteria

B CRISPR-Cas9 to edit DNA

C Fermentation to make beer and cheese

D Recombinant DNA technology to make transgenic organisms

Question 11

Which of the following is an ethical issue related to creating genetically modified crops for human consumption?

A The plants may have a faster growth rate.

B The plants may be more resistant to pests.

C Humans are interfering in the evolutionary process.

D Crops may be more resistant to herbicides and have less competition from weeds.

Question 12

Two people were debating the use of human embryonic stem (ES) cells in biotechnology. Person A believed that ES cells should never be used, because blastocysts are destroyed and this is the same as terminating a life. Person B believed that ES cell research should be allowed, because blastocysts are only a collection of cells and are not a human life. What type of issue is this debate an example of?

A Ethical

B Political

C Scientific

D Economic

Genetic technologies

Question 13

Which of the following statements is **not true** about artificial insemination and artificial pollination?

A They both reduce genetic variation.

B They both select for desirable traits.

C They are both forms of assisted reproduction.

D They both use recombinant DNA technology.

Question 14

What is the name of the enzyme that cuts DNA to allow for the insertion of a gene into a plasmid?

A DNA ligase

B DNA polymerase

C Reverse transcriptase

D Restriction endonuclease

Question 15

A gene inserted into a plasmid is called

A recombinant DNA.

B recombinant mRNA.

C recombinant protein.

D recombinant polypeptide.

Question 16

Which of these options is the correct order of the steps to make genetically modified bacteria?

A Digestion – ligation – transformation – selection

B Ligation – selection – transformation – digestion

C Selection – ligation – transformation – digestion

D Transformation – ligation – digestion – selection

Question 17

Golden Rice is a transgenic plant that has had two beta-carotene biosynthesis genes inserted into its genome. Which of the following is a social benefit of this biotechnology?

A The technology is patented.

B The rice plants are more resistant to disease.

C The rice plants have an increased growth rate.

D The rice grains reduce malnutrition in developing countries.

Question 18

Which of the following is a possible disadvantage of a transgenic organism?

A An increase in crop yields

B An increase in the growth rate of an animal

C A reduction in the amount of insecticide used on a farm

D Cross-fertilisation with closely related species in the wild

Question 19 ©NESA 2020 SI Q10

A farmer intends to artificially inseminate cows with semen from a bull that has been chosen based on characteristics of colour and muscle mass. The farmer does not know that the bull is heterozygous for a rare recessive allele not previously present in the farmer's cow population. The introduction of this recessive allele to the population of cows is an example of

A gene flow.

B genetic drift.

C natural selection.

D selective breeding.

Question 20 ©NESA 2019 SI Q20

The diagram shows how CRISPR/Cas9 can be used as a new tool for genetic engineering. This technology has dramatically improved scientists' ability to successfully modify genomes.

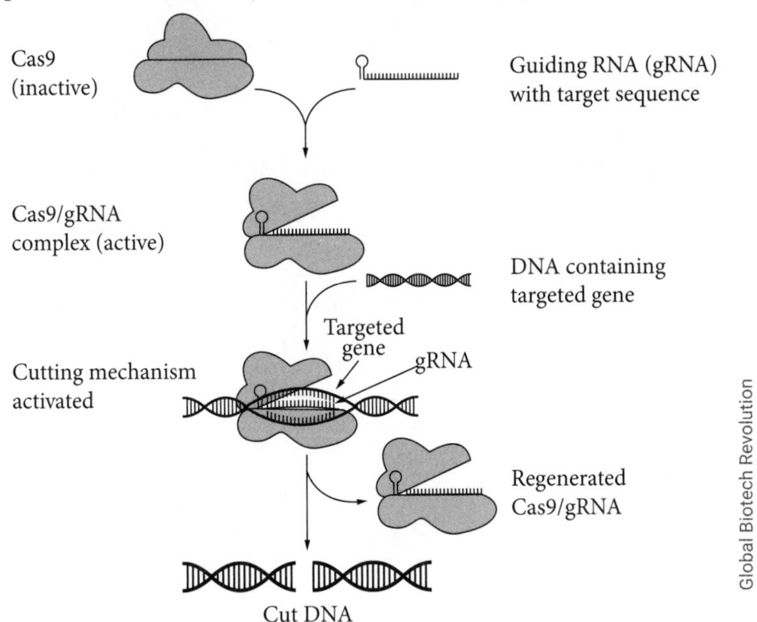

Scientists have been able to use biotechnology to 'cut and paste' DNA for decades. Why would the new CRISPR/Cas9 technology have improved the scientists' success in cutting DNA of specific genes?

A Cas9 is able to combine with specific DNA.

B Cas9 has an active site that cuts target DNA.

C gRNA has the same nucleotides as the target DNA.

D gRNA has nucleotides complementary to the target DNA.

Short-answer questions

Solutions start on page 256.

Mutations

Question 21 (4 marks)

a What is a point mutation? 1 mark

b Describe the cause and effect of **one** type of point mutation. 3 marks

Question 22 (5 marks)

The diagram below shows the coding DNA sequence of a section of an exon, and the same sequence with a genetic mutation (indicated by the arrow).

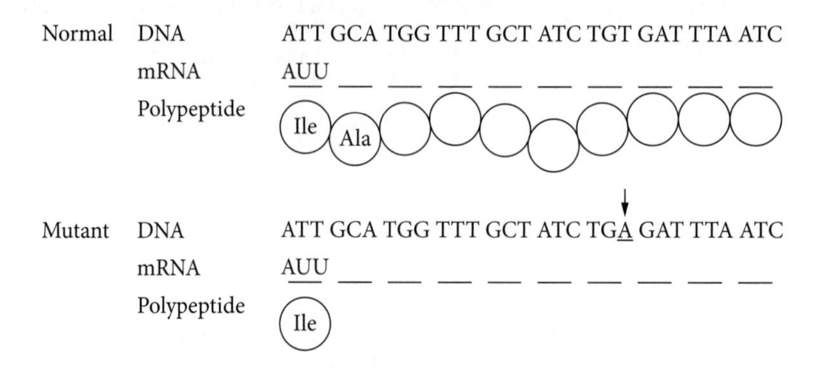

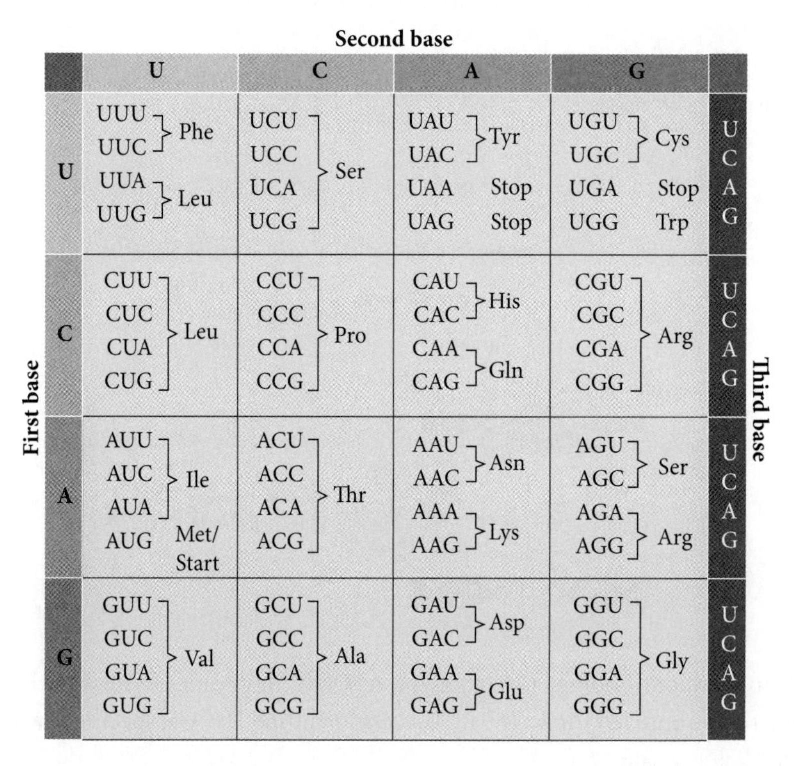

First base	Second base: U	Second base: C	Second base: A	Second base: G	Third base
U	UUU Phe UUC Phe UUA Leu UUG Leu	UCU Ser UCC Ser UCA Ser UCG Ser	UAU Tyr UAC Tyr UAA Stop UAG Stop	UGU Cys UGC Cys UGA Stop UGG Trp	U C A G
C	CUU Leu CUC Leu CUA Leu CUG Leu	CCU Pro CCC Pro CCA Pro CCG Pro	CAU His CAC His CAA Gln CAG Gln	CGU Arg CGC Arg CGA Arg CGG Arg	U C A G
A	AUU Ile AUC Ile AUA Ile AUG Met/Start	ACU Thr ACC Thr ACA Thr ACG Thr	AAU Asn AAC Asn AAA Lys AAG Lys	AGU Ser AGC Ser AGA Arg AGG Arg	U C A G
G	GUU Val GUC Val GUA Val GUG Val	GCU Ala GCC Ala GCA Ala GCG Ala	GAU Asp GAC Asp GAA Glu GAG Glu	GGU Gly GGC Gly GGA Gly GGG Gly	U C A G

a Use the amino acid table to complete the transcribed mRNA and the polypeptide sequence for the normal and mutated section of the gene. 2 marks

b What is the name of this type of mutation? 1 mark

c Describe what effect the mutation may have on the polypeptide. 2 marks

Question 23 (6 marks)

A student designed a computer program to model how genetic drift affects the frequency of an allele (p) in different population sizes. The simulation included three populations with a size of 5, 20 and 100 individuals (*N*). The frequency of the p allele was measured over 10 generations. The student presented the data in the table below.

Generation	Frequency of p allele		
	Population 1 (*N* = 5)	Population 2 (*N* = 20)	Population 3 (*N* = 100)
0	0.5	0.5	0.5
1	0.8	0.5	0.6
2	1.0	0.8	0.5
3	1.0	0.4	0.7
4	1.0	0.2	0.8
5	1.0	0.3	0.6
6	1.0	0.2	0.5
7	1.0	0	0.4
8	1.0	0	0.6
9	1.0	0	0.5
10	1.0	0	0.6

a Draw a graph of the data for the three populations. 2 marks

b In which generation did the p allele become 'fixed' in population 1? 1 mark

c Use the graph to explain why smaller populations are more likely to experience more extreme effects of genetic drift. 3 marks

Question 24 (7 marks)

The Chernobyl disaster was a nuclear accident that happened in 1986 in the former Soviet Union. The graph below shows the incidence of thyroid cancer in children, adolescents and young adults in the region in the decade after the accident.

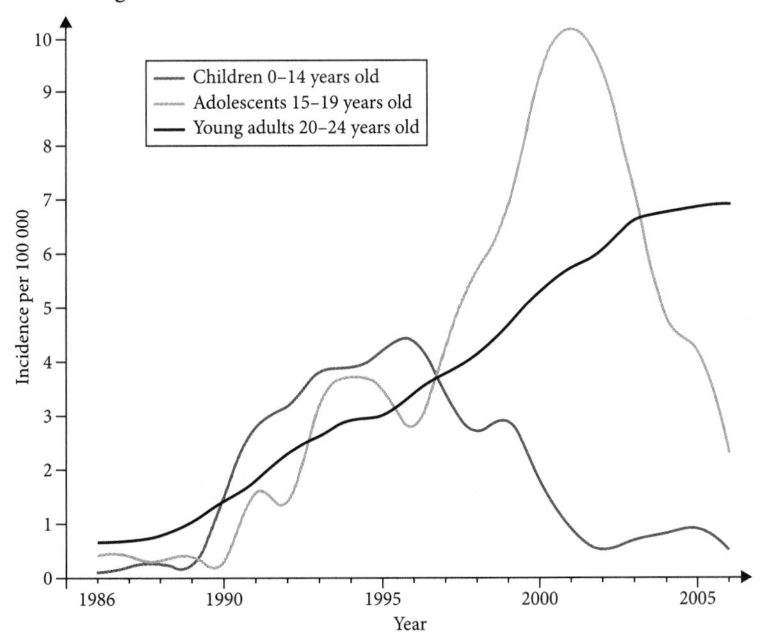

a What type of mutagen caused the cancer? 1 mark

b Describe how the mutagen caused genetic mutations that resulted in cancer. 2 marks

c Explain the trend in cancer incidence in children (0–14 years old). 4 marks

Question 25 (2 marks)

a Give **one** example of a coding and a non-coding region of the genome. 1 mark

b Describe the possible effect of a mutation in an intron. 1 mark

Question 26 (5 marks) ©NESA 2019 SII Q26

The map shows the percentage of adult indigenous populations able to digest lactose.

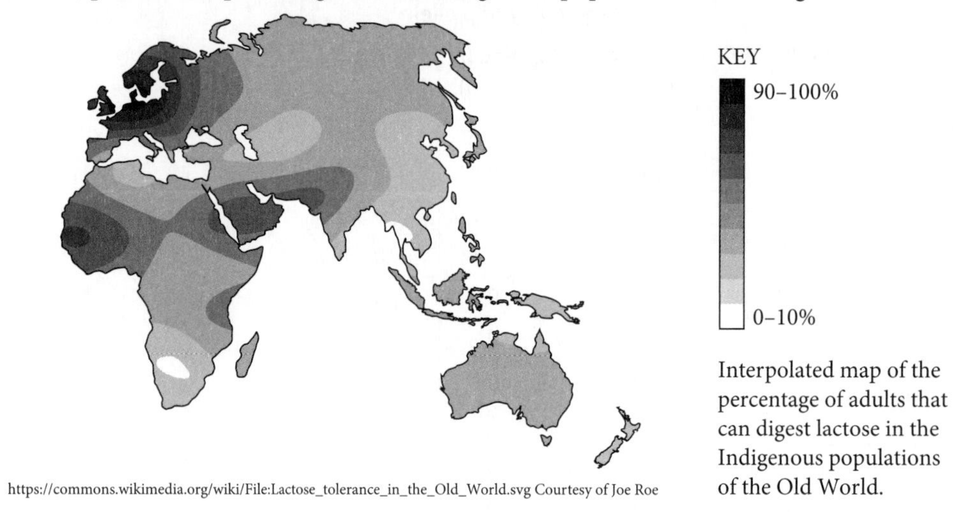

Interpolated map of the percentage of adults that can digest lactose in the Indigenous populations of the Old World.

https://commons.wikimedia.org/wiki/File:Lactose_tolerance_in_the_Old_World.svg Courtesy of Joe Roe

The ability to digest lactose is due to the presence of an enzyme (lactase) which can metabolise the sugar (lactose) present in milk. The gene responsible for producing lactase is usually permanently switched off at some time between the ages of 2 and 5 years. However, some people remain able to digest lactose throughout their lives. With reference to evolution and DNA, provide possible reasons for the distribution shown in the map.

Question 27 (5 marks) ©NESA 2021 SII Q24b

The diagram shows the early stages of embryonic development from a fertilised egg. The developing ball of cells has split and monozygotic (identical) twins have formed. Mutations can occur at different times during embryonic development, for example Mutation *A* would result in both twins having the mutation in all their cells.

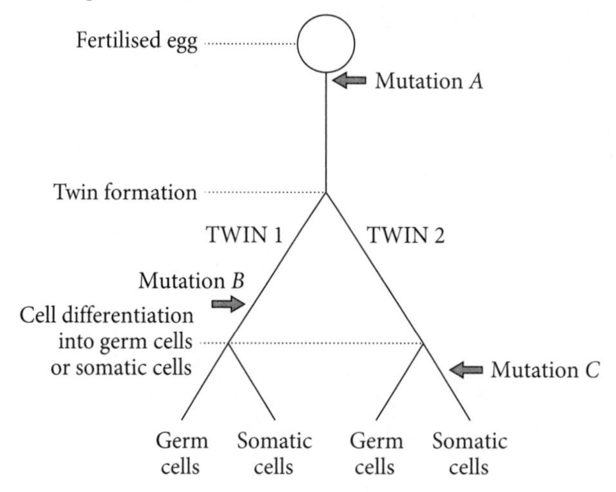

Explain the effects of Mutation *B* and Mutation *C* on each twin and on any offspring that they may have.

Question 28 (3 marks)

The sickle cell phenotype provides a selective advantage against malaria infection. In a population of 10 000 people in Africa, 16% of people are homozygous for the recessive allele (t). Use the following two equations to calculate the values for the table below.

$P + q = 1$

where P is the frequency of the dominant allele (T) and q is the frequency of the recessive allele (t).

Hardy–Weinberg equation:

$P^2 + 2Pq + q^2 = 1$

where P is the frequency of the dominant allele and q is the frequency of the recessive allele.

Genotype	Genotype frequency	Number of individuals
Homozygous dominant (TT)		
Heterozygous (Tt)		
Homozygous recessive (tt)		

Question 29 (5 marks) ©NESA 2020 SII Q29

Explain how **two** processes that affect the gene pool of populations can lead to evolution.

Biotechnology

Question 30 (4 marks)

Read the following statement to answer the questions that follow.

> **Xenotransplantation** is the transplantation of cells, tissues or organs from one species into another. In the 1990s, medical research began into the use of organs from genetically modified animals for organ transplants into humans. This research was driven partly by a worldwide shortage of donor organs from humans. Pigs have been the animal of choice in this type of research, because the size and anatomy of their organs is similar to those of humans. However, pig organs were originally rejected by the human immune system and there are concerns that this procedure could result in pig retroviruses infecting humans, possibly causing illness and disease pandemics. Research into xenotransplantation has slowed because of these serious concerns, but advances have been made in reducing immunological incompatibility between pigs and humans. In 2022, the first genetically modified pig heart was successfully transplanted into a human.
>
> (paraphrased from Clark J & Whitelaw B (2003), 'A future for transgenic livestock', *Nature Reviews Genetics* 4: 825–33, accessed at https://www.nature.com/articles/nrg1183)

a Identify a potential social benefit of this research. 1 mark

b Describe a negative social implication of performing xenotransplantation. 1 mark

c Provide **one** example of an ethical issue for, and **one** example of an ethical issue against, the use of genetically modified pigs for xenotransplantation. 2 marks

Question 31 (5 marks)

Use **one** example of a future application of biotechnology to answer the following questions.

a What is the name of the technology? 1 mark

b Describe the process used in the application. 2 marks

c Explain how the technology could benefit society. 2 marks

Question 32 (15 marks) ©NESA 2021 SII Q33a & c (ADAPTED) ●●●

Genetically engineered Atlantic salmon have been produced and approved for aquaculture in the USA. These salmon have a transgene that includes a protein-coding sequence from a Chinook salmon's growth hormone gene and the promoter region of an Ocean Pout's antifreeze protein gene. This makes the transgenic fish grow more quickly than wild-type salmon. The following diagram provides an overview of the production of the transgenic salmon.

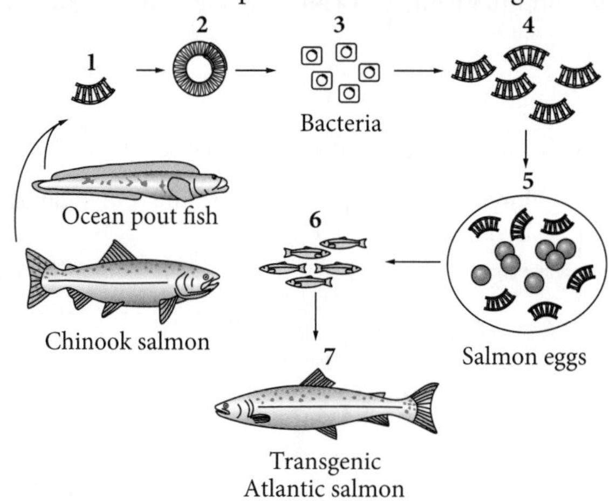

a Explain the processes shown in steps 1–5. 6 marks

b Transgenic fish can reproduce and pass on the dominant transgene (T).

Reproduction for aquaculture is strictly controlled using a variety of techniques in order to protect and preserve biodiversity.

Some of these techniques are outlined below.

1. Homozygous (TT) female (XX) breeding stock are kept in quarantine.
2. The female fish undergo hormone treatment that results in sex reversal and the development of male sex organs and sperm.
3. The sperm produced is collected and used to fertilise eggs obtained from wild-type, non-transgenic salmon.
4. The eggs are treated with pressure shock to prevent the completion of meiosis II. As a result, offspring are triploid (three copies of each chromosome). All offspring are transgenic female fish and have XXX (XXX fish cannot develop sex organs).
5. Offspring are transported to inland aquaculture tanks to be grown to market size.

Analyse how these techniques protect and preserve biodiversity. 9 marks

Genetic technologies

Question 33 (6 marks) ©NESA 2020 SII Q28 ●●

A student drew a diagram to model part of the process of meiosis.

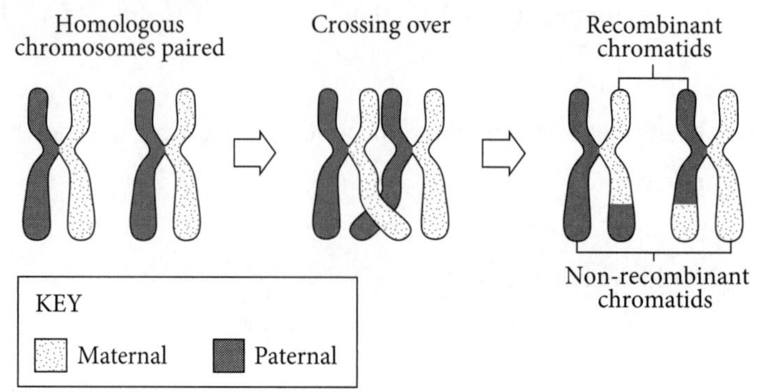

a Explain the misunderstanding of meiosis shown in this model. 3 marks

b Explain the effect of meiosis on genetic variation. 3 marks

Question 34 (5 marks)

The woolly mammoth (*Mammuthus primigenius*) became extinct in the Pleistocene epoch about 10 000 years ago. Mammoths lived in cold habitats in the northern hemisphere and their bodies were often well preserved in very cold permafrost. The mammoth's closest living relative is the Asian elephant (*Elephas maximus*). Some scientists believe whole-animal cloning can be used to bring extinct animals like mammoths back to life.

Using this information, complete the flow diagram below to show how a woolly mammoth could be cloned.

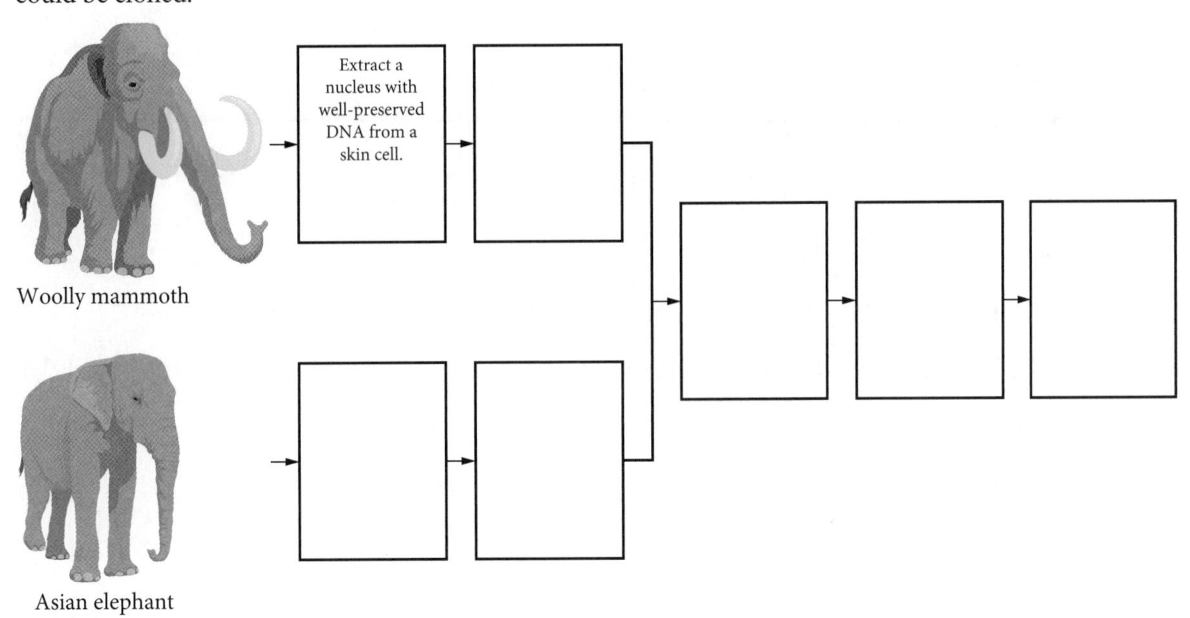

Question 35 (3 marks)

The following column graph shows the efficiency of whole-animal cloning in a variety of species. The data has been pooled from experiments performed by many research groups around the world.

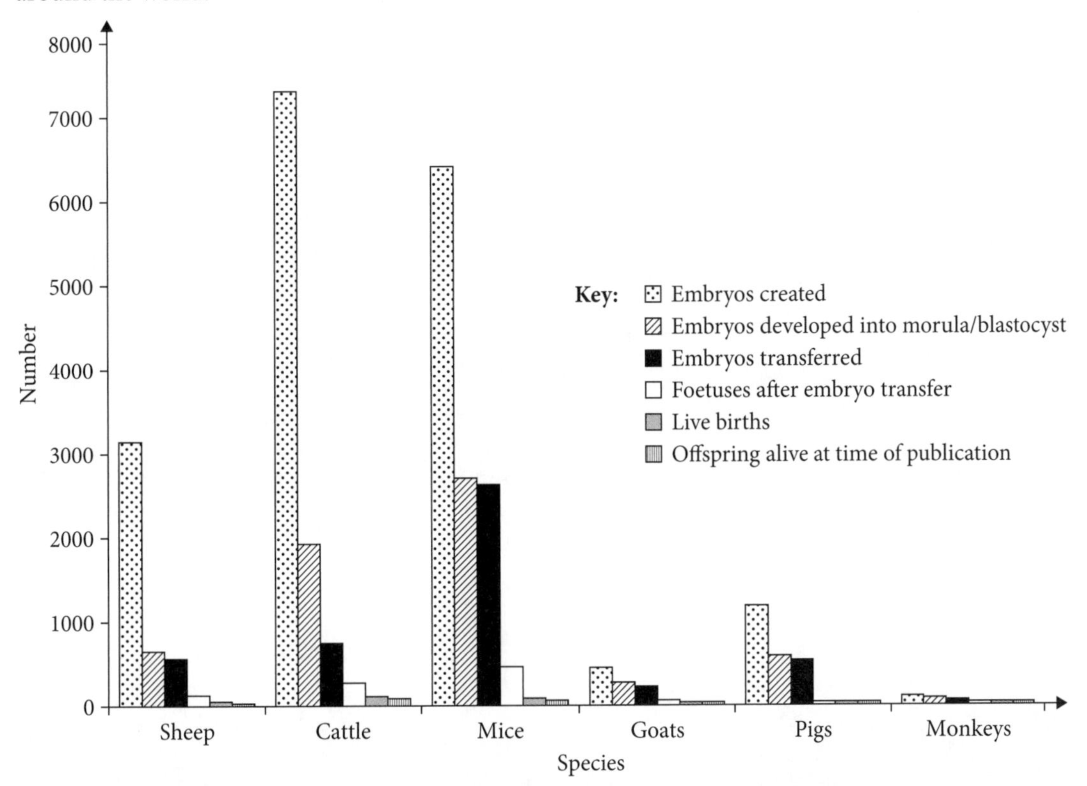

a Which species has had the most embryos created? 1 mark

b Describe the general pattern in the success rate of each stage in the cloning process. 2 marks

CHAPTER 3 MODULE 7: INFECTIOUS DISEASE

Chapter 3
Module 7: Infectious disease

Module summary

Module 7 explored the different pathogens that cause infectious diseases, and the historical and current developments and practices associated with their **treatment**, **prevention** and **control**. The challenges associated with controlling disease spread during epidemics and pandemics was also addressed. Module 7 focuses on the impact of infectious diseases in agriculture and how animals and plants respond to exposure to pathogens. Finally, there is a strong emphasis on understanding how the innate and adaptive immune systems protect the human body against invading pathogens.

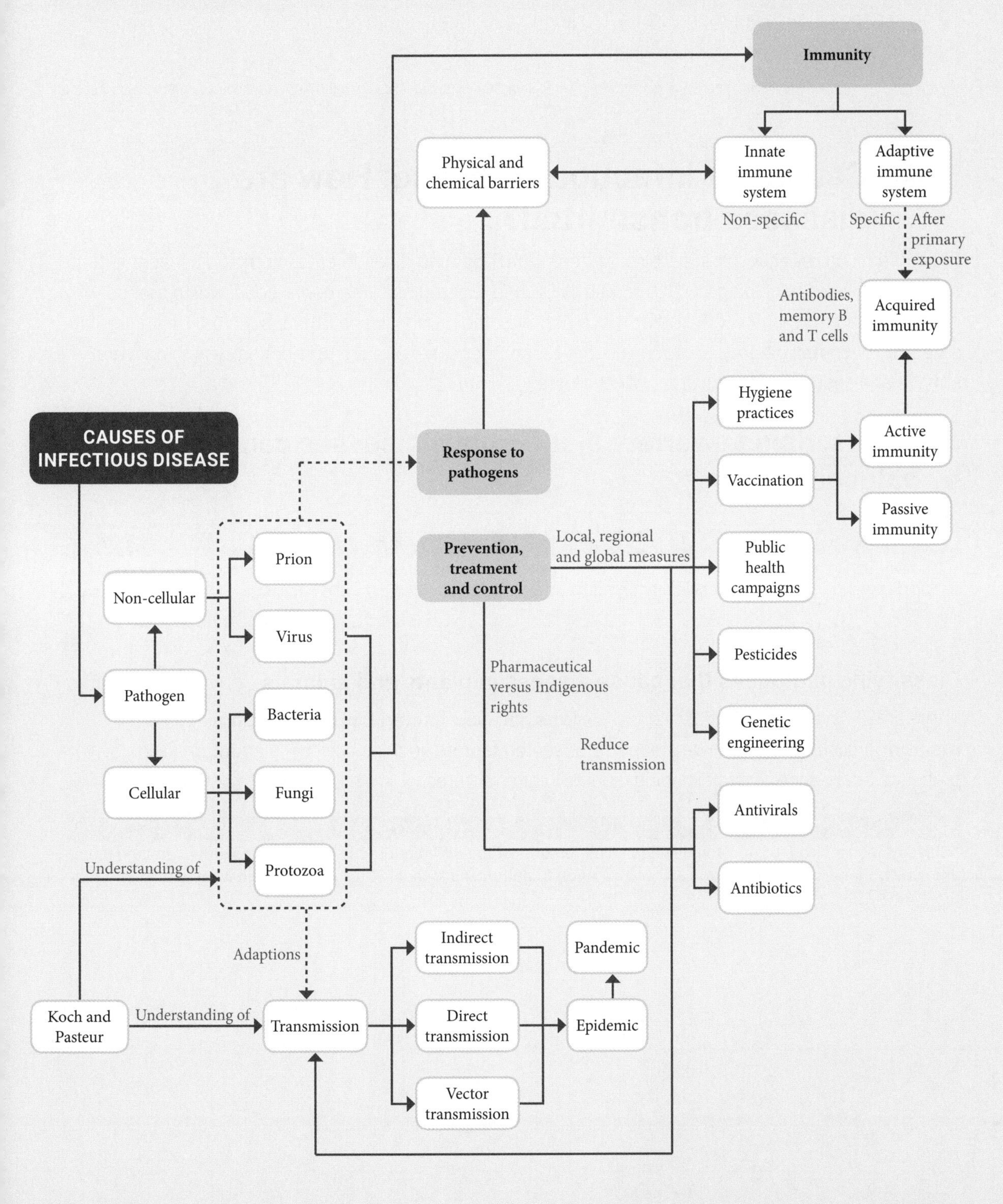

Outcomes

On completing this module, you should be able to:

- analyse infectious disease in terms of cause, transmission, management and the organism's response, including the human immune system.

Key science skills

In this module, you are required to demonstrate the following key science skills:

- develop and evaluate questions and hypotheses for scientific investigation
- design and evaluate investigations in order to obtain primary and secondary data and information
- conduct investigations to collect valid and reliable primary and secondary data and information
- select and process appropriate qualitative and quantitative data and information using a range of appropriate media.

3.1 Causes of infectious disease: How are diseases transmitted?

Each infectious **disease** is caused by a different **pathogen**: a disease-causing biological agent. Many of these diseases, including COVID-19, HIV/AIDS and tuberculosis, are contagious, which means they can be spread from an infected person to a non-infected person, while others such as malaria require a **vector** for **transmission**. In order to survive, pathogens use a range of mechanisms to gain entry, survive and ensure transmission between hosts.

3.1.1 Describing a variety of infectious diseases caused by pathogens

Note

This is an 'including' dot point, meaning you may be asked questions on microorganisms, macroorganisms or non-cellular pathogens, and on collecting primary and secondary-sourced data and information relating to disease transmission.

Classifying pathogens that cause disease in plants and animals

There are six main groups of pathogens: **prions**, **viruses**, **bacteria**, **protozoa**, **fungi** and **macroorganisms**. They can be classified based on factors such as size, cellular nature (living or not), presence or absence of nucleic acids, and cellular structures.

Note

'Classify' requires you to know the distinguishing features of the different groups of pathogens and to arrange them into categories. You could be given any combination of pathogens in the HSC exam. Focus on what sets them apart from each other.

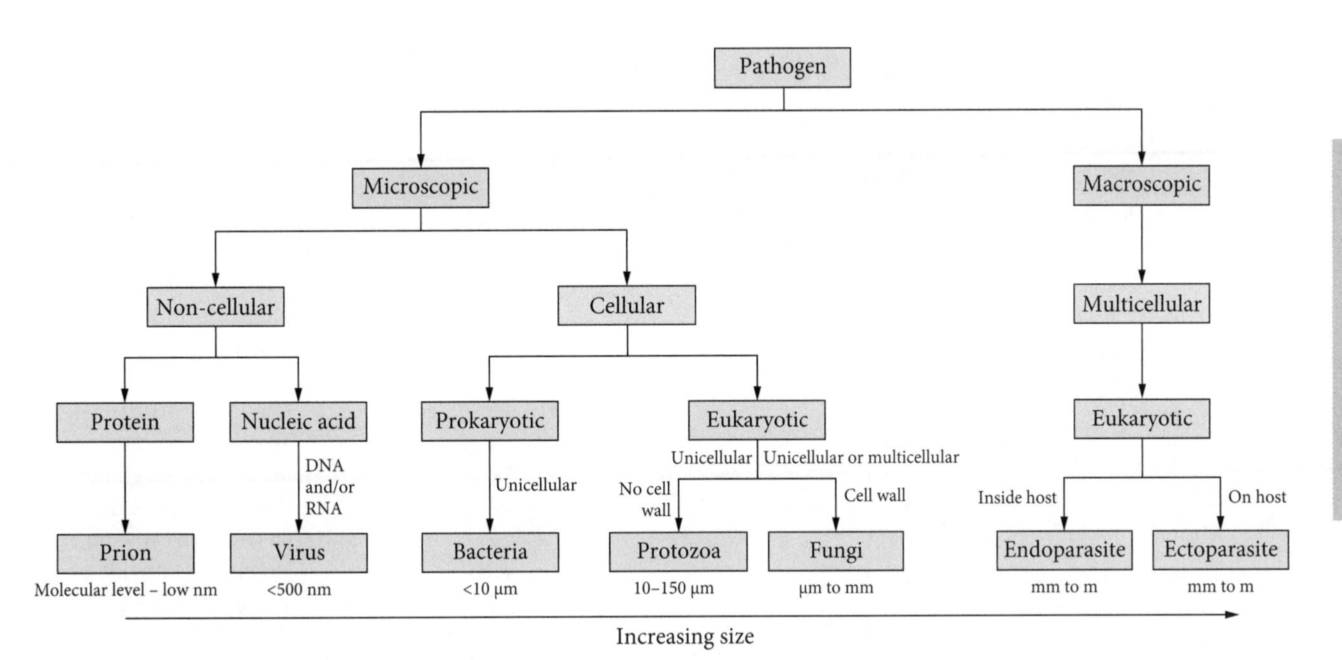

FIGURE 3.1 Classification of pathogens

TABLE 3.1 Features of the main pathogen groups

Pathogen type	Distinguishing features	Disease and causative pathogen (animal and plant)	Mode of transmission	Description of disease
Prions	• Non-living; contain no nucleic acids (DNA or RNA) • Modified/ misfolded proteins of 200–250 amino acids long • Only seen with electron microscope	• Creutzfeldt-Jakob disease (CJD) in humans	• Eating infected meat • Contaminated medical equipment • Receiving organ transplant from an infected individual	• Prion replicates by triggering conversion of normal PrP^{C} (cellular prion protein) on neurons into PrP^{Res} (resistant prion protein), which is resistant to protease enzymes responsible for breaking down incorrectly folded proteins. Brain tissue is destroyed, and infected individuals lose neurological function. No treatment or cure.
		• No known plant examples		
Viruses	• Non-cellular (non-living) • Have nucleic acids (DNA and/or RNA) • Unable to reproduce without host cell • 20–500 nm; only visible with electron microscope	• Acquired immunodeficiency syndrome (AIDS) caused by human immunodeficiency virus (HIV)	• Exchange of bodily fluids – primarily unprotected sex with infected person or sharing of contaminated needle	• Targets the immune system and weakens the body's ability to fight off infectious diseases. This leads to AIDS, which is defined by the onset of several cancers and other life-threatening infections. Over 40 million deaths recorded.
		• Named based on the symptoms caused in the host plant, e.g. tobacco mosaic virus disease caused by TMV	• Insect vector • Contact with infected plant or contaminated equipment	• Discolouration and mottled browning of leaves

››

TABLE 3.1 cont.

››

Pathogen type	Distinguishing features	Disease and causative pathogen (animal and plant)	Mode of transmission	Description of disease
Bacteria	• Prokaryotic • Smallest of the cellular pathogens, 1–10 µm • Visible with light microscope	• Tuberculosis (TB) caused by *Mycobacterium tuberculosis*	• Inhalation of shed bacteria	• Primarily affects lung tissue. When active, highly contagious and presents with persistent coughing and leads to significant tissue damage. Has caused over 1 billion deaths (more than any other infectious disease in history). Growing concerns with rise of antibiotic-resistant strains.
		• Fire blight caused by *Erwinia amylovora*	• Insect vector • Contact with infected plant or contaminated equipment	• Affects all parts of pome fruit (apples and pears) plants, and makes the plant look burnt
Protozoa	• Single-celled, eukaryotic, no chloroplasts or cell wall • 10–150 µm • Visible with light microscope • Typically, complex life cycle that includes multiple hosts	• Malaria caused by five species of *Plasmodium* genus. *P. falciparum* is most dangerous	• Bite from infected female *Anopheles* mosquito	• High fever, shaking chills and flu-like illness caused by rupture of red blood cells and damage to blood vessels. Most significant protozoan disease, with 2.3 million new cases per year and estimated 400 000 deaths.
		• Phloem necrosis in coffee plants caused by *Phytomonas leptovasorum*	• Insect vector feeds from infected plant and transmits to uninfected plant	• Yellowing of leaves, eventual root dieback and death of the plant
Fungi	• Eukaryotic, heterotrophic, have a cell wall	• Body ringworm (tinea corporis) caused by *Epidermophyton floccosum*	• Skin-to-skin contact • Contact with contaminated towels, clothes or flooring	• Red, raised, itchy and cracked skin. Symptoms are caused by digestive enzymes released by the fungus.
		• Powdery mildew • Not caused by a single pathogen but by many hundreds of species that produce the same symptoms	• Wind- or water-borne • Contact with infected plant or contaminated equipment	• Powdery mildew produces a white powdery appearance on the plant due to large numbers of spores.

››

TABLE 3.1 cont.

Pathogen type	Distinguishing features	Disease and causative pathogen (animal and plant)	Mode of transmission	Description of disease
Macro-organisms	• Multicellular, eukaryotic, visible to naked eye • Endoparasites or ectoparasites • Cause disease directly or as vectors that transmit other pathogens	**Endoparasite** • Taeniasis (a range of abdominal and digestive complaints) caused by *Taenia saginata* (beef tapeworm)	• Contracted by cattle after ingesting pasture mites infected with tapeworm larvae	• Abdominal pain, loss of appetite and weight loss. Tapeworm segments are passed in faeces. • Rash and flu-like symptoms often associated with joint pain and weakness in limbs
		Ectoparasite • Lyme disease, although caused by bacteria from genus *Borrelia*, is contracted from bite of an infected paralysis tick (*Ixodes holocyclus*) • The same tick species is responsible for tick paralysis in a range of animal species, including humans	• Contracted by humans after ingesting undercooked meat that contains tapeworm eggs or larvae • Bite from paralysis tick	• Paralysis tick secretes a neurotoxin as it feeds. Results in loss of muscle control, difficulty breathing, heavy salivation and unconsciousness if the tick is not removed. In animals, will lead to respiratory or heart failure if untreated.
		• Plant diseases linked to macroorganisms are typically the result of the organism acting as a vector • Mosaic virus diseases	• Aphid vector	• Aphids are sap-sucking insects. They transfer the virus between plants by eating from an infected plant before moving to a non-infected plant. All mosaic viruses cause irregular colouring and shaping of leaves.

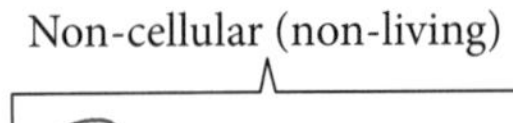

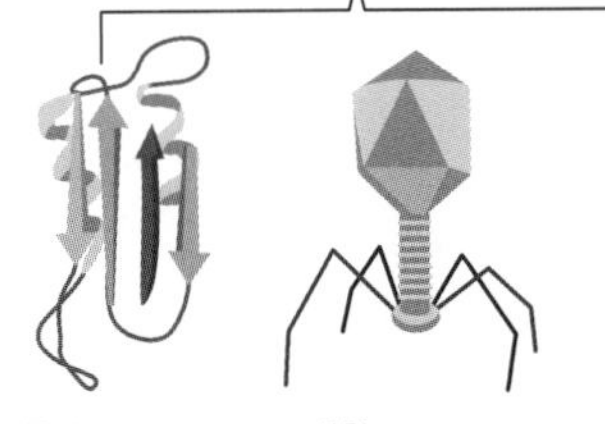

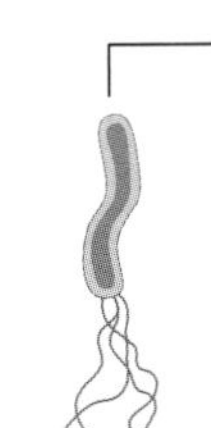

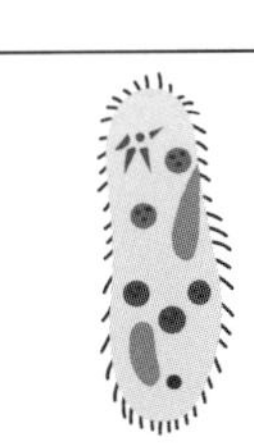

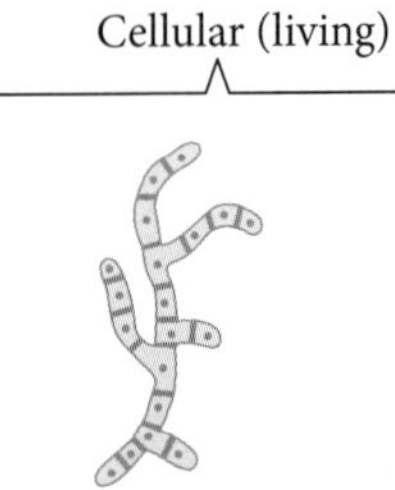

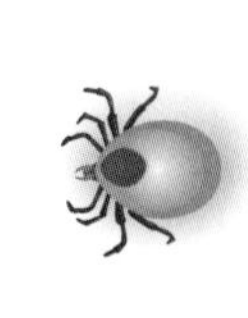

FIGURE 3.2 Cellular and non-cellular pathogens and examples of diseases they cause

Note

Make sure you know the difference between the pathogen and the disease. You will lose marks if asked for a disease and you give the causative pathogen instead.

Investigating transmission of a disease during an epidemic

An **epidemic** is an outbreak of an infectious disease that spreads rapidly among a large number of individuals within a defined geographical area. In contrast, a **pandemic** is typically the spread of a new disease, or a new variant of a disease, across a wider geographical area, often worldwide. The World Health Organization (WHO) monitors many diseases closely due to concerns about the risk of possible epidemics and pandemics. These include Ebola virus disease (EVD), cholera, **zoonotic** influenzas (bird flu and swine flu) and several diseases caused by coronaviruses (SARS, MERS and COVID-19).

CASE STUDY 1

ZIKA VIRUS EPIDEMIC IN BRAZIL

Zika disease is caused by the Zika virus. Symptoms of the disease are typically very mild and include fever, rash and headache or, in many cases, no symptoms at all. However, the disease is very problematic for pregnant females as the virus can pass through the placenta to the baby. It can result in miscarriage, preterm birth and microcephaly, a significant birth defect.

Transmission

Zika virus is a zoonotic pathogen transmitted primarily by the bite of an infected *Aedes aegypti* mosquito. Secondary modes of transmission include from mother to foetus during pregnancy, or sexual contact. Zika was first identified in a sick monkey in the Zika Forest of Uganda in 1947. In 1952, the first reported human case was confirmed, and it has subsequently been detected in nearly 90 countries.

In 2016, the WHO declared a Zika virus epidemic in Brazil when scientific studies demonstrated a link between a more virulent strain of the virus and increasing numbers of birth defects. An estimated 1–1.5 million infections occurred during the epidemic, with several thousand confirmed cases of congenital Zika syndrome. By mid-2017, cases of the disease dropped to zero in Brazil. While there is no definitive explanation for the disappearance of the virus, scientists believe it was due to a combination of seasonal weather patterns that impaired the mosquito vector, vigilant control measures and a degree of herd immunity that resulted from the high levels of infections.

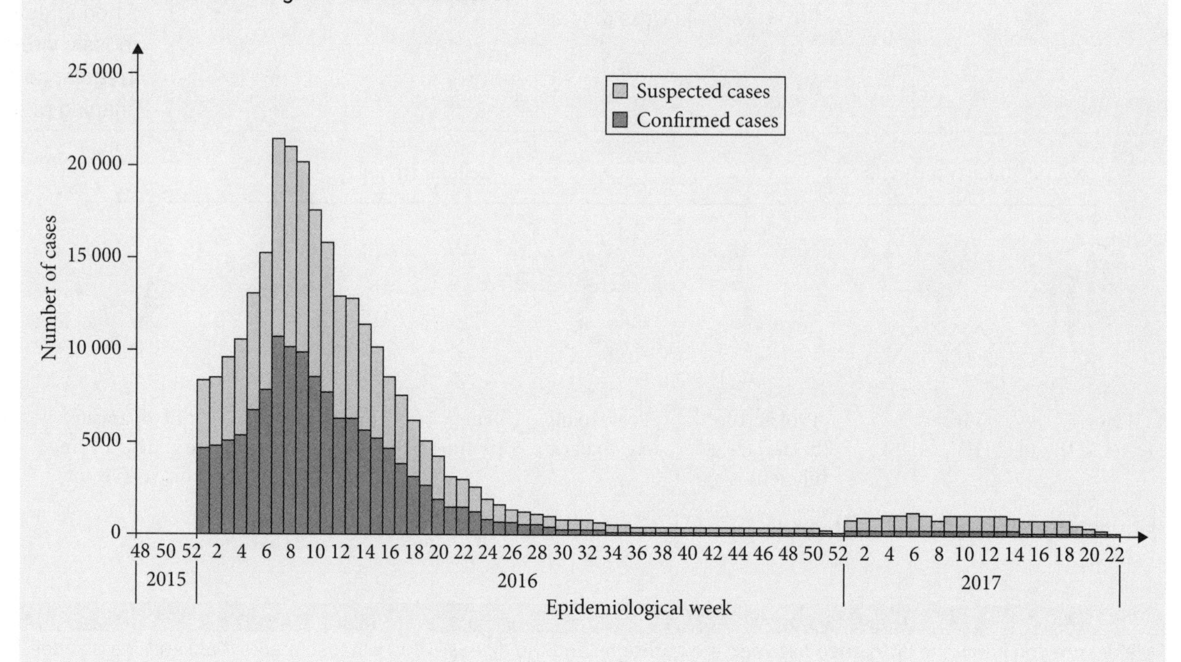

FIGURE 3.3 Suspected and confirmed cases of Zika virus in Brazil 2015–2017

Designing and conducting a practical investigation relating to the microbial testing of water or food samples

This dot point highlights the application of scientific skills linked to knowledge and understanding outcomes, specifically WS12-2 and WS12-3. There are multiple forms your practical investigation could take. For example, a simple practical investigation could be designed to identify the presence of **microbes (microorganisms)** in a sample of water. A more complicated experiment could determine the effect of an identified independent variable on a dependent variable. This latter example would require explicit controlled (constant) variables.

In all cases there should be an experimental control to ensure that any observed microbes are the result of the samples being tested, a risk assessment given the hazards associated with microbes, and a methodology that incorporates sterile technique to avoid cross-contamination.

Note

This dot point, and similar methods-based dot points, will be assessed by asking you to either write a method or analyse a method provided to you. In both cases, the focus will be on your ability to demonstrate a clear knowledge of the different variables, experimental controls, **validity** and **reliability**.

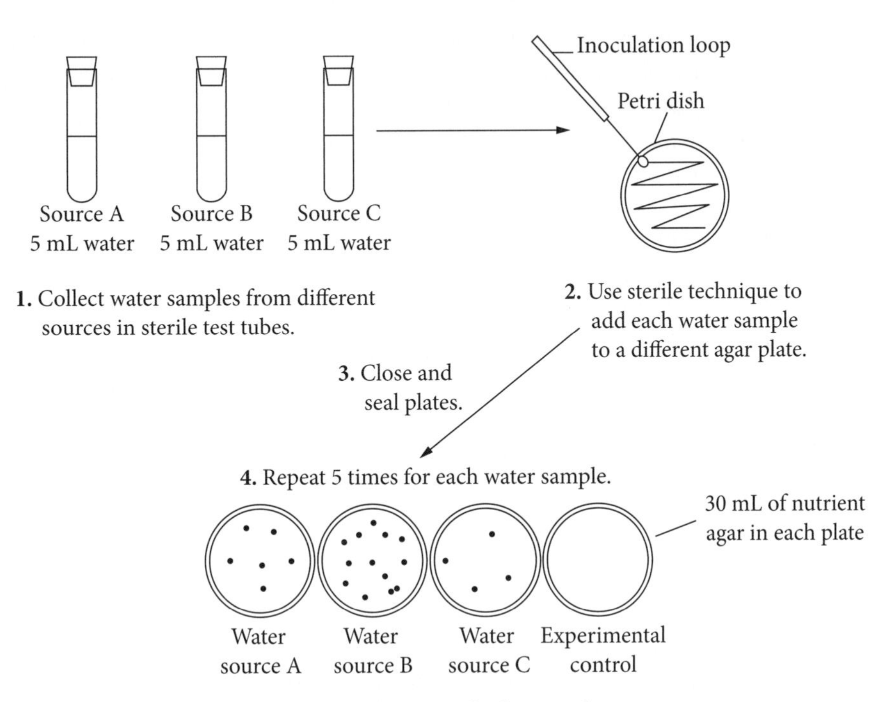

- **Independent variable:** source of water
- **Dependent variable:** number of colonies recorded or percentage of agar covered by colonies
- For **validity**: control the variables, e.g. equal volumes collected, same temperature and incubation time
- For **reliability**: repetition and averaging of consistent results after removing outliers
- To improve **accuracy**: could include technology to better count microbial colonies or determine percentage of agar covered by colonies

6. Record:

- *Quantitative data*: number of microbial colonies on each plate, or percentage of agar covered by colonies. Determine an average for each sample.
- *Qualitative data*: general appearance, diameter and colour of colonies

FIGURE 3.4 The experimental set-up to test for microbes in water samples

Investigating modes of transmission of infectious diseases, including direct contact, indirect contact and vector transmission

Understanding the mechanism by which pathogens are transmitted between hosts is essential to controlling the spread of infectious diseases. Modes of transmission are broadly categorised as direct, indirect or vector-borne. Many diseases (e.g. measles and influenza) can be transmitted by both direct and indirect contact.

TABLE 3.2 Modes of transmission of infectious diseases

Mode of transmission	Details	Examples
Direct contact	• Transfer of pathogen upon physical contact between infected and non-infected individuals • Exchange of bodily fluids during kissing or sexual intercourse • Inhalation of airborne droplets that have been coughed or sneezed into the surrounding environment by an infected person	• COVID-19 (SARS-CoV-2) • AIDS (HIV) • Tuberculosis (*Mycobacterium tuberculosis*) • Measles (measles virus) • Influenza (influenza virus)
Indirect contact	• Transfer of pathogen that does not involve direct human-to-human contact • Can include transmission of airborne droplets, where pathogens remain viable outside the host body for many hours • New individuals become infected when they touch a contaminated object and then transfer the pathogen to one of their mucous membranes • Consumption of contaminated food or water is also a common mode of indirect transmission	• Cholera (*Vibrio cholerae*) • Food poisoning (*Escherichia coli*) • Measles (measles virus) • Influenza (influenza virus)
Vector-borne	• Pathogen is transmitted typically by an **arthropod** vector • Responsible for nearly 20% of all infectious diseases, killing an estimated 700 000 people annually • Malaria causes an estimated 400 000 deaths annually • Dengue fever is the most significant viral vector-borne disease, with 96 million cases per annum, 40 000 deaths and over half the world's population are at risk of contracting the disease	• Malaria (*Plasmodium* transmitted by the bite of an infected female *Anopheles* mosquito) • Dengue fever (Dengue virus transmitted by the bite of an infected female *Aedes* mosquito)

3.1.2 Investigating the work of Robert Koch and Louis Pasteur, to explain the causes and transmission of infectious diseases

In the late 1800s, Louis Pasteur (1822–1895) and Robert Koch (1843–1910) made some of the most significant discoveries in human history regarding the causes of infectious diseases and the mechanisms by which they spread. Their respective contributions are central to improved health outcomes and have allowed us to keep ahead of the evolutionary 'arms race' we are currently in with many pathogens.

Note

Analyse every dot point carefully. If it is in the syllabus, it could be in your HSC exam. In this case, know the work of Koch *and* Pasteur in detail and how each contributed to our understanding of cause *and* transmission of infectious diseases. Be ready to clearly articulate the difference between cause and transmission.

Prior to the work of Pasteur and Koch, **spontaneous generation** was a widely held theory used to explain how new organisms came into existence directly from non-living matter. For example, it was believed that rats came from garbage and maggots came from rotting meat.

Koch's postulates

Robert Koch hypothesised that each infectious disease was *caused* by a specific pathogen. To test his theory, Koch worked extensively with the bacterium *Bacillus anthracis* and demonstrated what is now known as Koch's postulates, the basis behind identifying a causal relationship between a specific pathogen and a disease. Koch followed the same procedure to determine that *Mycobacterium tuberculosis* was the causative pathogen of tuberculosis and *Vibrio cholerae* was responsible for cholera. His work highlighted that *transmission* of an infectious disease occurs when an infected individual passes on the causative pathogen to a non-infected individual. Koch's work in identifying the causative pathogens of a disease was influenced by Pasteur's germ theory.

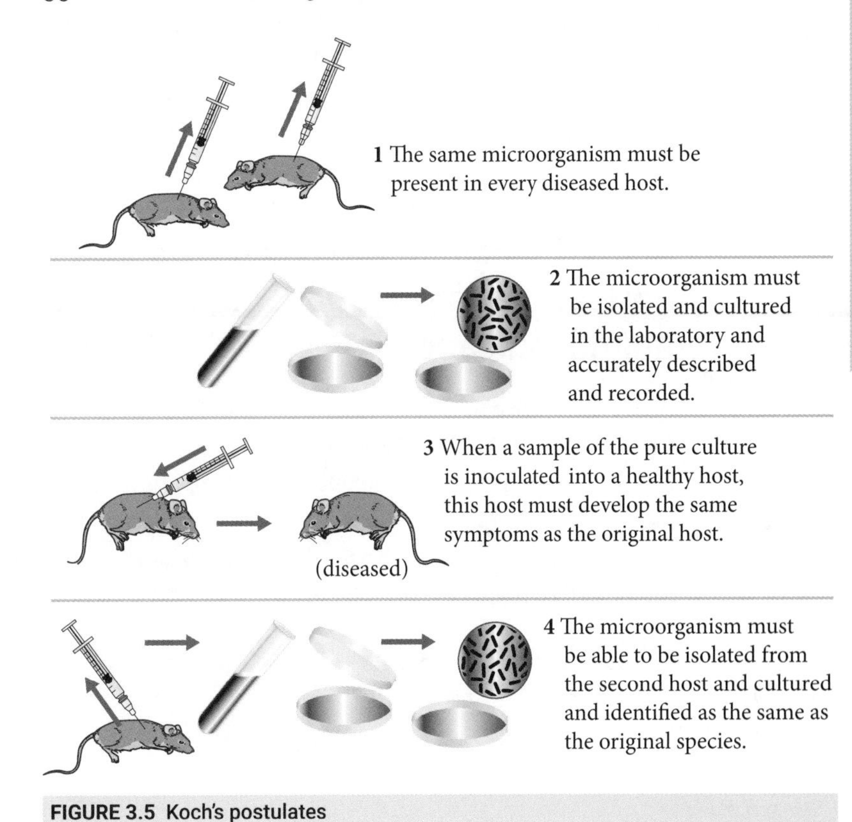

FIGURE 3.5 Koch's postulates

Pasteur's experiments on microbial contamination

Pasteur **hypothesised** that the long-held theory of spontaneous generation was incorrect. He believed that microbes were airborne and that food and liquids spoil when these microbes land and become active. In the early 1860s, Pasteur performed his famous swan-neck flask experiments, which disproved the theory of spontaneous generation and clearly demonstrated that microbes are airborne and could therefore be *transmitted* from an infected to a non-infected individual. Over 150 years later his 'control' flask is on display in the Pasteur Institute in Paris, and it is still microbe free!

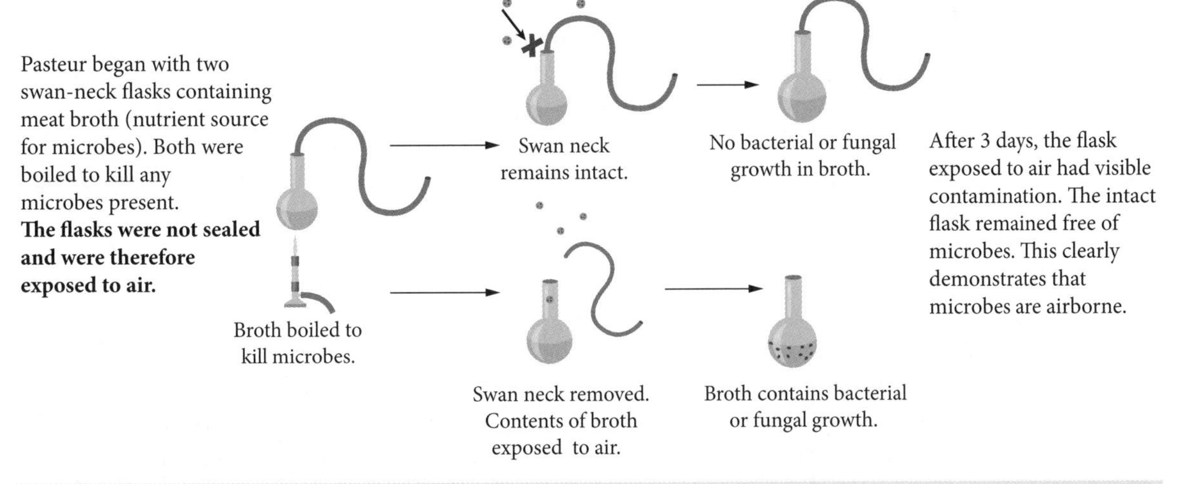

FIGURE 3.6 Pasteur's swan-neck flask experiment

Pasteur also worked extensively with cholera, anthrax and rabies. In 1861, his combined areas of research led him to propose the germ theory, which states that specific microbes are the *causative* agents of infectious diseases. His work was influenced by Koch's postulates, which assisted him in developing vaccines specific to chicken cholera in 1879, anthrax in 1881 and rabies in 1885. In developing the anthrax vaccine, he agreed to a public experiment. He took 50 sheep, and provided two doses of an **attenuated** vaccine to 25 while the other 25 received no vaccine. All 50 sheep were then exposed to the causative pathogen, *Bacillus anthracis*. After three days the vaccinated sheep remained healthy, while all 25 non-vaccinated sheep died.

3.1.3 Assessing the causes and effects of diseases on agricultural production, including but not limited to plant and animal diseases

Agricultural diseases have a significant impact on global production. An estimated 20–40% of global crop production is lost annually to pests and diseases, at a cost of over A$300 billion. Similarly, 20% of livestock is lost to disease, with many more animals affected by disease, which in turn reduces productivity. Collectively these losses limit the availability of food, fibre and dairy products, which could otherwise support those in need of these resources. With the world's growing population predicted to reach over 8.5 billion by 2030, a 50% increase in current food production is required.

Note

For this dot point you need to make a judgement about the cause *and* effect of animal *and* plant diseases on agricultural production, *and* be familiar with an additional example ('including but not limited to'). Consider structuring your response to clearly distinguish cause, effect and judgement for each of your chosen diseases. This could be done via a table or in a more formal written response with headings.

These growing food demands typically require a move to industrial-scale farming practices. While these practices produce higher yields of the desired product at a cheaper cost, large numbers of genetically similar animals and plants are kept in extremely close proximity to one another. With this comes an increased risk of disease. The ease and volume with which agricultural products are transported on the global economic market further increases the risk of disease transmission and outbreaks.

Plant diseases

CASE STUDY 2

PANAMA DISEASE

While there are over 80 wild species of bananas, which can reproduce sexually and asexually, only two are used in commercially grown cultivars: *Musa acuminata* and *Musa balbisiana*. Over time, selective breeding of mutant strains, chosen for their size, sweetness and seedless phenotype, has produced the commercially available fruit of today. The seedless mutation makes them unable to reproduce sexually. As a result, the significant majority of bananas grown worldwide have been propagated asexually and are therefore genetically identical. This, combined with the monoculture practices used to grow them, places the world's banana crops at significant risk of disease outbreaks.

In the 1950s, Panama disease (banana wilt) caused by the fungal pathogen *Fusarium oxysporum*, decimated the global population of the *M. acuminata* Gros Michel (Big Mike) cultivar that dominated the world market. The fungus enters the plant's vascular tissue from the soil, blocking the **xylem** and preventing water and nutrient transport, causing the plant to wilt and die.

In the late 1950s, a second sterile cultivar, *Musa acuminata* Cavendish, was found to be resistant to Panama disease while also producing higher yields. It replaced the Gros Michel banana and currently constitutes 50% of total world banana production and 99% of the world's export market. An estimated 50 billion tonnes of Cavendish bananas are produced globally each year, with an export value of over A$10 billion. In Australia, banana production primarily supplies the domestic market and is valued at A$450 million. In 1989, a new strain of *F. oxysporum* called Tropical Race 4 (TR4) was discovered in Taiwan in a Cavendish banana crop. Despite attempts to contain TR4 it has now spread around the world and has been detected in Australia and all major banana-growing countries. It is again anticipated that the world's banana crops will be decimated in the next decade if new alternatives are not developed, which will likely include genetically modified (GM) crops.

Animal diseases

CASE STUDY 3

NEWCASTLE DISEASE

Newcastle disease (ND) is a highly contagious disease of birds and is of particular concern to the poultry industry. The disease is made more problematic by the high animal density associated with battery cage farming of chickens.

The causative pathogen is the Newcastle disease virus (NDV), which is transmitted primarily through droppings and secretions from infected birds. The disease presents with a range of symptoms, including loss of appetite, shortness of breath, oral and nasal discharge, bright green diarrhoea and often significant neurological impairment.

While the disease can be fatal, there is also high **morbidity**, which has a significant impact on poultry meat and egg productivity and therefore income revenue. To protect against the disease, NDV-free countries, including Australia, place restrictions on the importation of animal products from regions where the disease is **endemic**. As a result, when outbreaks occur in countries dependent on their poultry export market, flocks are destroyed, including non-infected animals.

The Australian chicken egg and meat industries are valued at over A$825 million and A$2.5 billion per annum respectively. Because many countries only import poultry products from NDV-free countries, experts believe that even a small ND outbreak in Australia would impact significantly on the economic viability of the egg and meat industry. As a reference, a 2002–2003 outbreak of ND in the USA required the culling of over 3 million birds at an estimated cost of over A$230 million.

Fungal diseases

Note

The dry bubble disease case study addresses the 'including but not limited to' aspect of the dot point.

CASE STUDY 4

DRY BUBBLE DISEASE IN MUSHROOMS

Dry bubble disease is a fungal disease of commercially grown mushrooms (also fungi) caused by *Lecanicillium fungicola*. The pathogen's primary host is the white-button mushroom. Symptoms vary depending on the time of infection, with late infections causing brown necrotic lesions on the fruiting body, and early infections causing deformed masses of undifferentiated mushroom tissue. The disease is highly contagious, with spores spread easily through water and air.

Mushroom crops are grown asexually and in monoculture, which results in extremely low genetic variation in commercially grown populations of mushrooms. While this produces crops of great quality and quantity, they are also highly susceptible to disease outbreaks.

Australian growers produce over 70 000 tonnes of mushrooms per annum, valued at over A$450 million, while the global market is valued at over A$65 billion. Currently, dry bubble disease is estimated to cost the industry 5% of total annual revenue.

3.1.4 Comparing the adaptations of different pathogens that facilitate their entry into and transmission between hosts

For continuity of a pathogen species, the pathogen must gain entry, survive and reproduce within a host before being transmitted to a new host to continue the cycle. This requires specific adaptations to breach the defence mechanisms of the host, and other adaptations to survive and reproduce in the hostile environment inside the host. If the pathogen is eliminated before being transmitted to a new host, it runs the risk of extinction.

Note

Note the plural – *pathogens*. You must be familiar with at least two different pathogens. A comparison requires both similarities *and* differences. A table is recommended. For this dot point, clearly distinguish between adaptations that facilitate entry and those that facilitate transmission.

Specific adaptations of *Plasmodium falciparum*

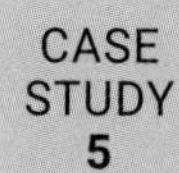

CASE STUDY 5

PLASMODIUM FALCIPARUM (MALARIA)

Recall that there are five species of protozoans from the *Plasmodium* genus that cause malaria, with *P. falciparum* the most significant in terms of human mortalities. All species have a complex life cycle consisting of distinct stages in different host organisms. They are unable to survive outside their host organism(s).

1. Female *Anopheles* mosquito takes a blood meal, inoculating human (or other animal) with **sporozoite** stage of *P. falciparum*.

2. Sporozoites migrate to liver and infect **hepatocytes**. Sporozoites reproduce asexually by multiple fission to produce **merozoites**. They cause host cells to rupture and release merozoites into bloodstream, where they infect red blood cells.

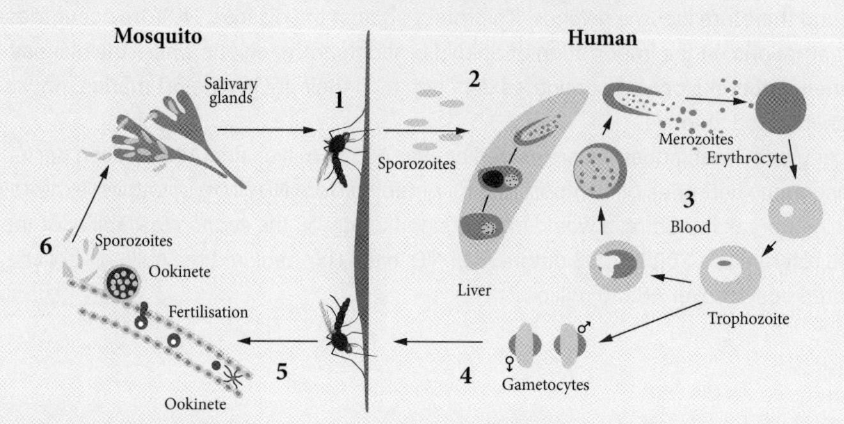

3. When in the RBCs the merozoites enter the next stage of the life cycle, called a **trophozoite**. The trophozoite takes one of two possible pathways.

First, they can reproduce via multiple fission to produce more merozoites, which cause the RBCs to rupture, releasing the merozoites into the bloodstream to repeat the cycle. This stage causes the typical symptoms of malaria: fever, chills, sweats and anaemia.

4. The second pathway involves merozoites differentiating into either male or female **gametocytes** (**microgametocytes** and **macrogametocytes** respectively).

5. The gametocytes are ingested by another female *Anopheles* mosquito during a blood meal where they mature in the digestive tract of the mosquito to a form either a male gamete (microgamete) or female gamete (macrogamete).

6. The gametes fuse to form a mobile diploid zygote (ookinete), which migrates through the midgut wall and develops into an **oocyst**.

FIGURE 3.7 The life cycle of *Plasmodium* species, highlighting adaptations of the pathogen to facilitate entry into and transmission between hosts

Adaptation of *Plasmodium falciparum* to facilitate entry into host

1 To enter a new human host, *P. falciparum* first requires the infected female *Anopheles* mosquito to bite and penetrate the skin. To obtain a blood meal, the mosquito releases an anticoagulant protein that prevents the blood from clotting. It is at this point that *P. falciparum* exits the mosquito's salivary gland.

Adaptations of *Plasmodium falciparum* to facilitate transmission between hosts

2 At each stage of the pathogen's life cycle, different antigen molecules are produced, preventing the host from initiating an effective immune response against the pathogen.

3 Once it has entered the red blood cells, *P. falciparum* produces adhesion proteins, which are presented on the surface of the cell. The proteins change the shape of the RBCs, slowing their movement through blood vessels. This aids the **merozoites** in exiting the cell.

Specific adaptations of influenza

CASE STUDY 6

INFLUENZA

There are four families of influenza viruses: A, B, C and D. Influenza C viral infections typically cause only mild illness in humans, while influenza D viruses do not cause illness in humans. In contrast, human influenza A and B viruses cause seasonal winter epidemics, while only influenza A has caused global pandemics. Symptoms of influenza A vary depending on the virulence of the virus variant and can include fever, chills, headache, sore throat, and coughing and sneezing.

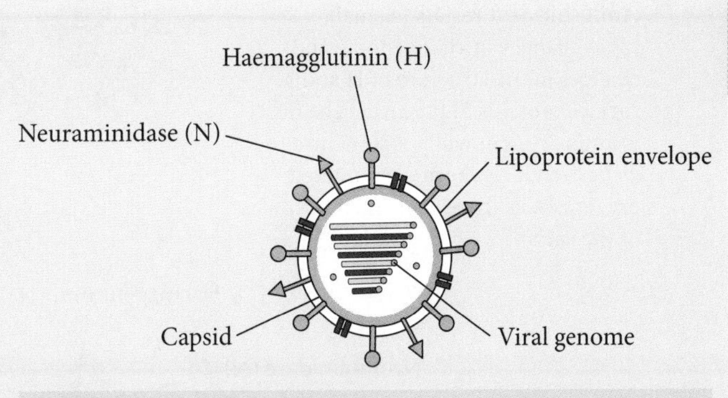

FIGURE 3.8 The influenza A virus

Influenza A viruses are categorised based on variations in two surface proteins: haemagglutinin (H) and neuraminidase (N). There are 18 haemagglutinin subtypes (H1–H18) and 11 neuraminidase subtypes (N1–N11). For example, the 1918 influenza pandemic, which killed an estimated 50 million people, was caused by the influenza A H1N1 variant.

Adaptation of influenza to facilitate entry into host

1 The H and N surface proteins of the influenza virus have evolved to bind to specific receptor proteins on target epithelial cells that line the upper and lower respiratory tract. Once inhaled, the virus quickly binds to host cells, where it is endocytosed before hijacking the cell's genetic machinery to replicate. While inside the host cell, it remains hidden from the host's innate and adaptive immune systems.

> **Note**
> It is important to remember coughing and sneezing, as these relate to transmission of the virus.

Adaptations of influenza to facilitate transmission between hosts

2 After the virus has reproduced within epithelial cells, it leaves the cell by pushing through the cell membrane. As it exits, the virus surrounds the RNA capsule with lipids and **glycans** from the host cell. By hijacking these host cell molecules, the virus is able to avoid detection by the immune system while outside host cells.

3 The influenza A virus has a high mutation rate, referred to as antigenic drift. Changes in the H and N surface proteins (antigens) mean the host's immune system will not recognise the virus even if it has been encountered previously, as there is no 'memory' of the variant. This results in ongoing transmission of the virus. Antigenic drift makes it necessary to develop annual vaccines against the two primary variants (H1N1 and H3N2) responsible for seasonal flu outbreaks.

4 Antigenic shift occurs when there is a new version of the H or N surface proteins introduced into the human population. Importantly, this can involve zoonotic strains exchanging genetic information with human influenza A strains, resulting in antigens not previously encountered by the human immune system. The lack of recognition by the immune system results in rapid transmission of the virus and is responsible for major influenza pandemics, including the 1918 pandemic and the more recent 2009 swine flu pandemic, both H1N1 variants.

5 The virus's main mode of transmission is through inhalation of shed virus in airborne droplets. The typical symptoms of coughing and sneezing are therefore highly effective at transmitting the virus. Furthermore, once shed, the virus can remain viable on some surfaces for up to 2 days outside a host. It is therefore highly effective at indirect transmission.

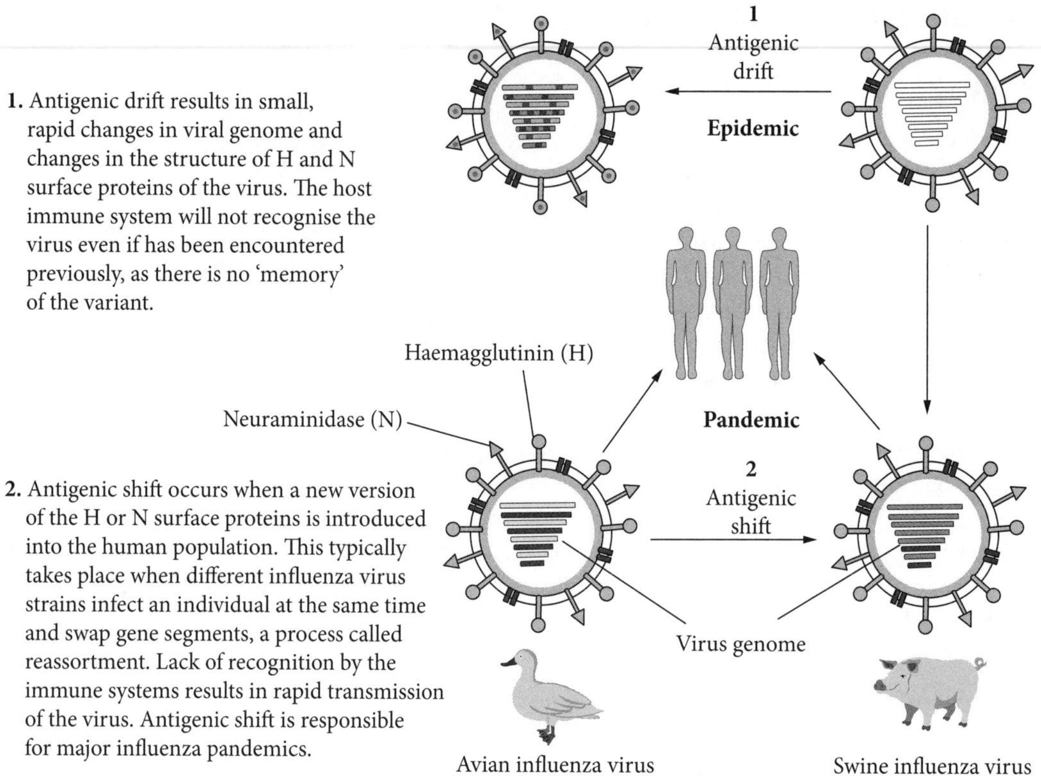

1
Antigenic drift
Epidemic

1. Antigenic drift results in small, rapid changes in viral genome and changes in the structure of H and N surface proteins of the virus. The host immune system will not recognise the virus even if has been encountered previously, as there is no 'memory' of the variant.

Haemagglutinin (H)
Neuraminidase (N)
Pandemic
2
Antigenic shift

2. Antigenic shift occurs when a new version of the H or N surface proteins is introduced into the human population. This typically takes place when different influenza virus strains infect an individual at the same time and swap gene segments, a process called reassortment. Lack of recognition by the immune systems results in rapid transmission of the virus. Antigenic shift is responsible for major influenza pandemics.

Virus genome
Avian influenza virus
Swine influenza virus

FIGURE 3.9 Antigenic drift and antigenic shift: adaptations of influenza A virus to facilitate entry into and transmission between hosts

> **Note**
> You do not need to remember all these adaptations. For each pathogen, choose one that facilitates entry and one that facilitates transmission.

3.2 Responses to pathogens: How does a plant or animal respond to infection?

Pathogens have the potential to cause significant damage, or even kill host organisms. Both plants and animals have a range of structures and mechanisms to prevent pathogen entry. However, if these barriers are breached, the host organism will respond to minimise the harm caused. This section focuses on diseases caused by fungal and viral pathogens in Australian plants and the physical and chemical changes that occur at the cellular and tissue level in host organisms in response to the presence of pathogens.

3.2.1 Investigating the response of a named Australian plant to a named pathogen through practical and/or secondary-sourced investigation; for example, fungal or viral pathogens

> **Note**
> You *must* know the name of a specific Australian plant *and* the respective pathogen. If you provide a plant that is not native to Australia, or provide a 'generic' name (e.g. fern), you will lose marks. Also, the dot point stipulates that a fungal or viral pathogen can be used as an example. This means you can choose either one of these, or a viable alternative.

The significant majority (nearly 90%) of plant diseases are caused by fungal pathogens. Airborne fungal pathogens penetrate plant tissues via natural openings (stomata and lenticels), while soil-borne fungal pathogens enter via the plant's root system.

All plants (and animals) have a range of **physical barriers** and **chemical barriers** that protect against pathogen entry and slow the spread if entry is achieved. Physical barriers are structural obstructions that protect against microbe entry; chemical barriers are any of the chemical characteristics of the organism that oppose colonisation by microbes.

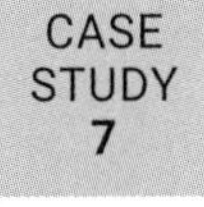

PHYTOPHTHORA DIEBACK

Phytophthora cinnamomi is a soil-borne fungal pathogen that affects a wide range of Australian plant species. Many thousands of potential host species are classified as 'at risk'. Between the 1940s and the 1970s a significant outbreak of *P. cinnamomi* caused the dieback of *Eucalyptus marginata* (jarrah) and other understorey plants over an area of 300 000 acres in Western Australia.

Physical and chemical defence mechanisms of *E. marginata* that prevent entry or inhibit the spread of *P. cinnamomi* include cellulose and lignin in the cell wall, which make the cell difficult to penetrate. The plant also secretes a range of antimicrobial chemicals. For example, like all eucalypts, *E. marginata* produces an oil, which is stored in sub-dermal secretory glands. The active ingredient of the oil is 1,8-cineole, which has antimicrobial properties that inhibit the spread of *P. cinnamomi* through the plant tissues.

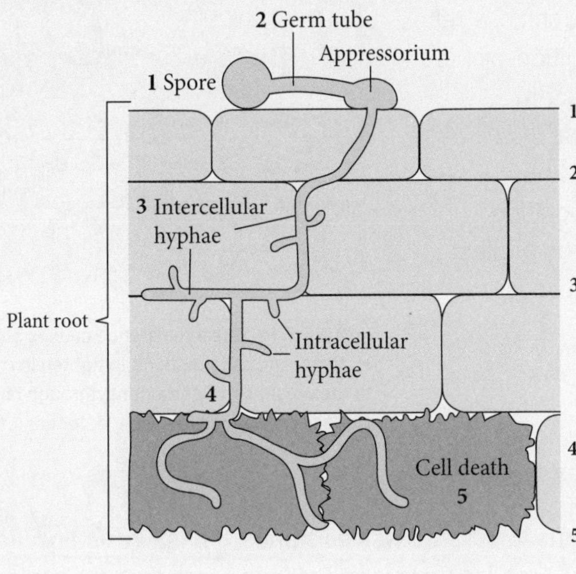

1. *P. cinnamomi* infects a new plant when spores come in contact with a root.
2. Spore produces a **germ tube** with a specialised cell (appressorium) that releases enzymes that breach the root's physical and chemical barriers.
3. Germ tube continues to differentiate by mitosis to produce the mycelium. Intercellular hyphae spread through the plant, while intracellular hyphae penetrate the plant's tissues and absorb nutrients and water.
4. Mycelium grows throughout the vascular tissue of the plant, blocking it and preventing transport of water and nutrients.
5. Leads to root and crown rot, chlorosis and eventual defoliation and death of the plant.

FIGURE 3.10 ***Phytophthora cinnamomi*** **spore entering root of** ***Eucalyptus marginata***

Cell- and chemical-mediated defences

If a pathogen breaches the protective barriers of a plant, the plant will respond with a combination of cell- and chemical-mediated defence mechanisms. These can be applied to the *E. marginata* and *P. cinnamomi* case study or an alternative of your choice.

Gene-for-gene resistance

Plants have resistance genes (R genes) that code for R proteins. Each R protein can bind directly with a specific avirulence (AVR) protein that is coded for by an AVR gene. AVR proteins are required to infect host cells. Once bound, the R-AVR complex prevents the pathogens from infecting new cells. Because closely related species of pathogens have conserved AVR proteins, an R gene offers broad-spectrum protection against any pathogen that produces the same AVR protein. If a plant does not have the relevant R gene, it is at significantly increased risk of infection.

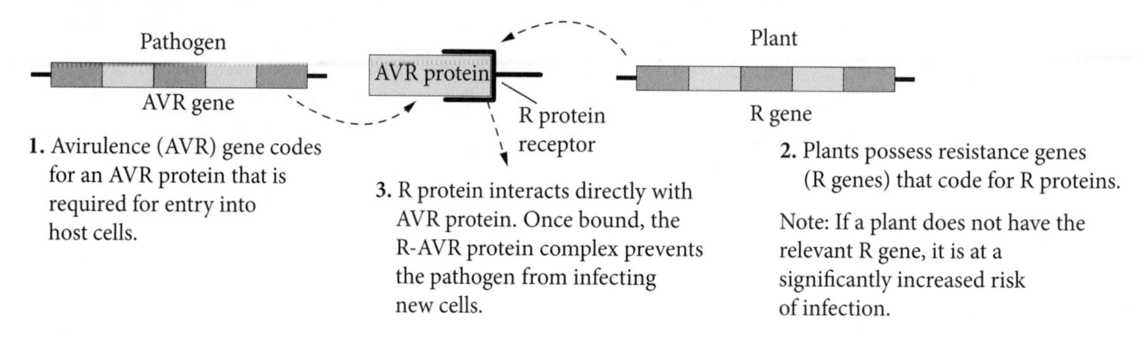

FIGURE 3.11 Gene-for-gene resistance

Basal resistance

Basal resistance is triggered by the recognition of **pathogen-associated molecular patterns (PAMPs)**, which are highly conserved molecules common to different groups of pathogens. They include chitin, a cell wall component in fungal pathogens. In addition, plants can sense 'self' **damage-associated molecular patterns (DAMPs)**, which include the release of cell wall components from infected cells. The PAMPs and DAMPs are recognised by host pattern recognition receptors (PRRs), which results in fortification of plant tissues. Cell junctions are closed, restricting access to invading pathogens.

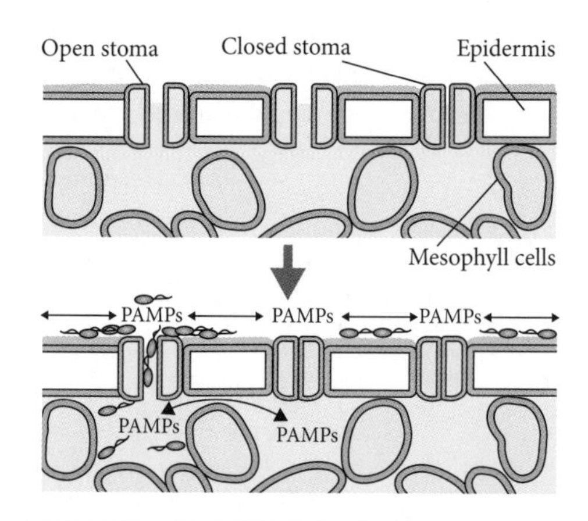

FIGURE 3.12 Basal resistance causes stomata to close, and cell junctions to tighten in response to the presence of pathogens, through chemical signalling methods involving detection of PAMPs.

Hypersensitive response

Many pathogens have developed mechanisms to inhibit basal resistance. If basal resistance fails, the localised hypersensitive response (HR) is activated to restrict and stop the pathogen from spreading to other parts of the plant. The plant cells in the region alter the cell wall chemistry, and the pathogen is trapped inside the host and dies with the cell during **apoptosis**. The HR is more effective if the R gene–AVR gene complex is present.

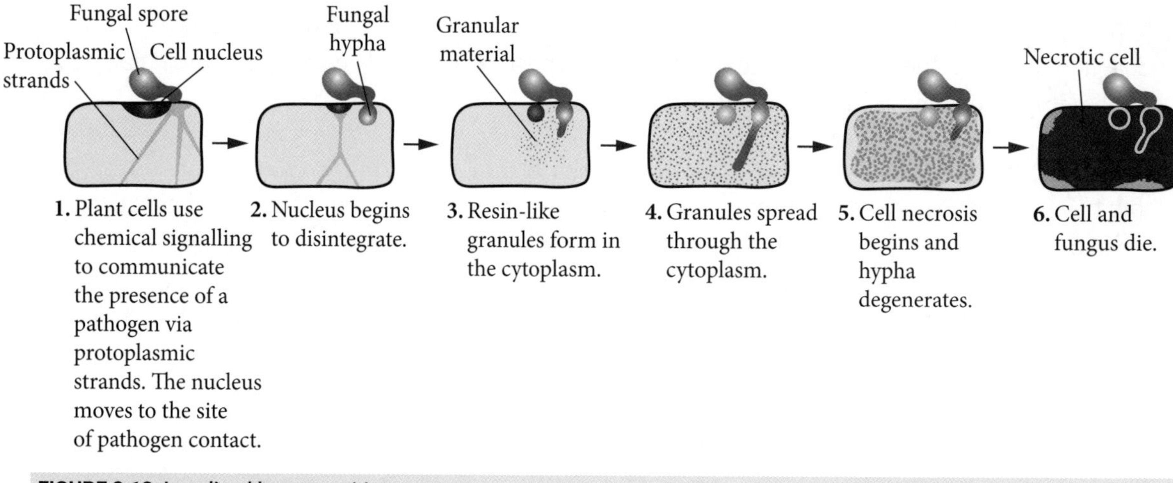

FIGURE 3.13 Localised hypersensitive response

Systemic acquired resistance

Systemic acquired resistance (SAR) is a non-specific, whole-plant response that occurs once the HR has been triggered. It results in plant tissues becoming highly resistant to a wide range of pathogens for an extended period. This systemic resistance is activated by the accumulation of salicylic acid (SA), which initiates a signalling pathway to other parts of the plant, causing them to produce a range of broad-spectrum **pathogenesis**-related proteins. These 'prep ' the plant for any pathogens that bypass the localised HR.

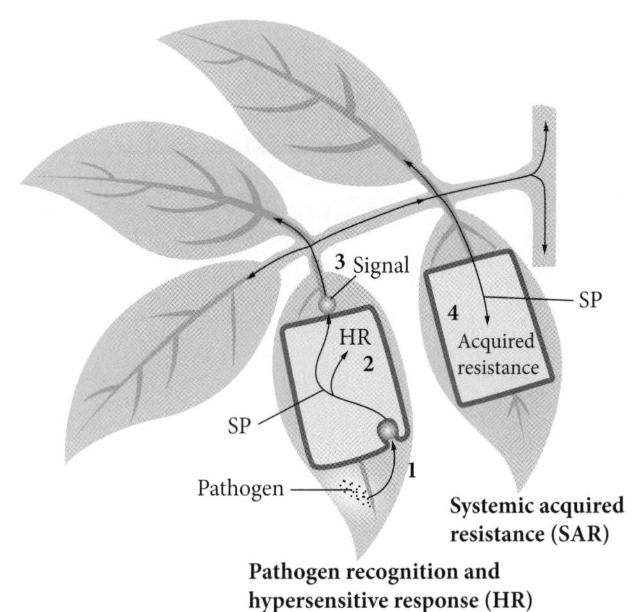

FIGURE 3.14 Hypersensitive response and systemic acquired resistance

3.2.2 Analyse responses to the presence of pathogens by assessing the physical and chemical changes that occur in the host animal's cells and tissues

The **innate immune system** can be broadly divided into two categories: the physical and chemical barriers that prevent the entry of pathogens into the body; and the cellular and chemical responses that occur if a pathogen breaches these barriers and enters the body. Both are non-specific.

> **Note**
> This dot point requires both an 'analysis' *and* an 'assessment' of animal responses to pathogens. You need to know a range of physical and chemical responses and how they interact with each other (the relationship). You should be able to explain the outcomes of these responses, understand the implications if they are impaired and, based on this knowledge, provide a judgement as to the importance of each response.

Innate immune system: Physical and chemical barriers

The body has a range of non-specific physical and chemical barriers that work together to prevent entry of microbes into the body. Physical barriers include the skin, the mucous membrane and epithelial layers of the respiratory, gastrointestinal and urinary tracts, and motile cilia of the lungs and respiratory tract. Chemical barriers include microenvironmental factors such as pH and a range of innate antimicrobial molecules that can aid in the destruction of pathogens.

Table 3.3 highlights the main physical and chemical barriers of the body. It is important to note that, while they have been classified here as either physical or chemical, there are often significant overlaps.

> **Note**
> This dot point refers to animals, not specifically humans. However, because the same principles apply to the significant majority of animals, the human context has been discussed for familiarity.

TABLE 3.3 The body's physical and chemical barriers against pathogens

Barrier	Description and location	
Physical		
Skin	• Upper *epidermis* contains keratin, a protein that helps bind the skin's cellular matrix, making it impenetrable to microbes. • Middle *dermis* contains hair follicles, sweat and sebaceous glands. Sebaceous glands produce sebum, which seals the hair follicle, preventing access by pathogens. • *Hypodermis* is the bottom fatty layer of the skin and contains blood and lymph vessels. These vessels deliver cells involved in the innate immune response if the skin is breached.	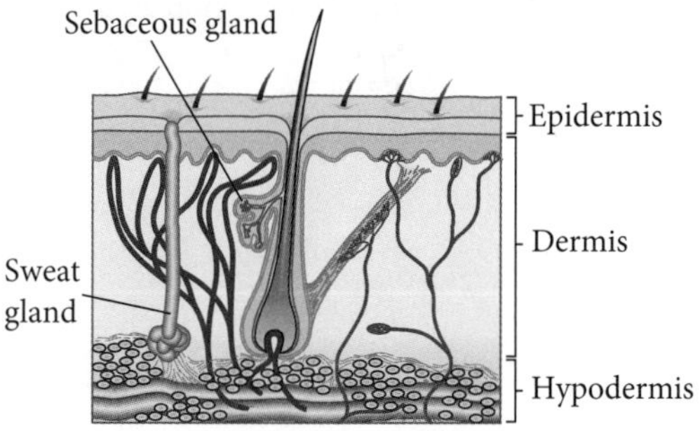
Epithelial lining of mucous membrane	• Tight cell junctions block microbe entry. • Contain motile **macrophages** called Langerhans cells, which lie ready for any breaches of the layer.	
Cilia	• Ciliated respiratory epithelial cells work in conjunction with mucous membranes to direct trapped microbes out and away from the body. • Microbes are removed or destroyed via coughing or sneezing, or when they come in contact with stomach acid after being swallowed.	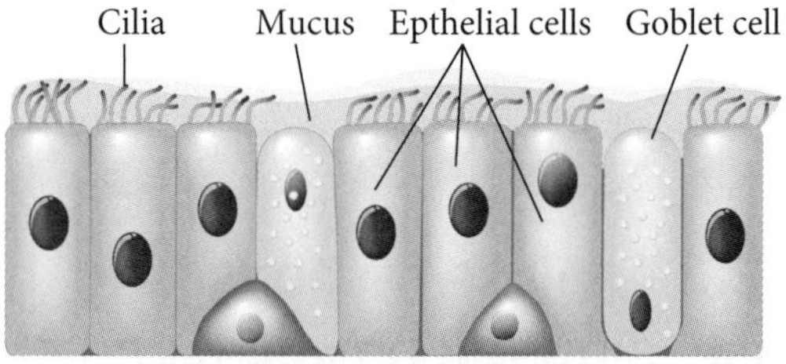
Mucous membranes	• Line the respiratory, urinary and digestive tracts. Specialised epithelial cells (goblet cells) secrete mucus, which lubricates the membrane and traps any microbes that enter the body. Microbes are then removed via various methods depending on their location: sneezing, coughing, defecation or urination.	
Peristalsis	• Peristaltic waves of the digestive tract mix microbes with stomach acid and mucous membranes, preventing microbial colonies from settling and causing infection.	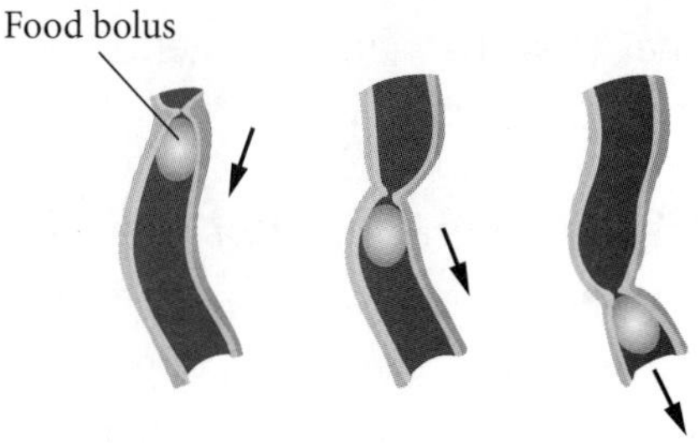
Urine flow	• Urine is passed under pressure during urination, which removes any microbes present in the urinary tract. • Urine is slightly acidic (pH 6), which kills many microbes.	

››

TABLE 3.3 cont.

Barrier	Description and location
Chemical	
Acidic and alkaline secretions	• Specialised stomach epithelial cells (parietal cells) produce hydrochloric acid (pH 1–2), which kills any microbes that enter via food or water. • The duodenum has a pH of 7–8.5. Bile made in the liver is secreted into the duodenum, where it neutralises acid entering from the stomach and creates alkaline conditions. The change in pH is usually enough to kill any microbes that survive the acidic environment of the stomach.
Sebum and sweat	• See skin. Sebum has acidic pH (4–6), making it an effective bactericide. • Sweat has a slightly acidic pH (6.5), which hinders bacterial growth. Sweat also contains a protein called dermcidin, which binds to bacteria and disrupts their function.
Cerumen (earwax)	• Earwax is high in fatty acids, which lowers the pH to 4, which kills most microbes.
Lysozymes	• Saliva, tears, sebum and sweat contain enzymes called lysozymes, which act on the cell wall of bacterial microbes, causing them to **lyse**.

Innate immune system: Cellular and chemical responses

If a microbe breaches the physical and chemical barriers of the body, **antigen** molecules on the surface of the microbes are recognised by the host as 'non-self', which initiates the innate immune response, including inflammation and phagocytosis. The activation of the innate immune system in turn initiates the **adaptive immune response** as required (e.g. production of B and T lymphocytes and antibodies).

Innate cellular responses

Leukocytes are a diverse group of cells, commonly called white blood cells, that are involved in both the innate and adaptive immune responses. Collectively their function is to fight infections. Leukocytes involved in the innate immune response are categorised as either granulocytes or monocytes, which are then further subdivided into specific cell types based on their function. Granulocytes (**neutrophils**, **eosinophils** and **basophils**) contain enzyme-filled granules that damage or digest pathogens after phagocytosis. Monocytes differentiate into either macrophages or **dendritic cells**, both of which are phagocytotic, meaning they engulf foreign bodies to break them down internally.

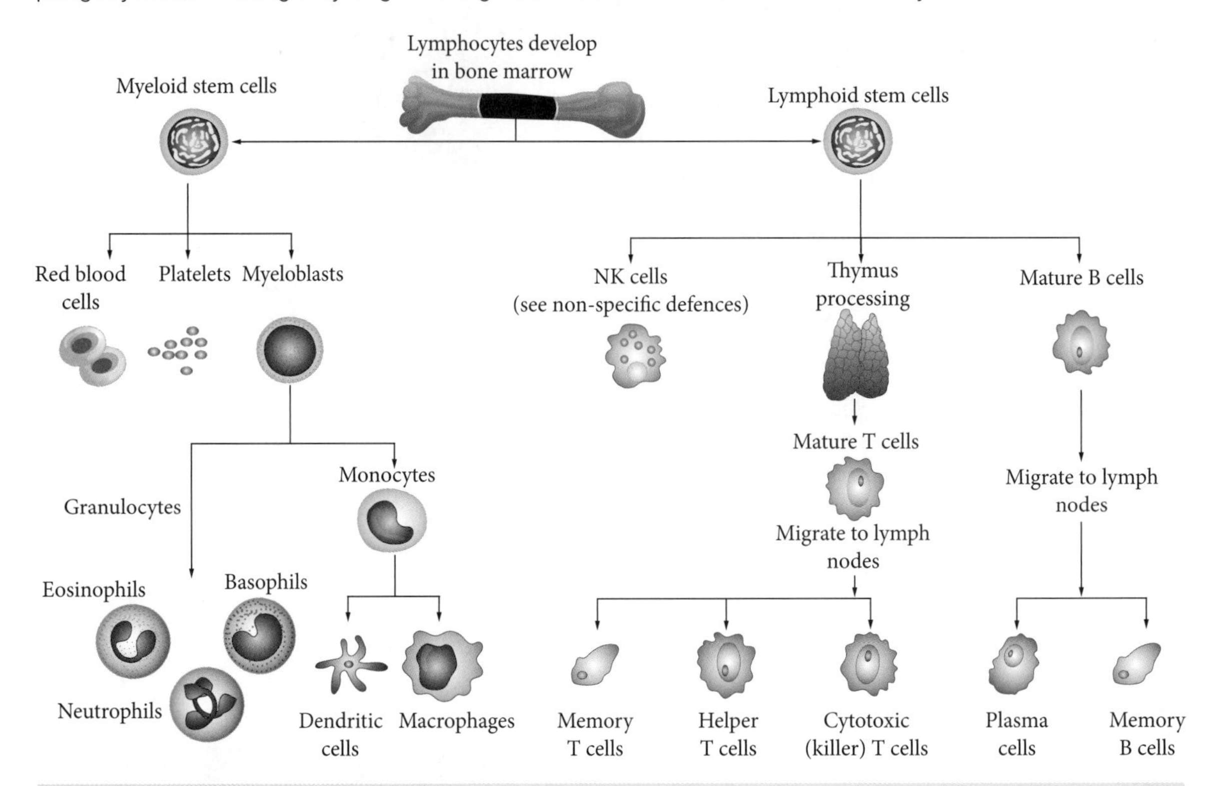

FIGURE 3.15 Important cells in the innate and adaptive immune responses

TABLE 3.4 A summary of leukocytes involved in the innate immune response

Leukocyte	Category	Role	Image
Macrophage	Monocyte	• Act primarily as phagocytes. Also release **chemoattractants**, which activate other immune cells and increase their migration to an infected site. • Fixed macrophages are located in tissues susceptible to pathogen invasions (skin, lungs and intestine). • Mobile macrophages circulate in the bloodstream and lymph system, and are attracted to an infected site by the chemoattractant molecules released by **mast cells**.	Scott Camazine / Alamy Stock Photo
Dendritic cell	Monocyte	• Phagocytic antigen-presenting cells (APCs) that express both major histocompatibility complex (MHC) class I and II molecules on their surface (see section 3.3.2). Bridge the innate immune system with the specific adaptive immune system. • Circulate in the bloodstream, where they capture and phagocytose pathogens directly; or are attracted to an infection site upon detecting chemoattractants released by other immune cells.	Science Photo Library / Alamy Stock Photo

››

TABLE 3.4 cont.

Leukocyte	Category	Role	Image
Mast cell	Granulocyte	• In all vascularised tissue. Stimulate **vasodilation** and vessel permeability through release of histamine, which leads to increased migration of macrophages and neutrophils into infected area.	Science Photo Library / Alamy Stock Photo
Neutrophil	Granulocyte	• Phagocytic cells that make up 50–60% of all leukocytes. • Defend against smaller pathogens, including bacteria, viruses and fungi. The death of neutrophils and tissue surrounding an infection site is visible as pus.	David M. Phillips / Science Photo Library
Eosinophil	Granulocyte	• Defend against parasitic infections that are too large to be phagocytosed directly. Instead, they attach en masse to the parasite and release digestive enzymes from their granules to destroy the pathogen.	Science Photo Library / Alamy Stock Photo
Basophil	Granulocyte	• Can phagocytose, but their primary role is the release of histamine and heparin. Histamine causes further vasodilation and subsequent inflammation. Heparin prevents clotting from occurring too quickly, which would otherwise inhibit the immune response.	Science Photo Library / Alamy Stock Photo
Natural killer (NK) cell	Granulocyte (lymphocyte)	• Do not require prior activation and therefore play a central role in the innate immune system. • Continually patrol the body and can distinguish, via the MHC I receptor pathway, between healthy cells and pathogen-infected cells.	Science History Images / Alamy Stock Photo

Note

Don't confuse the terms 'leukocytes' and 'lymphocytes'. All leukocytes are collectively called white blood cells and are involved in both the innate and adaptive immune responses. Lymphocytes are a type of leukocyte involved only in the adaptive immune response (B and T lymphocytes).

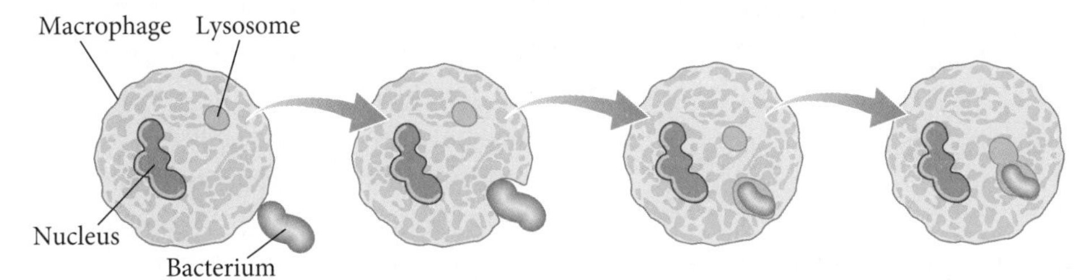

1. Bacterium comes into contact with macrophage.
2. Plasma membrane of macrophage envelopes the bacterium and engulfs it.
3. The bacterium is encased in a vesicle called a phagosome.
4. The phagosome fuses with the lysosome to become the phagolysosome, which now has low pH and contains digestive enzymes that destroy the bacterium.

FIGURE 3.16 The process of phagocytosis

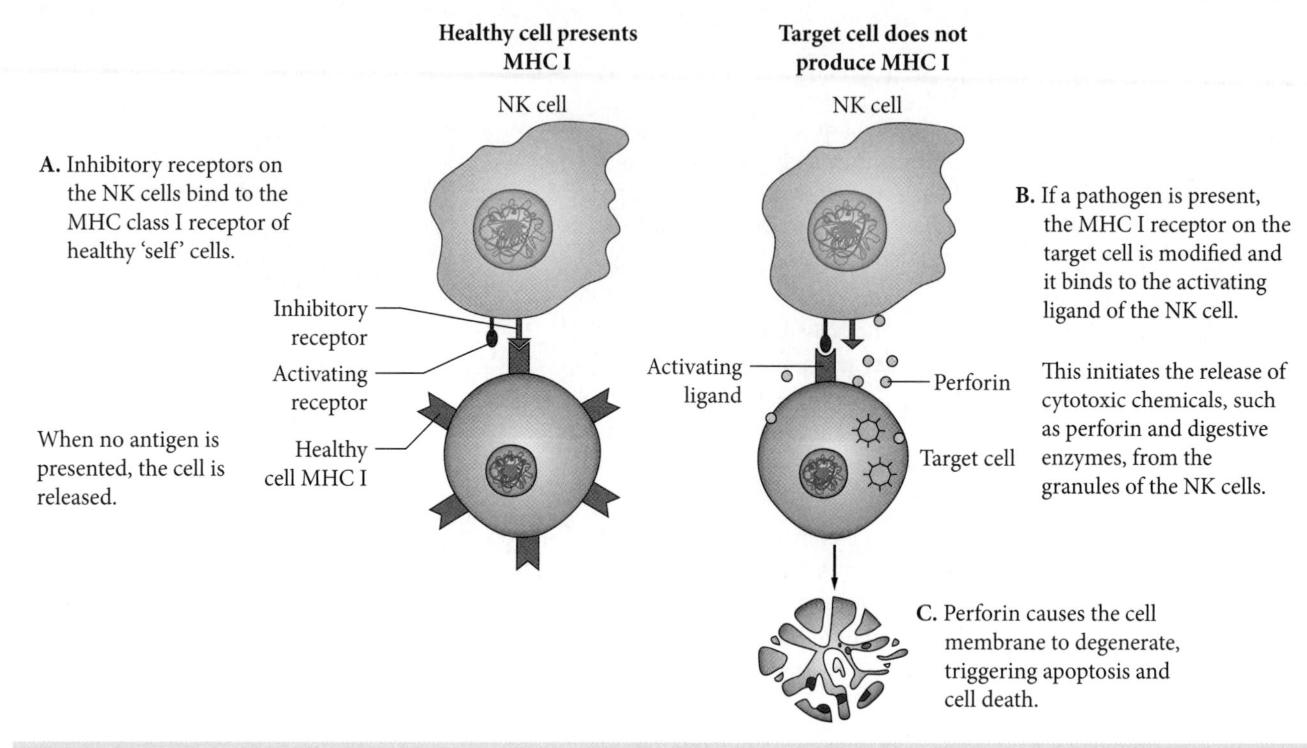

FIGURE 3.17 Natural killer cells recognise infected cells and induce apoptosis.

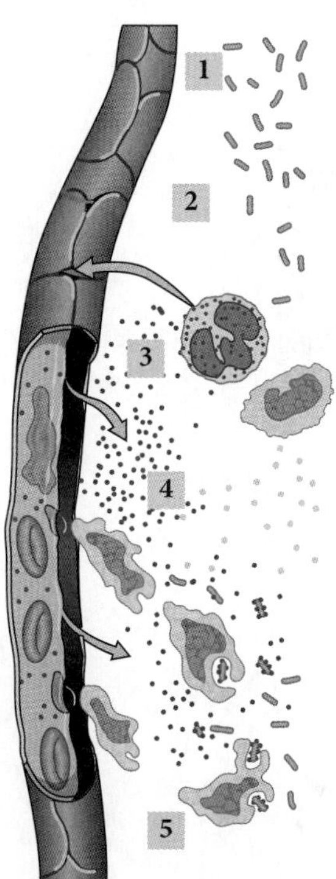

FIGURE 3.18 Inflammation allows phagocytes entry to the site of infection.

Innate chemical responses

Inflammation response

Inflammation occurs at the site of an infection. Mast cells are activated when their pattern recognition receptors (PRRs) detect the presence of pathogen-associated molecular patterns (PAMPs) on the pathogen's surface, which causes the degranulation of the mast cells and the release of **proteases** and histamine. The mast cells and fixed macrophages also release **cytokines**, which attract mobile phagocytes to the site of infection.

Cytokines, including interleukins and interferons, coordinate the innate inflammation response and also the adaptive immune response. Interleukins act as messenger molecules between immune cells, while interferons are released by infected cells, triggering surrounding cells to become more resistant to the invading pathogen. Interferons also signal phagocytes and NK cells to target and destroy infected cells.

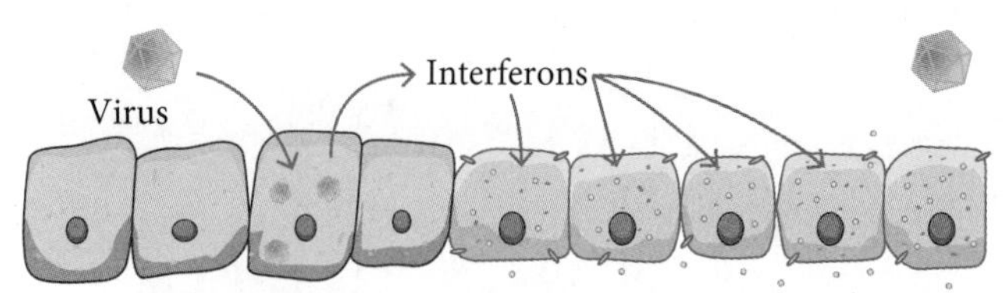

FIGURE 3.19 Interferons released by infected cells trigger resistance by surrounding cells.

Complement system

The complement system includes a group of up to 50 proteins that become activated when they encounter a pathogen. Activation triggers a cascade that coats a pathogen in complement proteins, which enhances (complements) the binding of antibodies to the pathogen and subsequent phagocytosis. The complement system also increases inflammation, and some proteins act directly on the cell membrane of pathogens to destroy them.

3.3 Immunity: How does the human immune system respond to exposure to a pathogen?

The human body has a range of strategies to minimise the impact of pathogens if they enter the body. This section explores the interaction of the innate and adaptive immune systems, and focuses on the different **leucocytes** involved in the immune response. This includes the response of the immune system after primary exposure to pathogens, the specificity and interaction of B and T lymphocytes, and the immunological memory induced by the exposure. It is the production of memory B and T lymphocytes after the primary exposure to a pathogen that provides **acquired immunity**.

3.3.1 Investigating and modelling the innate and adaptive immune systems in the human body

In this context, a model is a representation that provides a simplified explanation of the different cell types (and their relationships) involved in the innate and adaptive immune systems. While the choice of *model* is yours to make, a diagram, flow chart or computer simulation would be ideal to highlight the different cell types involved, the interactions between them and the specificity of the process.

The information covered in sections 3.2.2 and 3.3.2 will provide you with the details required to construct a model and explain how the innate and adaptive immune systems respond after primary exposure to a pathogen. To demonstrate your knowledge, you should be comfortable in comparing the antibody- and cell-mediated responses (see Figure 3.26 on page 145).

Note

This dot point would typically be assessed in the HSC by asking you to identify the correct sequence of a series of images, or you may be asked to comment on the limitations of a provided image.

3.3.2 Explaining how the immune system responds after primary exposure to a pathogen, including innate and acquired immunity

The innate immune system was covered in detail in section 3.2.2. This section focuses on the adaptive immune system and the development of acquired immunity.

Adaptive immune system

The adaptive immune system is specific and produces an immunological memory of antigens that have been encountered. Antigens are any molecules, usually peptides or proteins, that a host recognises as foreign or 'non-self' when the physical and chemical barriers of the body have been breached. Adaptive immunity offers a rapid and highly effective response to specific antigens that have been encountered previously.

Two broad groups of lymphocytes are involved in the adaptive immune system.

- **B cells** are produced and mature in the bone marrow (thus 'B' cell). They are responsible for antibody-mediated immunity (or humoral immunity).
- **T cells** mature in the thymus (thus 'T' cell). They are responsible for cell-mediated immunity.

B and T cells can exist in three states.

- Naive cells have not been exposed to an antigen.
- Effector cells have encountered a specific antigen and are actively involved in its removal. Once the pathogen is defeated, proliferation stops and the cells die.

9780170465267

- Memory cells continue to circulate after the antigen has been eliminated. It is the memory cells that initiate a quick immune response if the antigen is encountered again.

Note
It is only after a primary exposure to a pathogen that acquired immunity is achieved.

Activating the adaptive immune system

All nucleated cells in the body express proteins on their surface, known as major histocompatibility complexes class I (MHC I) molecules. Each person has unique MHC I molecules, and this allows the cells of the innate and adaptive immune systems to identify 'self' from 'non-self' antigens. If a 'self' cell has been infected by a pathogen, it presents the foreign antigen on its surface via the MHC I molecules. When an immune cell binds with the MHC I molecule, the foreign antigen is recognised and an immune response will result.

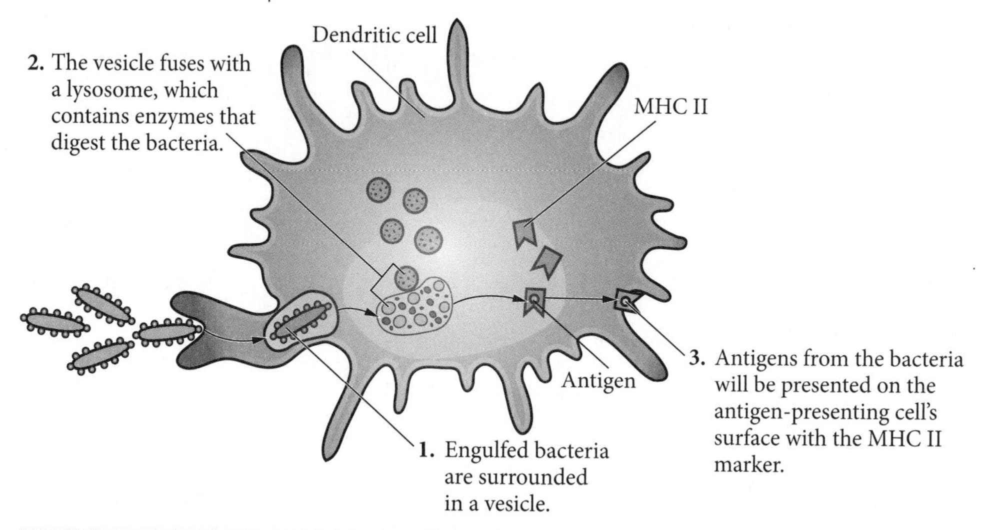

FIGURE 3.20 Presenting antigens via the MHC II receptor

In addition to the MHC I molecules, antigen-presenting cells (APCs: macrophages, dendritic cells and B cells) also express MHC class II (MHC II) molecules. After phagocytosis, a pathogen is digested, and a portion of the pathogen is expressed on the surface of the phagocyte via the MHC II molecule. The APC then migrates to the lymph system, where it presents the specific antigen to helper T (T_h) cells via the MHC II molecule or to killer (cytotoxic) T (T_c) cells via the MHC I molecule. It is at this point that the adaptive immune system is activated. Once activated, both T_h and T_c cells mature and proliferate.

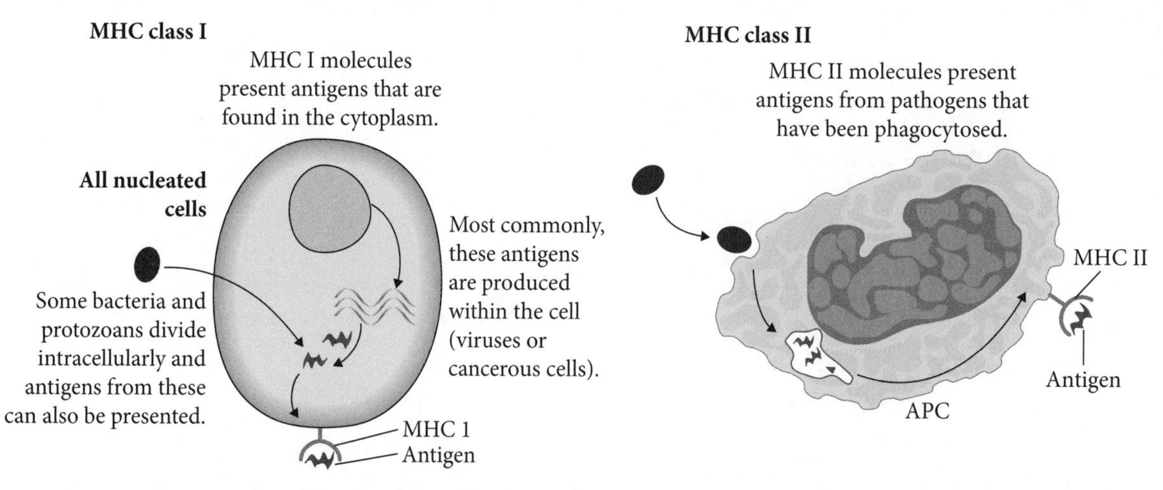

FIGURE 3.21 Major histocompatibility complexes I and II presenting antigens

Antibody-mediated immunity

Helper T cells activate the antibody-mediated response. The T_h cells bind (via their T cell receptors) to the MHC II molecule of naive B cells that are expressing the same specific antigen that activated the

T_h cells. This, combined with the release of cytokines by the T_h cells, promotes rapid clonal expansion of the B cell population specifically linked to the identified antigen.

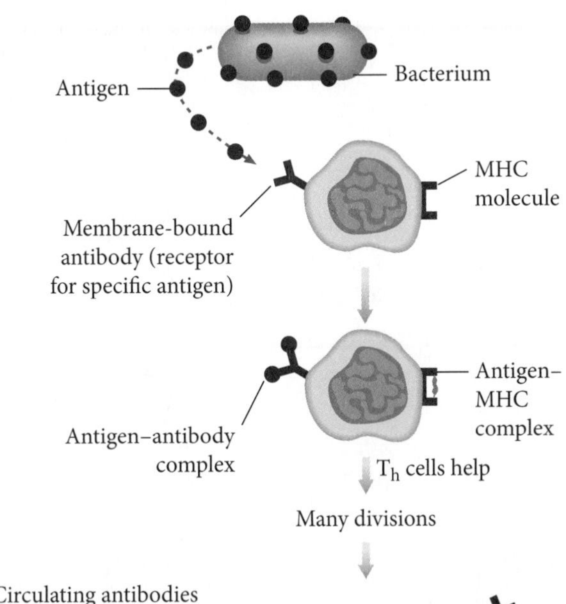

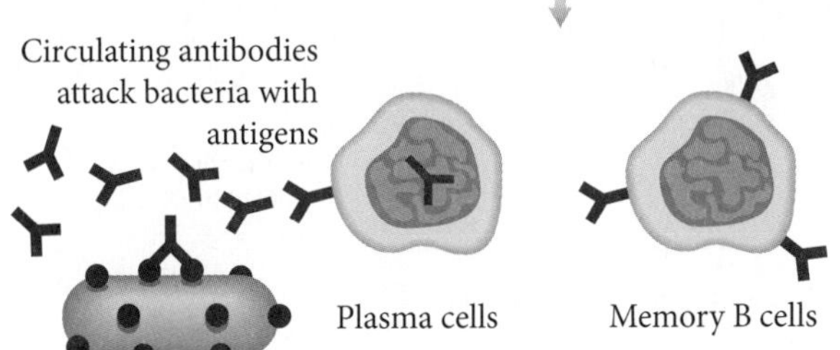

FIGURE 3.22 B cell activation by a helper T cell, resulting in clonal expansion

Each naive B cell contains a unique B cell receptor (BCR) capable of recognising a specific antigen. The antigen is processed and presented on its cell membrane via MHC II molecules. The activated B cells differentiate into memory B cells or plasma cells, which act as antibody factories. The circulating antibodies bind directly with antigens to form an **antibody–antigen complex**, preventing them from causing further infection and making them easier to recognise and be destroyed by phagocytes. The memory B cells (and memory T cells) are the basis of acquired immunity. They provide ongoing immunity due to their receptors, which are specific to a previously encountered antigen.

Antibodies

Antibodies (immunoglobulins) are quaternary proteins consisting of four polypeptides: two heavy chains and two light chains joined to form a Y-shaped molecule. Most of the heavy chain and the lower end of the light chain are conserved in all antibodies. In contrast, the ends of the heavy and light chains form variable regions that are unique to each antibody and act as the antigen-binding site. Antibodies are naturally produced by B cells in response to antigen exposure and therefore play a critical role in the immune system's defence against infection and disease.

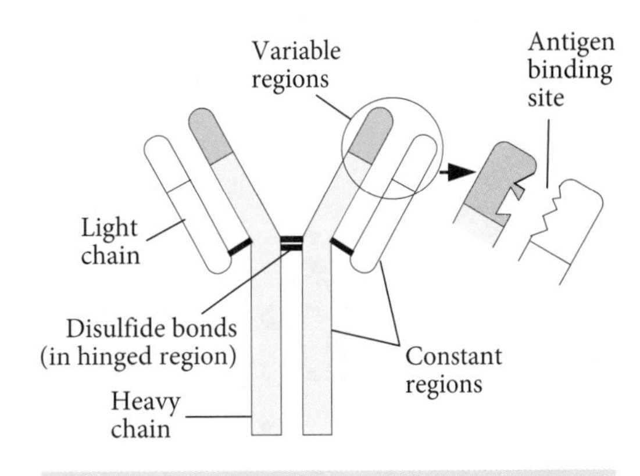

FIGURE 3.23 The general structure of an antibody

Antibodies function in one of several ways, including **neutralisation**, **agglutination**, **opsonisation** and **complement activation**. In all cases, pathogen mobility and reproduction are inhibited, allowing other components of the immune system to destroy the pathogen.

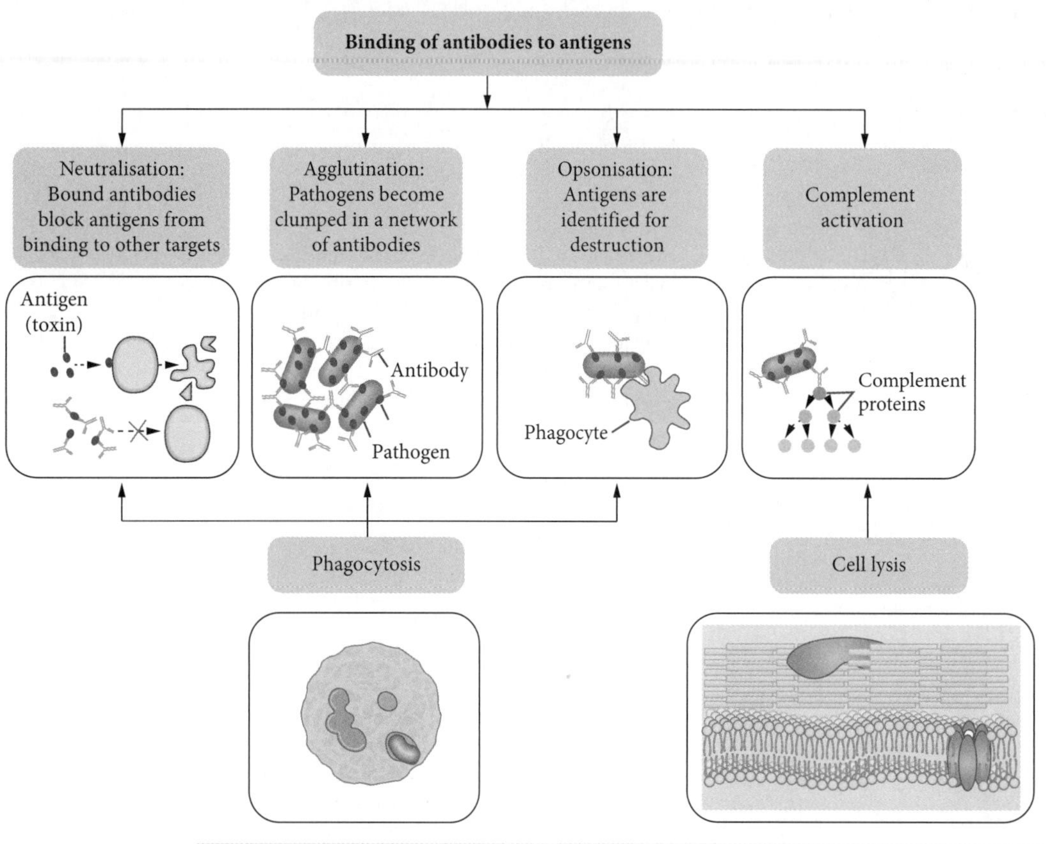

Neutralisation: Antibodies bind directly to antigen molecules on the surface of individual pathogens, preventing them from entering or damaging new host cells.

Agglutination: Antibodies bind directly to antigen molecules on the surface of multiple pathogens, causing them to clump together, where they are phagocytosed altogether.

Opsonisation: Antibodies bind to the antigens expressed on the surface of infected host cells. The antibody-bound cells are detected by innate phagocytes or granular NK cells, which release cytotoxic chemicals to lyse the host cell, killing any pathogens contained within it.

Complement activity: Protein cascade binds pathogens, causing cell lysis.

FIGURE 3.24 Protective mechanisms and consequences of antibody–antigen binding

Cell-mediated immunity

Antibodies are unable to bind antigens within a cell. As a result, the cell-mediated immune response is required to target pathogens that have already infected host cells. This process is regulated by T cells, which only recognise antigens when they are expressed on the surface of an APC (via MHC II molecules) or an infected cell (via MHC I molecules).

All T cells express a T cell receptor – **cluster of differentiation** 3 (TCR–CD3) receptor complex, which is required to bind both MHC I and MHC II molecules. The CD3 receptor is conserved between the cell types, while the TCR is highly variable, to account for different antigens. The different types of T cells are classified based on their additional surface CD receptors. For example, T_h cells express CD4, and T_c cells express CD8, which are also required to bind MHC II and MHC I molecules of APCs, respectively.

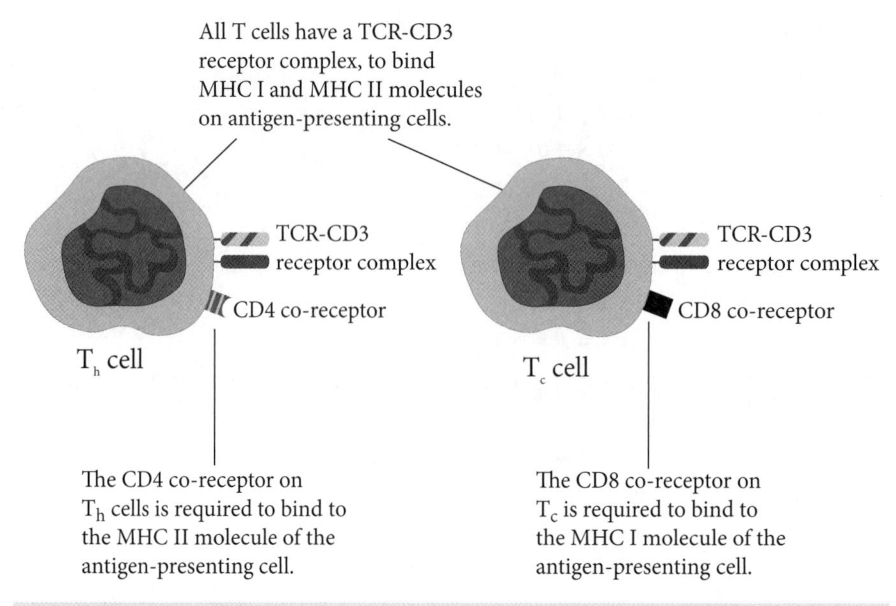

FIGURE 3.25 Different receptors of T_h and T_c cells for binding to MHC II and MHC I molecules

The cell-mediated response is activated when an APC presents a specific antigen via an MHC II molecule to a complementary T_h cell. The T_h cells then release cytokines to activate the mass cloning of T_c cells and memory T cells with receptors specific to the antigen.

Activated T_c cells exit the lymph system and migrate to the site of infection, where they bind via their specific receptor to infected cells presenting the antigen via their MHC I receptors. The T_c cells degranulate, releasing cytotoxins including perforin, which lyse the cell, killing any pathogens within it. T_c cells perform a similar function to NK cells, with one key difference: they require activation and, as such, form part of the adaptive immune response.

Once an infection has been defeated, regulatory T (T_{reg}) cells are produced. T_{reg} cells release cytokines that prevent further production of T_c cells and B cells. Most circulating immune cells linked to the defeated pathogen die off, except for a population of memory T and B cells. These memory cells can remain circulating or accumulate in the spleen and lymph system, where they will rapidly proliferate to very large numbers if they are exposed to an antigen they have encountered before. Acquired immunity has been achieved.

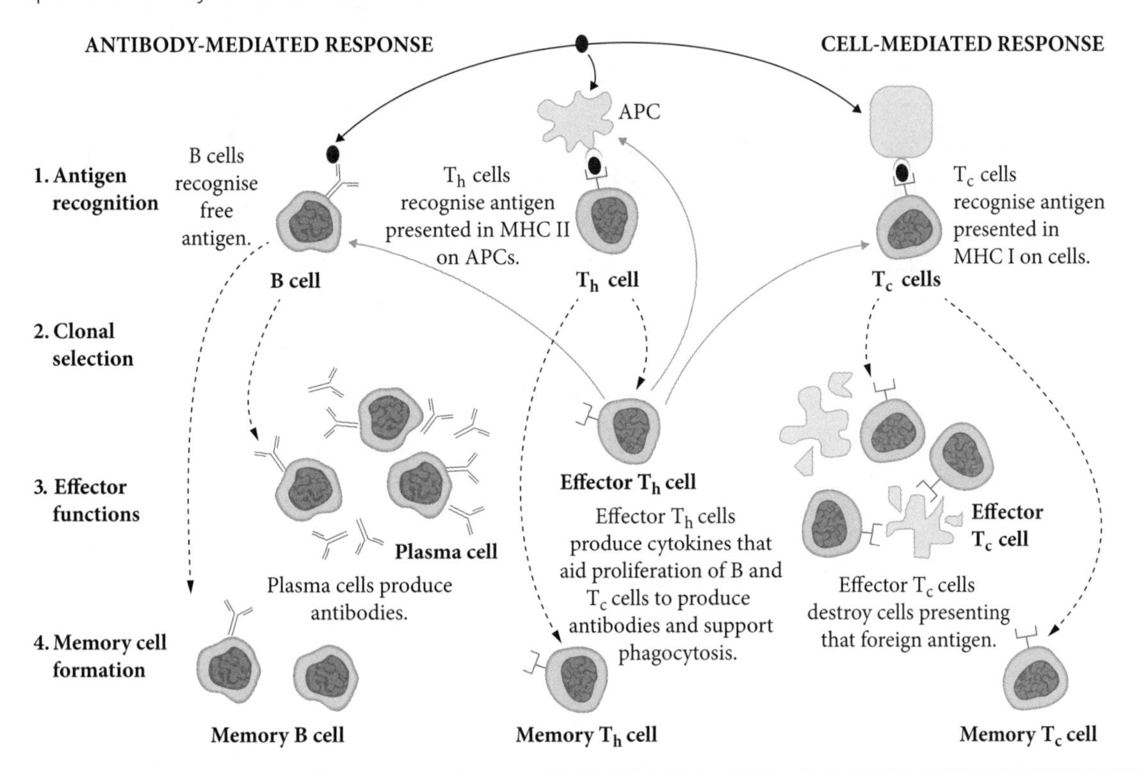

FIGURE 3.26 A comparison of the antibody-mediated and cell-mediated responses

3.4 Prevention, treatment and control: How can the spread of infectious diseases be controlled?

Knowing the cause of an infectious disease and the mechanisms by which it is transferred allows government and health authorities to put in place appropriate preventative and control measures to minimise the impact of the disease. While health decisions and legislation occur at the regional and global level, it is at the local level that the most specific and enforceable measures can be taken to minimise disease spread.

3.4.1 Investigating and analysing the wide range of interrelated factors involved in limiting local, regional and global spread of a named infectious disease

In order to limit the spread of an infectious disease, contact between infected and non-infected individuals must be kept to a minimum. Key considerations about what measures need to be implemented to reduce the spread include:

- the **virulence** of the pathogen involved and its incubation period
- population density and mobility.

Note

For this dot point, you need to focus on a specific disease. The choice is yours, but you could consider COVID-19. It is topical and you will already be aware of many of the local, regional and global measures introduced to limit the spread of the disease. It is important to clearly distinguish between the different levels of intervention and to understand how the range of strategies complement each other to limit the spread of the disease.

Consider COVID-19, which is caused by the viral pathogen SARS-CoV-2. The pandemic has highlighted the ease and speed with which an infectious disease can spread around the world.

In January 2020, the WHO identified a 'novel' virus that caused pneumonia-like symptoms in fewer than 100 individuals in Wuhan, China. Soon after the disease was identified, the number of new infections and deaths started to grow exponentially. Chinese authorities placed Wuhan and surrounding regions into strict lockdown, effectively quarantining 20 million people. But it was too late. By late January, the virus had been detected in Europe, Asia and the USA.

On 11 March, only two months after COVID-19 was first identified, the WHO declared that COVID-19 met the three criteria to be categorised as a pandemic:

- illness resulting in death
- sustained person-to-person spread
- worldwide spread.

In under two years, the number of global infections reached 350 million. The official death toll is already over 5.5 million, although the actual number is probably much higher.

Table 3.5 highlights some of the local, regional and global strategies introduced to limit the spread of COVID-19. Strategies either reduce contact between infected and non-infected individuals or limit the risk of transmission when they do come in contact.

TABLE 3.4 A summary of leukocytes involved in the innate immune response

Level of intervention	Details of intervention strategies
Local	• Forced two-week quarantine for individuals returning from overseas; fines for breaches • Forced lockdowns, mandatory mask wearing in all public spaces, nighttime curfews and travel bubbles where enforceable • 'Pop up' testing and tracking facilities, central to limiting the spread of COVID-19 • Vaccine targets linked to eased restrictions with mandatory **vaccination** required for certain professions. Proof of vaccination required to enter businesses and public facilities While introduced at the regional level, most of these measures were monitored and enforced at the local level.
Regional	• 'Hard' border lockdowns to restrict movement within and between states • Extended lockdowns, with heavy restrictions on the liberties and movement of residents. Government approval and permits required to travel into regional areas • National Cabinet formed to address the specific health and economic needs of each state and territory. Included a staged approach to the vaccination rollout, with elderly and vulnerable populations prioritised, and a range of financial incentives provided to support people forced into unemployment by the lockdowns • 'No jab, no play' and 'No jab, no pay' legislation

››

TABLE 3.5 cont.

Level of intervention	Details of intervention strategies
Global	The WHO is responsible for the global governance of health and disease. It communicates directly with policymakers at the national and regional levels to inform them of trending disease patterns. It also recommends and supports specific strategies to limit disease spread. For COVID-19 this included: • mathematical modelling to track and predict spread of the different virus variants. This allowed regions to be proactive rather than reactive to outbreaks • a coordinated approach involving thousands of scientists and public health authorities, to produce and distribute vaccines. Within 18 months of the disease first being identified, multiple companies had produced vaccines • where necessary, the planning and building of emergency operations centres, training of health workers and financing of vaccination programs.

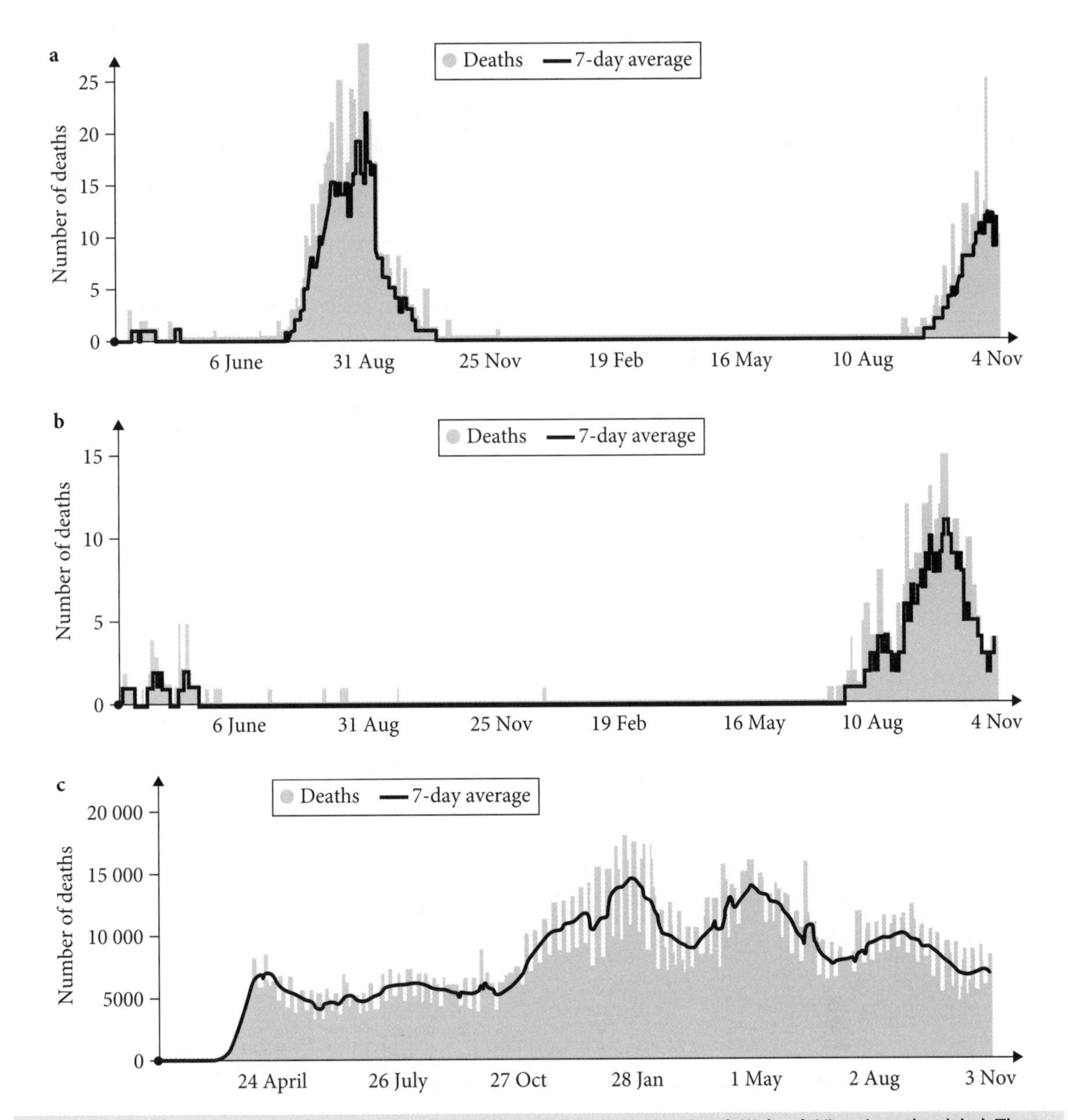

FIGURE 3.27 The number of daily deaths from COVID-19 in 2020-2021: **a** New South Wales, **b** Victoria and **c** global. The relative numbers and timing of peak case numbers highlight the need for different local, regional and global responses.

3.4.2 Investigating procedures that can be employed to prevent the spread of disease

Note

When investigating the different procedures that prevent disease spread, it is essential that you *explicitly* mention how they prevent the spread. For example, washing hands with soapy water kills microbes, preventing them from spreading. Also note the 'including but not limited to'. As always, this mean you must be familiar with the examples provided in the dot point and have additional examples.

Hygiene practices

Hygiene is defined as the conditions and practices that assist in maintaining health and prevent the spread of infectious disease. Hygiene practices can be broadly categorised as personal, domestic or community. The practices in each category either kill microbes or reduce the likelihood of an infected person passing on the pathogen.

TABLE 3.6 Hygiene practices

Level of hygiene	Details	Examples
Personal	Practices that an individual can do to maintain good health and minimise the risk of becoming infected or infecting others when sick	• Handwashing with soap and water after defecating and before eating or handling food • Frequent bathing • Correct respiratory hygiene when coughing and sneezing • Using protection during sexual intercourse
Domestic	Practices done at the household level to minimise disease spread and focusing on food storage and preparation	• Keeping raw and cooked foods separate • Cooking food for the appropriate time and at the appropriate temperature • Storing food at the recommended temperature • Using clean water for cooking
Community	Large-scale infrastructure programs that focus on providing society with access to clean water and food, and sanitary living conditions	• Water treatment and purification (filtration and chlorination) • Sewage removal and treatment • Correct animal husbandry techniques • Maintenance of hygiene in markets

Quarantine

Quarantine involves a period of strict isolation to prevent the introduction and spread of diseases or unwanted animals, plants and pests into the country or across state borders. The quarantine period is typically for the period in which a pathogen is known to be **communicable**. Quarantine, if effective, prevents infected individuals from encountering non-infected individuals. As an island nation, Australia has a greater opportunity to prevent the entry of diseases into the country. Combined with some of the strictest international and interstate quarantine regimes in the world, Australia remains free of many significant human and agricultural diseases.

Other biosecurity strategies that complement quarantine include border inspections, enforced destruction of diseased animals and plants, and banning the importation of organic products from most countries. Effective quarantine and biosecurity practices are essential in protecting Australia from serious communicable diseases and economic loss.

Vaccination, including passive and active immunity

Vaccination is the process of making people (or animals) resistant to diseases caused by specific pathogens. The immunity provided is either active or passive, depending on the source of the immunity.

Active immunity

Active immunity occurs when activated B cells differentiate into plasma cells and antibodies are produced. This can occur *naturally* when an individual is exposed to a pathogen or *artificially* through vaccination.

Vaccines that provide active immunity do so by activating the adaptive immune system, leading to the production of antibodies and memory B and T cells specific to the antigen. They work by exposing the immune system to a compromised version of a pathogen, which induces a primary immune response without the associated symptoms of the disease. This exposure provides active immunity with long-term protection from the disease.

If a secondary exposure occurs, through either a booster shot or exposure to the actual pathogen, the immune response is faster and stronger, due to the presence of circulating antibodies and memory B and T cells generated from the primary exposure. This leads to more effective elimination of the pathogen and results in higher levels of antibodies, which remain circulating for longer periods.

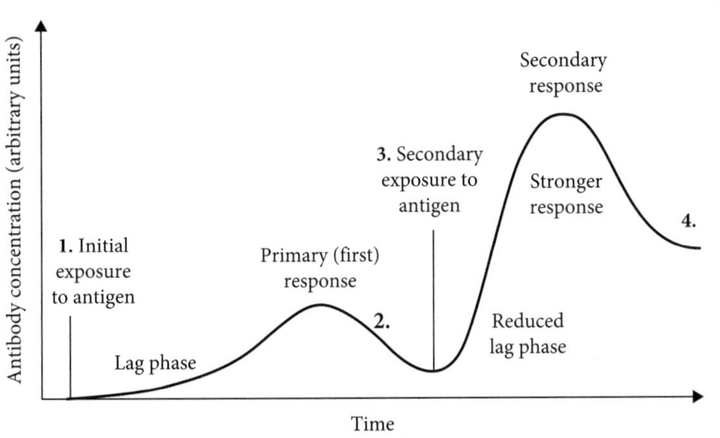

1. After initial exposure, there is a lag phase, when antibodies are not produced until activated B cells differentiate into plasma cells. With natural exposure, this delayed primary response provides time for pathogens to reproduce, which leads to presentation of disease symptoms.
2. Once an infection has been defeated, antibody levels slowly decline and a small number of memory B and T cells remain circulating.
3. Because the primary immune response generates memory B and T cells, secondary exposure to the same pathogen occurs more quickly and is much stronger.
4. Secondary exposure results in higher levels of antibodies, which remain circulating for a longer time.

FIGURE 3.28 Active immunity: primary and secondary exposure to an antigen

The most common active vaccines contain one of the following:

- a live attenuated (weakened) version of the pathogen (e.g. measles, mumps and rubella (MMR) and chickenpox vaccines)
- an inactivated (dead) version of the pathogen (e.g. poliovirus (IPV) and hepatitis A virus vaccines)
- viral messenger RNA (mRNA) (e.g. some COVID-19 vaccines).

Passive immunity

Passive immunity involves an individual receiving antibodies not produced by their own immune system. This can occur *naturally*, from mother to foetus via the placenta, or from mother to newborn during breastfeeding. It can also be achieved *artificially* through vaccination, where antibodies are injected directly into an individual (e.g. certain diphtheria and tetanus vaccines).

Passive immunity bypasses the adaptive immune system and provides immediate protection against a disease. However, because no memory B or T cells are produced, the protection is short-lived, and booster shots must be given if an individual is deemed to have been at risk of exposure.

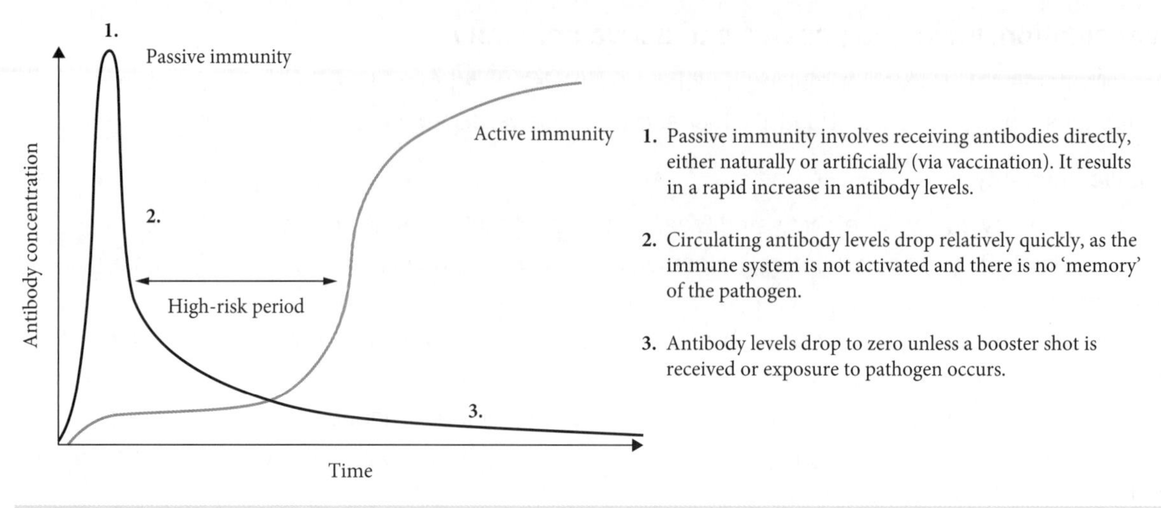

FIGURE 3.29 A comparison of passive and active immunity after primary exposure to an antigen

Herd immunity

Some individuals cannot be vaccinated, usually for age-related or health reasons. When a significant majority of individuals are vaccinated, usually around 90%, unvaccinated individuals are also protected as there are fewer hosts available for the disease to spread between.

> **Note**
> Herd immunity addresses the 'including but not limited to' component of the dot point.

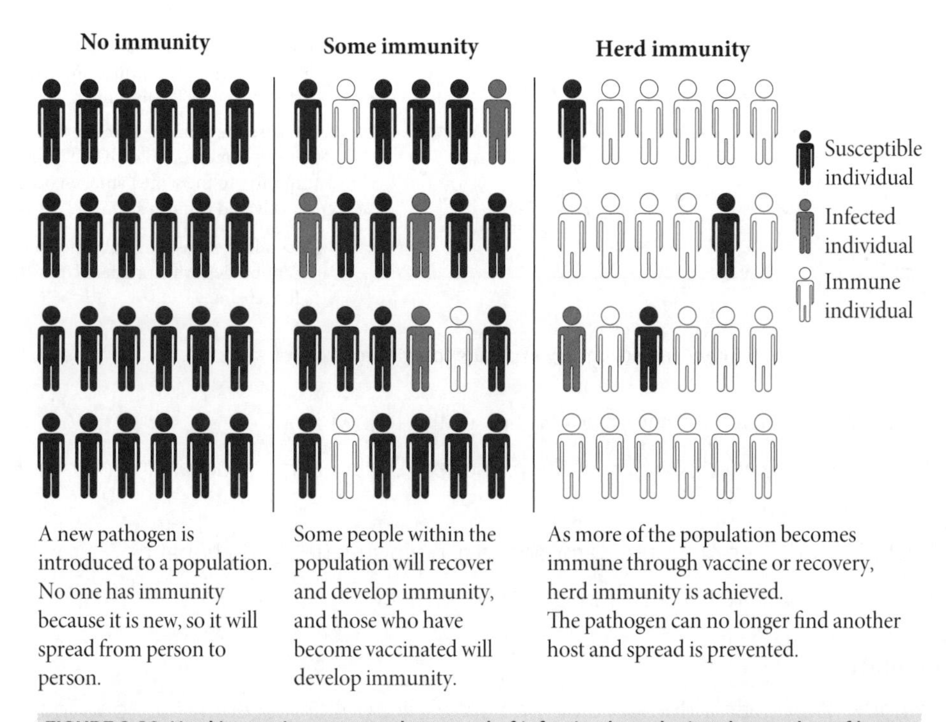

FIGURE 3.30 Herd immunity prevents the spread of infection by reducing the number of hosts available to the pathogen.

Public health campaigns

Public health campaigns seek to provide community members with the necessary information to have them engage in behaviours that improve their health and prevent the spread of infectious diseases. Campaigns use media platforms to disseminate information and reach target audiences.

The 1980s public health campaign to raise awareness about HIV/AIDS included TV commercials in which the Grim Reaper killed men, women and children with a bowling ball. The campaign was graphic and highly effective, and in conjunction with many other initiatives, resulted in the incidence of new cases falling in subsequent decades.

More recently, a national COVID-19 public health campaign saturated all forms of media. This included daily updates on the number of new infections and deaths, health experts highlighting the importance of vaccination, and television advertisements showing people who had contracted the disease, to promote uptake of the vaccination program. Collectively, the strategies contributed to around 90% of eligible Australians receiving double vaccination.

Use of pesticides

Pesticides (insecticides, herbicides and fungicides) are another strategy to prevent the spread of infectious animal and plant diseases. These chemicals target pathogens and insect vectors that carry pathogens.

While there is still widespread dependence on pesticides, there is also growing awareness of the impact they can have on people's health and the environment. Furthermore, because of the widespread use of pesticides and the rapid life cycle of target species, genetic resistance has developed in most cases. As a result, new pesticides must be continually developed, or alternative pest control measures used.

Genetic engineering

Genetic engineering involves the intentional modification of an organism's genome to alter its phenotype. While there are legitimate concerns associated with genetic engineering, it is particularly valuable in agriculture because of the growing resistance of pathogens and insect vectors to traditional chemical treatments. A familiar example is Bt crops (e.g. cotton, corn, potato), which are created by inserting a trans-gene from the soil bacterium *Bacillus thuringiensis* (Bt) into a plant's genome. When the trans-gene is expressed, the plant produces a protein toxin that kills any insect pests that attempt to eat it.

More recently, CRISPR-Cas9 has been used to genetically sterilise male *Anopheles* mosquitoes. When the males are released and allowed to mate, this causes the wild-type female to become infertile. After several reproductive cycles, the number of insect vectors drops significantly. *Anopheles* mosquitoes are known vectors for several pathogens, including malaria.

3.4.3 Investigating and assessing the effectiveness of pharmaceuticals as treatment strategies for the control of infectious disease

In his 1945 Nobel lecture on discovering penicillin (in 1928), Alexander Fleming predicted the rise of the 'superbug' when he said: 'The time may come when penicillin can be bought by anyone in the shops. Then there is the danger that the ignorant man may easily underdose himself and by exposing his microbes to non-lethal quantities of the drug make them resistant'. Nearly 80 years later, antibiotic resistance is commonplace.

> **Note**
> For this dot point, you need to assess the effectiveness of antivirals and antibiotics. To do this, you should know the strengths and limitations of each pharmaceutical and make a judgement of their effectiveness. Failing to provide the necessary depth required for an 'explain', 'assess' or 'evaluate' is one of the main reasons students lose 'easy' marks.

Antivirals

Recall that a virus is a non-living pathogen. It consists of a nucleic acid (either DNA or RNA) encased in a protein coat called a capsid. Some viruses are also enclosed in a second protective layer, called an envelope. The envelope consists of lipids derived from the host cell's membrane and viral proteins that play a role in infected new cells. Diseases caused by DNA viruses include hepatitis B and herpes. HIV and COVID-19 are caused by RNA viruses.

Because viruses require a host cell to reproduce, there are significant challenges to developing antiviral drugs. Any attempt to target the virus inevitably causes death or damage to the host cells.

As a result, antiviral drugs *never* provide a cure. Instead, they suppress the rate of viral replication by a range of mechanisms, including:

- inactivation of the capsid or envelope proteins, which inhibits viral attachment to the host cell
- inhibition of transcription and translation during viral protein synthesis
- inhibition of new viruses being released from host cells.

The effectiveness of **antivirals** is further challenged by the high mutation rate of viruses, particularly that of RNA viruses, which can be up to 1 million times higher than that of the host. This inevitably results in rapid development of drug resistance. In summary, antiviral medications are effective at reducing symptoms and in some cases extending the life of infected individuals. A decreased viral load also reduces the risk of transmission. However, the specificity of antiviral drugs and the high mutation rate of viruses make most drug development options economically non-viable.

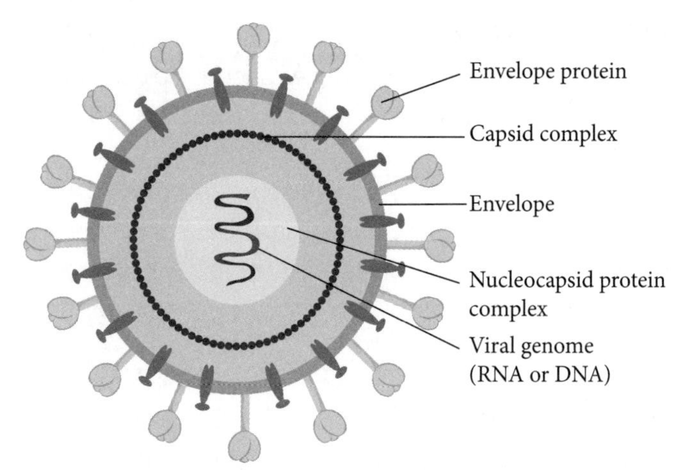

FIGURE 3.31 The general structure of a virus

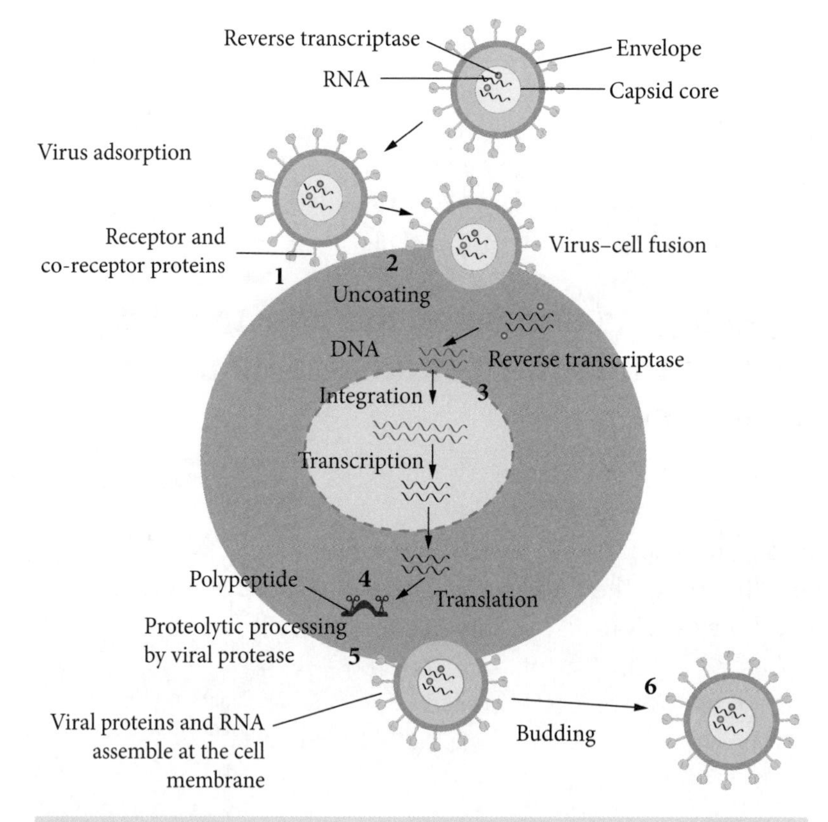

FIGURE 3.32 Antiviral target points: 1 cell attachment, 2 uncoating, 3 nucleotide replication, 4 capsid protein synthesis, 5 virus assembly, 6 release from cell

Antibiotics

There are more than 100 natural or synthetic **antibiotics** in use. They are categorised as either bactericidal or bacteriostatic.

Bactericidal antibiotics kill bacteria by interfering with the formation of the cell membrane, cell wall or cell contents. Examples are penicillin and vancomycin. **Bacteriostatic** antibiotics inhibit bacterial growth by interfering with DNA replication and protein synthesis. They do not kill the bacteria and therefore rely on the immune system of the host to clear the infection. Examples are amoxicillin and clindamycin.

Both bactericidal and bacteriostatic antibiotic groups have broad-spectrum and narrow-spectrum examples. Broad-spectrum antibiotics are effective against a wide range of bacterial species and are the most commonly prescribed. This is because they can be used when the causative pathogen is suspected of being bacterial without the actual species being identified. In contrast, each narrow-spectrum antibiotic is highly specific and acts only on a very small number of bacterial species.

One of the key strengths of antibiotics compared to antivirals is that they target cell structures and molecules not found in humans (and other animals), such as the cell wall or prokaryotic ribosomes. They do not affect host cells and therefore have very few side effects.

However, the overuse and misuse of antibiotics has led to the evolution of 'superbugs' – bacteria that are resistant to antibiotics. Examples are drug-resistant forms of tuberculosis, gonorrhoea, and infections caused by *Staphylococcus aureus*. Without the development of new drugs, diseases that have been long controlled by antibiotics will evolve to cause higher mortalities.

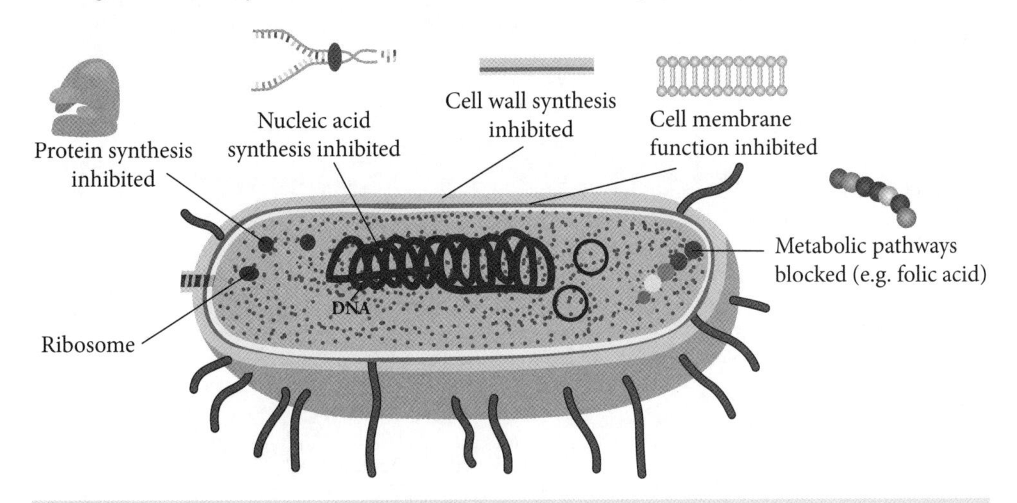

FIGURE 3.33 Mechanisms of antibiotic function

3.4.4 Investigating and evaluating environmental management and quarantine methods used to control an epidemic or pandemic

Recall that an epidemic is an outbreak of an infectious disease within a defined geographical area that spreads at a rate above what is normally expected. A pandemic is typically the spread of a new disease, or a new variant of a disease, across a wider geographical area, often worldwide.

The primary principle of controlling the spread of disease during epidemics and pandemics is the same: prevent contact between infected and non-infected individuals. To achieve this, the optimal approach will *always* involve a combination of both environmental management and quarantine methods. Effective environmental management includes the provision of clean water and food, appropriate sanitation and PPE, and maintaining good air quality. Quarantine methods that isolate infected or exposed individuals for the communicable period, combined with environmental management, offer the best opportunity to control disease spread.

> **Note**
> You need to know the strengths and limitations of different environmental management *and* quarantine methods, *and* provide a point of view as to their effectiveness. It is strongly suggested that your view incorporate a combined environmental management and quarantine approach to reducing disease spread.

CASE STUDY 8

COVID-19 (SARS-COV-2)

It is not a simple matter of saying that the combined measures were, or were not, effective at controlling the spread of COVID-19. For example, although the spread of COVID-19 in Australia was initially prevented through a range of environmental management and quarantine methods, flaws in biosecurity ultimately led to significant COVID-19 outbreaks in New South Wales and Victoria.

Once COVID-19 was in the community, environmental management methods, including 'pop-up' testing facilities to test, track and isolate infected individuals, forced lockdowns, mandatory mask wearing and social distancing, became central to slowing the spread of the virus until the vaccine rollout began in early 2021. These methods were highly effective, as highlighted by the drop to near zero new cases in Victoria after lockdowns 1 and 2. One environmental method, the wearing of masks in public spaces, has been scientifically shown to reduce the spread of exhaled aerosols by up to 80%. Combined with social distancing, this increases to 90%.

Of the nearly 2000 COVID-19 deaths in Australia between 2019 and 2021, the significant majority were unvaccinated, and of the 11.7% who had been double vaccinated, all were elderly or had underlying health issues. Furthermore, the comparison of double-vaccinated versus partially vaccinated highlights the cumulative protection offered by vaccination. While individuals can still contract COVID-19 after being vaccinated, the risk of becoming seriously ill is reduced significantly, with only 1.9% of intensive care unit (ICU) patients being double vaccinated.

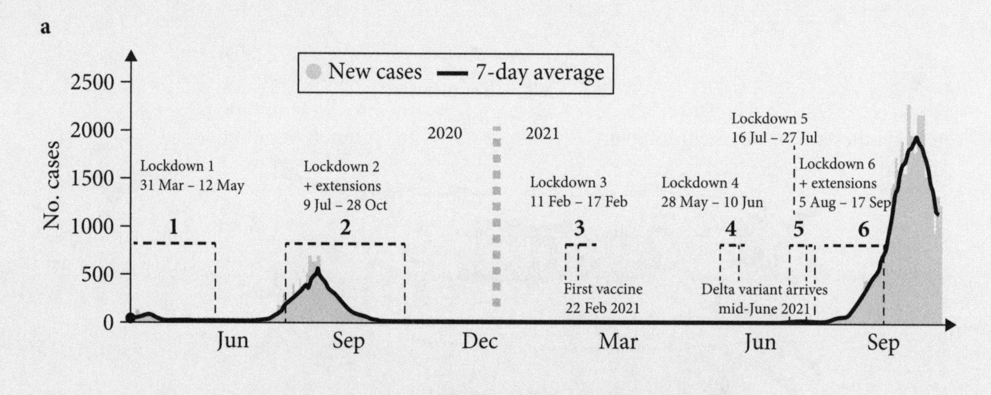

b

	ICU (%)	Deaths (%)
Fully vaccinated	1.9	11.7
Partially vaccinated	15	19.4
Unvaccinated	63.1	63.9
Not stated	19.9	5

FIGURE 3.34 a The number of Victorian COVID-19 cases and lockdown periods. b The proportion of vaccinated versus unvaccinated patients in ICU and deaths.

2066 Locally acquired in last 24 hours	**0** Overseas acquired in last 24 hours	**711** Under investigation in last 24 hours
19 240 Active cases (estimated)	**835** Currently hospitalised	**122 423** Tests in last 24 hours
181 578 Total cases	**1827** Total deaths	**44 582 410** Total tests

FIGURE 3.35 An example of Australia's COVID-19 summary statistics

Other considerations when evaluating the effectiveness of environmental management and quarantine methods relate to the impact they have on different stakeholders. For example, consider the impact of lockdowns and movement restrictions on the economy and on the mental health of the individuals involved. There are also concerns about civil liberties and whether forced lockdowns, restricting movement of citizens and mandatory vaccinations infringe upon individuals' rights. This is weighed against the concept of the 'greater good' and the burden that unvaccinated individuals will place on the health care system if they contract COVID-19 and require hospitalisation. This may prevent vaccinated individuals from receiving medical treatment for other often serious health conditions.

3.4.5 Interpreting data relating to the incidence and prevalence of infectious disease in populations

Incidence

In its simplest form, **incidence** is the number of new cases of a disease in a population in a specified time period. For example, in the data set on page 154, there were 2066 new cases of COVID-19 in Australia over 24 hours. However, incidence is often presented as a percentage, or as number of new cases per 100 000 of the population. If Australia's population is 25 900 000, the incidence is:

> **Note**
> You must be able to distinguish clearly between incidence and prevalence. This dot point has a heavy skill focus. Expect to be given graphs or data sets and asked to analyse the information in them.

$$\frac{\text{Number of new cases during a specific time}}{\text{Size of population at start of monitoring period}} \times 100$$

$$\frac{2066}{25900000} \times 100 = 0.008\%$$

The risk of a person contracting COVID-19 was 0.008% per day or 8 new cases per 100 000 individuals. Incidence data can be used to examine how quickly a disease is spreading through a population. Governments and health authorities can use this data to make decisions on how best to intervene and reduce transmission.

Prevalence

Prevalence is the total number of disease cases in a population in a given time period. Prevalence data can be used to determine the likely burden an outbreak will cause for health facilities and society in general. As highlighted in the 'epidemiologist's bathtub' analogy, prevalence is influenced by the number of new cases (incidence), the time people remain infected for before they recover, and **mortality**.

From the data set, there was an estimated prevalence of 19 240 active cases. Therefore:

$$\frac{\text{Total number of current cases during a specific time}}{\text{Size of population at start of monitoring period}} \times 100$$

$$\frac{19240}{25900000} \times 100 = 0.074\%$$

During this period of the outbreak, 0.074% of the population or 74 individuals per 100 000 would be infected with COVID-19 at any given time.

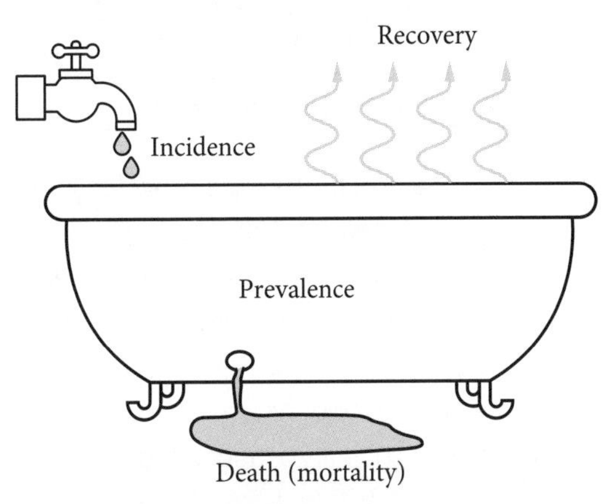

FIGURE 3.36 The 'epidemiologist's bathtub' is used to help distinguish between incidence, prevalence, mortality and recovery.

Mobility

Mobility of a population is an important consideration when assessing the risk of disease transmission. This includes movement within a region and between states, and international travel. All must be considered in combination with incidence and prevalence rates to determine what limitations need to be imposed on a population to reduce transmission.

The impact of reducing the mobility of individuals during a disease outbreak is highlighted in Figure 3.37. Prior to the introduction of vaccines in February 2021, the reduction in incidence and mortality must be attributed to forced lockdowns and restrictions that limited people's mobility.

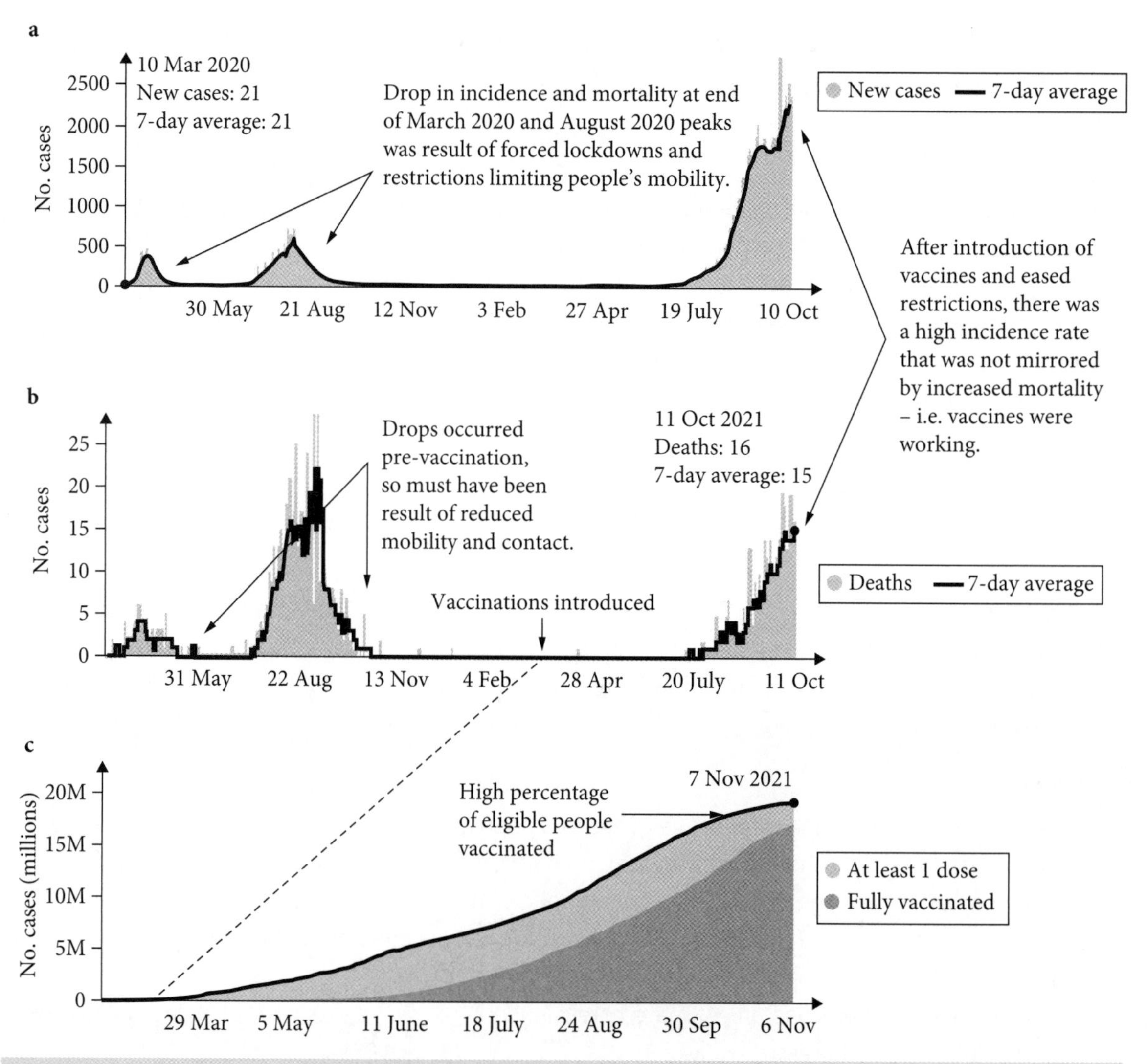

FIGURE 3.37 COVID-19: a comparison of a incidence, b mortality and c vaccination rate in Australia

Analysis of the data provided in the graphs in Figure 3.37 highlights the important collective role that reduced mobility and high vaccination rates have in reducing disease transmission and deaths. The vaccination rollout commenced in February 2021. By the end of 2021, the national average of individuals who were fully vaccinated was over 90%. This came with a significant easing of the restrictions that had limited people's mobility. With the easing of restrictions there was a significant rise in the incidence of new COVID-19 cases. This was not mirrored in the mortality rate– that is, limiting mobility reduced the spread of the disease, while vaccinations reduced the number of deaths.

Malaria or dengue fever in South-East Asia

Malaria in South-East Asia

The WHO estimates that the global prevalence of new malaria cases was 200–250 million per annum for the past two decades. While fluctuations occurred, there was an overall downward trend in case numbers. This was mirrored in mortality rates, with a fall from 736 000 in 2000 to 409 000 in 2019. Of the 87 countries where malaria is endemic, nine are in the South-East Asia region, with cases contributing to approximately 3% of the global malaria burden. The countries are Bangladesh, Bhutan, India, Indonesia, Maldives, Myanmar, Nepal, Sri Lanka and Thailand.

Note
You only need to be familiar with either malaria *or* dengue fever. Don't study both. The skills learned from analysing graphs and data sets from one disease can be applied to any context.

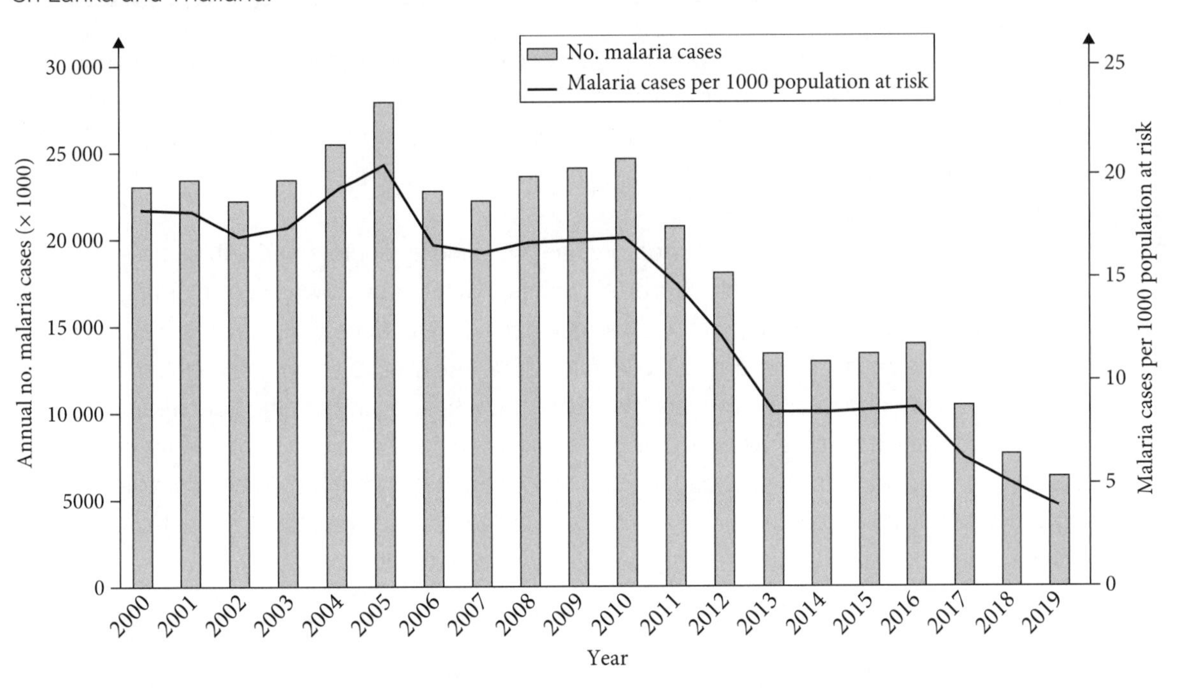

FIGURE 3.38 The annual incidence and prevalence of malaria in South-East Asia

Analysis of the graph highlights that between 2000 and 2019 the annual prevalence in this region dropped from 23 million to 6.3 million, while the incidence dropped from 18.1 to 3.9 cases per 1000 of at-risk population. While decreases in prevalence and incidence of malaria occurred in all six WHO regions, the drop was most significant in South-East Asia. With 25% of the world's population in South-East Asia, the WHO has actively targeted the region with financial support and a range of national malaria elimination programs.

3.4.6 Evaluating historical, culturally diverse and current strategies to predict and control the spread of disease

Prior to the germ theory, predicting and controlling disease spread was largely dependent on observed correlations between specific behaviours or environmental conditions and disease outbreaks. In response, practices were introduced to minimise contact between infected and non-infected people. After the development of germ theory, specific pathogens and their mode of transmission could be linked to a particular disease. This led to significant improvements in strategies used to predict and control disease spread.

Note
Your choices of historical, cultural and current strategies do not need to relate to the same disease. An HSC question targeting this dot point would likely require you to explain how and why predicting *and* controlling disease spread has changed over time, with specific examples from each category provided to support your argument.

Historical strategies

Historically, a lack of scientific knowledge meant that religion influenced cultural values and beliefs more so than it does today. For example, the bubonic plague, which killed a minimum of 25 million people during the Middle Ages (500–1450 CE), was attributed to the wrath of God punishing sinful behaviour. This early historical view of disease spread resulted in limited success in predicting and controlling outbreaks.

In the late 14th century, a link was made between the plague and movement of symptomatic people between towns and cities. In response, the concept of quarantine was first introduced in European ports to reduce transmission of the disease. Ships were required to remain at anchor for 40 days before crew were permitted to leave.

> **Note**
> When structuring your response, consider whether your examples are pre- or post germ theory, as this influences the success of the strategies in predicting and controlling disease spread.

In the mid-1840s, Ignaz Semmelweis, a Hungarian doctor, pioneered early antiseptic procedures, including handwashing prior to performing medical procedures. Despite society having no specific knowledge of pathogens, Semmelweis made the connection between women dying during childbirth from puerperal sepsis, a bacterial infection of the reproductive tract, and doctors delivering babies after performing autopsies on cadavers.

He *predicted* that the introduction of mandatory handwashing by doctors and antiseptic procedures prior to childbirth would reduce mortality rates. His hypothesis was correct. Prior to the introduction of his proposed hygiene practices, women had a 20% chance of dying during childbirth. After the procedures were instituted, this fell to less than 2%. The transmission had been *controlled*.

In 1854, British doctor John Snow mapped the number of cholera cases in the Soho region of London. He determined that the cause of the outbreak was a contaminated public water well. His discovery was significant, as it provided the first indication that cholera could be *controlled* by providing and maintaining a clean water source. Snow's collection and analysis of the data is one of the first recorded examples of an epidemiological study (refer to Module 8).

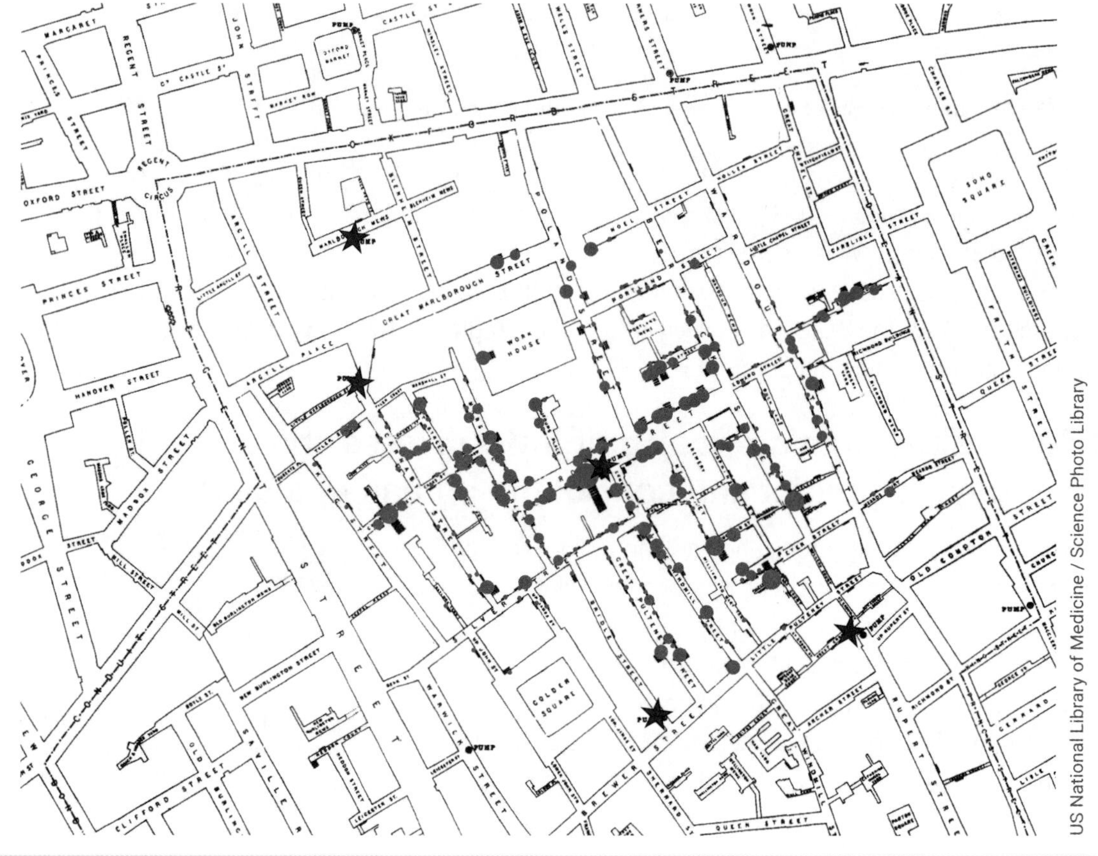

US National Library of Medicine / Science Photo Library

FIGURE 3.39 John Snow's 1854 cholera map of Soho, London. Snow determined that the cause of the outbreak was contaminated water from a public well. The pump was removed, and the number of new infections dropped immediately.

Cultural strategies

Culture is broadly defined as the ideas, customs and behaviours of a group of people. In some cases, cultural beliefs contribute to the prevention and control of disease spread. For example, for over 4000 years, Greek, Roman and Chinese cultures disinfected water by either boiling or charcoal filtration. While they had no concept of the germ theory, the link had been made between consumption of dirty water and disease.

In contrast, some cultural practices can actually promote, and therefore *predict*, the spread of infectious diseases. For example, in West African countries, burial ceremonies involve direct contact with the deceased individual. This practice proved to be one of the main modes of transmission of the Ebola virus in several recent epidemics in the region. The virus is transmitted through bodily fluids and is highly contagious, with a mortality rate as high as 90%. The epidemics were bought under *control* through a range of initiatives including improved public health and education campaigns, contact tracing and the enforcement of safe burials.

Current strategies

The WHO uses satellite imagery and global climate patterns to develop models that can be used to track and *predict* the spread of seasonally linked diseases such as cholera, malaria, dengue fever and Zika virus. This allows *control* measures to be implemented prior to the outbreak, to reduce transmission.

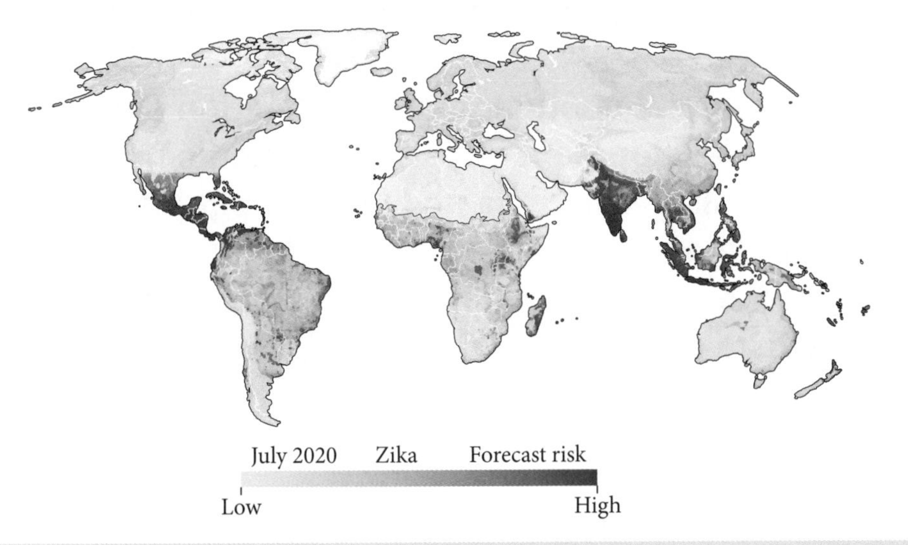

FIGURE 3.40 Global modelling predicting 2020 incidence of Zika virus disease. The data is based on NASA satellite imagery, socioeconomic status, prevalence of the mosquito vector *Aedes aegypti* and global climate modelling.

More recent control measures also focus on the R_0 (R-nought) value, which is an indicator of how contagious an infectious disease is. R_0 is determined by the infectious period, contact rate and mode of transmission. The value represents the number of new individuals a contagious person will infect in a 24-hour period. It is used to predict whether an epidemic or pandemic is spreading and whether control measures are working. The goal is to reduce R_0 to less than 1.

TABLE 3.7 R_0 value and disease status

R_0 value	Disease status
<1	In decline and will eventually die out, as each existing infectious person will cause less than one new infection
1	Stable within a population. Each existing infectious person will cause one new infection
>1	Infectious and spreading

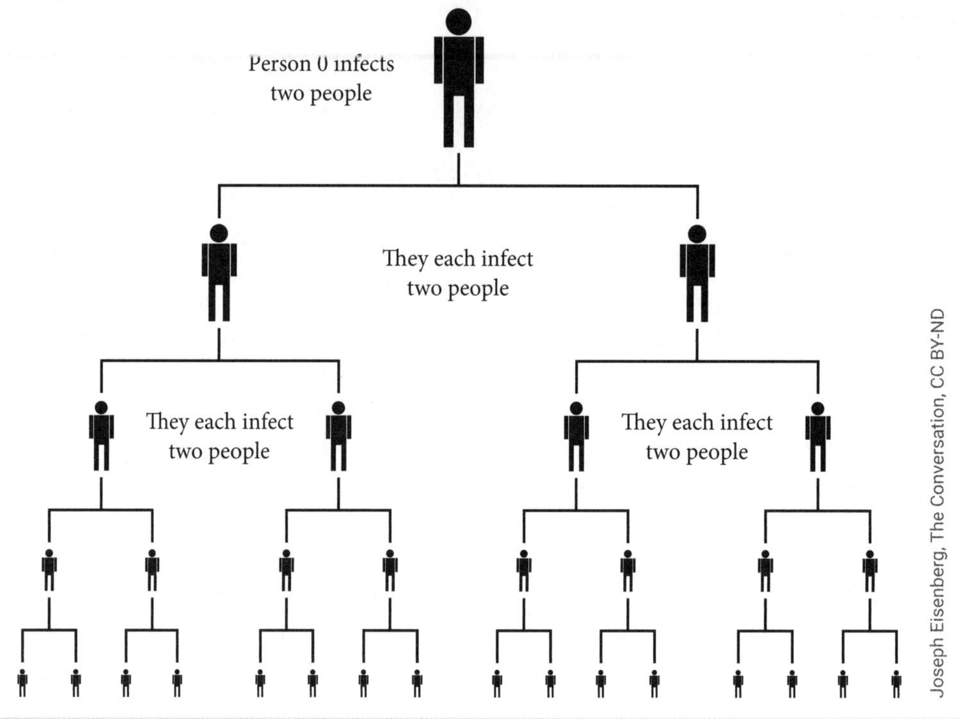

FIGURE 3.41 The spread of a disease with an R_0 of 2, similar to cholera at its most infectious period

3.4.7 Investigating the contemporary application of Aboriginal protocols in the development of particular medicines and biological materials in Australia and how recognition and protection of Indigenous cultural and intellectual property is important, for example: bush medicine, smoke bush in Western Australia

Bush medicine can be defined as the historical and traditional use of native Australian plants for both physical and spiritual healing by Indigenous Australians. Many such practices date back more than 50 000 years.

More recently, research into a variety of bush medicines has discovered a range of active ingredients, including some with anti-cancer and antiviral properties. Once isolated, pharmaceutical companies typically seek to patent the active ingredient in order to ensure long-term financial benefit from their research. However, this raises ethical and legal considerations as to who has legitimate ownership of the intellectual property, given that Indigenous Australians have used the product for thousands of years, albeit without knowledge of the active ingredient.

Note

The inclusion of 'for example' in this dot point means you have choice. The focus is not specifically on bush medicine or smoke bush but instead using your chosen example to highlight the importance of recognising and protecting Indigenous cultural and intellectual property.

CASE STUDY 9

SMOKE BUSH

In the 1960s, the Western Australian government awarded a licence to the US National Cancer Institute to screen native Australian plants for compounds with anti-cancer effects. None were found, but the plants were tested again in the 1980s and smoke bush (*Conospermum stoechadis*) was found to contain an active ingredient, *conocurovone*, a compound that could destroy HIV in low concentrations. In the 1990s an Australian pharmaceutical company was awarded an exclusive licence to develop the patent for a *conocurovone*-based drug to treat HIV. The projected royalties, if successful, were estimated at $100 million per year. Indigenous Australian people were not consulted about the collection and testing of plants from Country and received no financial benefit or acknowledgement from the project, despite the cultural importance of the plant and their knowledge of its medicinal properties.

KAKADU PLUM

CASE STUDY 10

The Kakadu plum (*Terminalia ferdinandiana*) is endemic to eucalypt forests of northern Australia. Indigenous people local to the region have used the fruit of the Kakadu plum for over 40 000 years as a source of food and for its natural healing properties to treat colds, flu and headaches. Extracts from the fruit and tree were also used as an antiseptic balm to treat a variety of ailments, including rheumatoid arthritis and fungal and bacterial infections.

More recently, scientific analysis has identified the Kakadu plum as having the highest vitamin C content of any fruit. Vitamin C has strong antioxidant properties that support the immune system in fighting infections and is also necessary to prevent scurvy, a nutritional disease (Module 8).

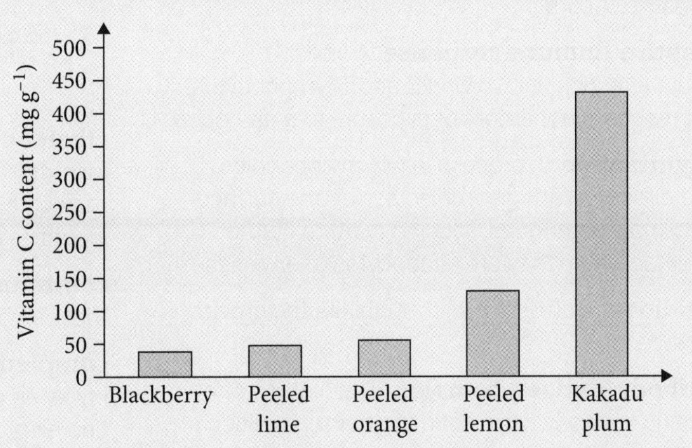

FIGURE 3.42 The vitamin content of Kakadu plum compared with other fruits

Studies also indicate that the fruit is a high source of gallic and ellagic acid, which have been linked to a range of health benefits, including anti-cancer, anti-inflammatory and antimicrobial effects. Several pharmaceutical and cosmetic companies have lodged patents to gain exclusive licensing rights to the research and any economic benefit derived from the fruit. For example, in 2010 a US company, Mary Kay, applied to patent the gallic and ellagic acid extracts from the Kakadu plum. The cosmetic company sought to use these active ingredients for their antioxidant properties and ability to repair damaged skin and reduce collagen breakdown. In 2011, Mary Kay withdrew the patent application due to concerns raised by local Indigenous Australian representatives regarding their ongoing access and use of the Kakadu plum. Another factor was tighter government regulations, which were introduced in response to a similar case that occurred with smoke bush in Western Australia.

These two case studies highlight why Indigenous Cultural and Intellectual Property (ICIP) laws are now in place to protect against the exploitation of traditional arts and culture of Aboriginal and Torres Strait Islander peoples. There must be an appropriate balance between competing interests. Any financial benefits gained from research into bush medicines must now be shared with any Aboriginal or Torres Strait Islander communities involved.

Glossary

acquired immunity The outcome of primary exposure to a pathogen where specific memory B and T lymphocytes are produced and cause a heightened immune response if secondary exposure occurs

adaptive immune response A specific immune response involving B and T lymphocytes that results in a memory of exposure to a pathogen

agglutination A process in which antibodies bind directly to antigen molecules on the surface of multiple pathogens, causing them to clump together, where they are phagocytosed en masse

antibiotic A drug that kills or inhibits the growth of bacteria

antibody–antigen complex The binding of a specific antibody to an antigen after recognition has occurred

antigen A molecule on the surface of microbes that is recognised by the host as 'non-self'; initiates the innate response, including inflammation and phagocytosis

antiviral A drug that inhibits the growth of viruses

apoptosis Programmed cell death; occurs in multicellular organisms

arthropod An invertebrate animal that has an exoskeleton, segmented body and paired jointed appendages (e.g. insects and arachnids)

attenuated Reduced in ability to perform a function

bacteria A prokaryotic organism and the smallest of the cellular pathogens; can be classified as either intracellular or extracellular; some are pathogenic

bactericidal Describes an antibiotic that kills bacteria by interfering with the formation of the cell membrane, cell wall or cell contents

bacteriostatic Describes an antibiotic that inhibits bacterial growth by interfering with DNA replication and protein synthesis

basophil A type of white blood cell; while phagocytic, the primary role is the release of histamine, which causes further vasodilation and inflammation

B cell A type of cell that is involved in the adaptive immune system; produced and matures in bone marrow

chemical barrier Any of the chemical characteristics of an organism's body that oppose colonisation by microbes

chemoattractant A substance that attracts motile cells

https://get.ga/aplus-hsc-bio-u34

A+ DIGITAL FLASHCARDS

Revise this topic's key terms and concepts by scanning the QR code or typing the URL into your browser.

cluster of differentiation A protocol used to identify cell surface proteins; each unique surface molecule is assigned a different number, which identifies cell phenotypes

communicable Describes a pathogen or disease that can be spread from person to person

complement activation A system of reactions involving proteins that bind to pathogen cells and eliminate them by lysis, phagocytosis and inflammation

control In relation to disease, the reduction of disease incidence, prevalence, morbidity or mortality to an acceptable level through the use or introduction of specific measures

cytokine A small signalling protein secreted by certain cells of the immune system; influences the behaviour of other cells

damage-associated molecular pattern (DAMP) A molecular maker expressed by 'self' cells that have been compromised by a pathogen; important component of the innate immune response

dendritic cell An antigen-presenting cell that activates the adaptive immune system

disease A condition that negatively affects the structure and function of an organism

endemic Describes a disease known to have a long-term presence in a specific region

eosinophil A granulocyte that releases toxins to defend against parasitic infections that are too large to be phagocytosed directly

epidemic An outbreak of an infectious disease that spreads rapidly among a large number of individuals within a defined geographical area

fungus A eukaryotic heterotrophic organism belonging to kingdom Fungi; a small number are pathogenic in humans, with most of these being opportunistic in immune-compromised individuals

genetic engineering The process of using recombinant DNA technology to change the genetic make-up of an organism

glycan A carbohydrate usually found attached to proteins and lipids in living organisms; released by some pathogens to help them evade the host's immune system

hypothesis A tentative explanation for an observed phenomenon expressed as a precise and unambiguous statement that can be supported or refuted by investigation

incidence The number of new cases of a disease in a population in a specified time period; often presented as a percentage, or number of new cases per 100 000 of the population

innate immusporone response A non-specific immune response that you are born with and does not have to be learned through exposure to a pathogen

leucocyte One of a broad group of cells collectively called white blood cells that circulate in the blood and lymph and form an essential part of the immune system

lyse Rupture

macroorganism An organism large enough to be seen by the naked eye; can live as a parasite inside a host (endoparasite) or on a host (ectoparasite)

macrophage A phagocytic cell that digests foreign pathogens

mast cell A granulocyte in connective tissue; releases histamine and other substances during inflammatory response

merozoite A stage of the malarial pathogen life cycle; infects red blood cells

microbe (microorganism) An organism that cannot be seen by the naked eye

mobility The degree of movement of a population

morbidity The associated health issues caused by a disease

mortality The number of deaths caused by a disease

natural killer (NK) cell A granular lymphocyte that does not require prior activation (innate immune response)

neutralisation A process in which antibodies bind directly to antigen molecules on the surface of individual pathogens, preventing them from entering or damaging new host cells

neutrophil A phagocytic cell that defends against smaller pathogens including bacteria, viruses and fungi

opsonisation A process in which antibodies bind to the antigens on the surface of infected host cells; the antibody-bound cells are detected by innate phagocytes or granular NK cells, which release cytotoxic chemicals to lyse the host cell, killing any pathogens in it

pandemic A new disease, or a new variant of a disease, that has spread across a wide geographical area, often worldwide

pathogen A cellular or non-cellular disease-causing agent

pathogen-associated molecular pattern (PAMP) A conserved molecular marker common to a group of closely related pathogens; an important aspect of recognition by the adaptive immune response

pathogenesis The nature and presentation of disease development

physical barrier A structural obstruction against microbe entry

prevalence The total number of disease cases in a population at a given time period

prevention The action of stopping a non-infected individual from becoming infected

prion A non-cellular pathogen consisting of misfolded protein; resistant to protease enzymes responsible for breaking down incorrectly folded proteins

protease An enzyme involved in breaking down proteins

protozoan An 'animal-like' protist; pathogenic species range in size from 10 to 150 μm and can be seen under a light microscope

reliability The extent to which repeated observations and/or measurements taken under identical circumstances will yield similar results

spontaneous generation A historical theory that living organisms develop from non-living matter

T cells A family of cells involved in the adaptive immune response; produced in bone marrow and mature in the thymus gland

transmission The passage of a pathogen between an infected and a non-infected individual

treatment The medical care given to an individual for an illness or injury

vaccination The process of making people (or animals) resistant to diseases caused by specific pathogens; the immunity provided is either active or passive, depending on the source of the immunity

validity The extent to which tests measure what was intended, and extent to which data, inferences and actions produced from tests and other processes are accurate

vasodilation The dilation of blood vessels

vector An organism that carries and transmits a pathogen

virulence The degree to which a pathogen can cause disease

virus A non-cellular pathogen that can only replicate inside the cells of a host organism

xylem Tissue in plants that transports water and minerals from the roots to the rest of the plant

zoonotic Describes a pathogen or disease that can be transferred between animal species

Exam practice

Multiple-choice questions

Solutions start on page 260.

Causes of infectious disease

Question 1

A cellular pathogen infecting a plant was found to have no membrane-bound organelles and a cell wall.

The pathogen can best be classified as a

A prion.

B fungus.

C bacterium.

D protozoan.

Question 2 ©NESA 2020 SI Q4 (ADAPTED)

Dengue fever is a disease in humans caused by the dengue virus. It is transmitted by female *Aedes aegypti* mosquitoes.

Which of the following is true of dengue fever?

A Both the dengue virus and the female *Aedes aegypti* mosquito are vectors.

B Both the dengue virus and the female *Aedes aegypti* mosquito are pathogens.

C Dengue virus is the pathogen and the female *Aedes aegypti* mosquito is the vector.

D The female *Aedes aegypti* mosquito is the pathogen and dengue virus is the vector.

Refer to the following information to answer Questions 3 and 4.

A student conducted a first-hand investigation using nutrient broth, beakers and an S-shaped delivery tube in an attempt to model one of Pasteur's experiments. The equipment and data collected are shown.

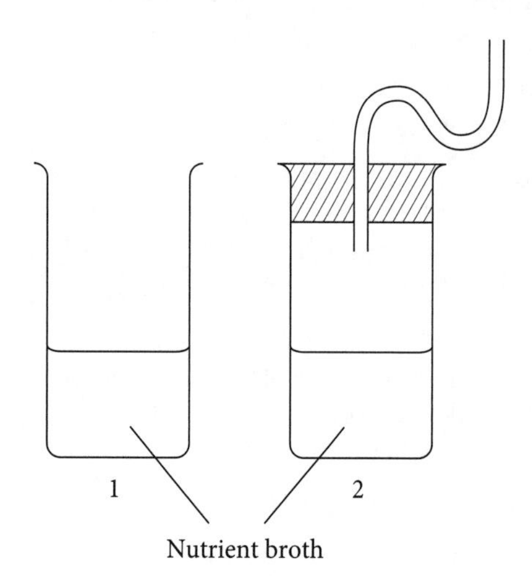

Test tube	Observation of nutrient broth Day 1	Observation of nutrient broth Day 14
1	Clear	Cloudy
2	Clear	Cloudy

Question 3 ©NESA 2014 SI Q12

The data collected by the student are

A quantitative because data were collected for fourteen days.

B qualitative because the appearance of the broth is described.

C qualitative because this is a model of a past scientific experiment.

D quantitative because the results were recorded for two different beakers.

Question 4 ©NESA 2014 SI Q13

The student's results were different from Pasteur's results. Which of the following provides the best explanation for the difference?

A The nutrient broth was different from Pasteur's.

B The nutrient broth always goes cloudy as it ages.

C The nutrient broth was not boiled thoroughly on Day 1.

D The nutrient broths were both exposed to oxygen from the outside air.

Question 5 ©NESA 2012 SI Q4

The diagram shows a pathogen called *Giardia*.

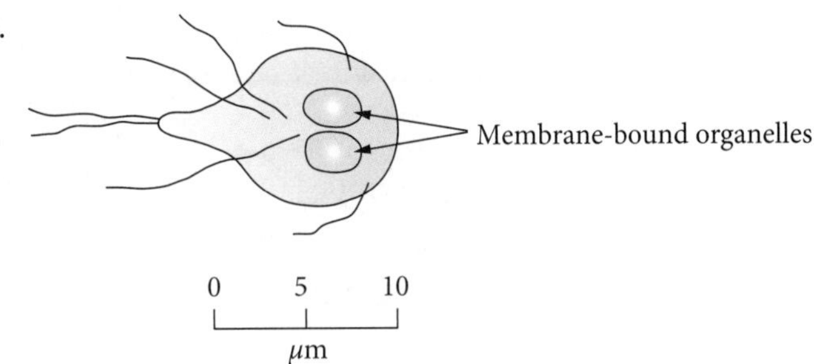

What type of pathogen is *Giardia*?

A Bacterium

B Prion

C Protozoan

D Virus

Responses to pathogens

Question 6

If a pathogen breaches the physical barriers of the innate immune system, mast cells in the surrounding tissue release histamines.

What is the role of the histamines?

A To lyse the pathogen

B To attract cytotoxic killer T cells to the area of infection

C To increase the permeability of blood vessels in the area of infection

D To stimulate the production of memory B cells that are specific to the pathogen

Question 7

Which of the following options correctly identifies a physical barrier and a chemical barrier to pathogens?

	Physical barrier	Chemical barrier
A	Intact skin	Stomach acid
B	Intact skin	Peristalsis
C	Mucous membrane	Cilia
D	Stomach acid	Lysozymes in saliva

Question 8

What process is shown in the following image?

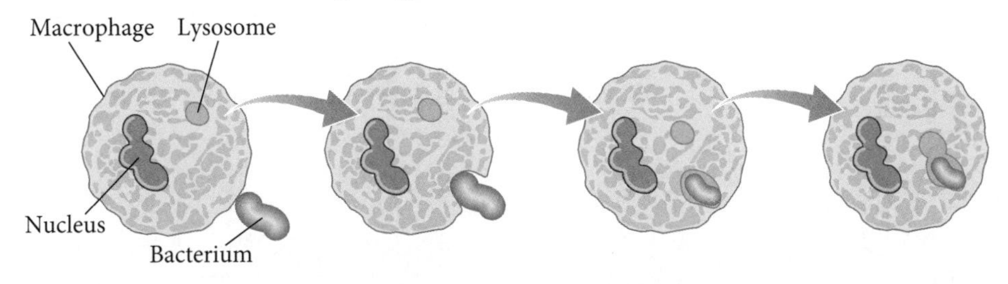

A Apoptosis

B Phagocytosis

C Inflammation

D Chemoattraction

Immunity

Question 9

Which type of cell involved in the adaptive immune response will destroy a virally infected cell?

A Phagocyte

B B lymphocyte

C Natural killer cell

D Killer T lymphocyte

Question 10 ©NESA 2010 SI Q20

The diagram illustrates an immune response.

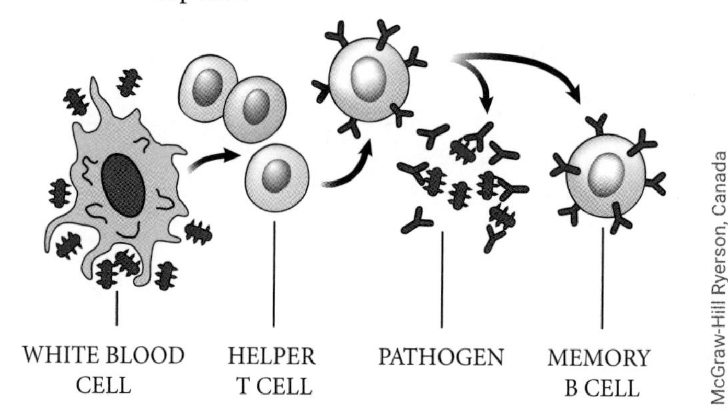

Below is a list of statements, each describing a step in the immune response.

1 Antibodies are produced to immobilise the pathogens.

2 B cell is activated by a helper T cell.

3 Helper T cells are activated by the white blood cell.

4 Memory B cell is ready to respond to further infections.

What is the correct sequence of events?

A 2, 3, 1, 4

B 2, 3, 4, 1

C 3, 2, 1, 4

D 3, 4, 2, 1

Question 11 ©NESA 2019 ADDITIONAL EXAM MOD 7 Q6

A student was vaccinated for rubella when they were 13. Three years later, they were exposed to the active rubella virus.

Which graph best represents the student's production of antibodies over time?

A

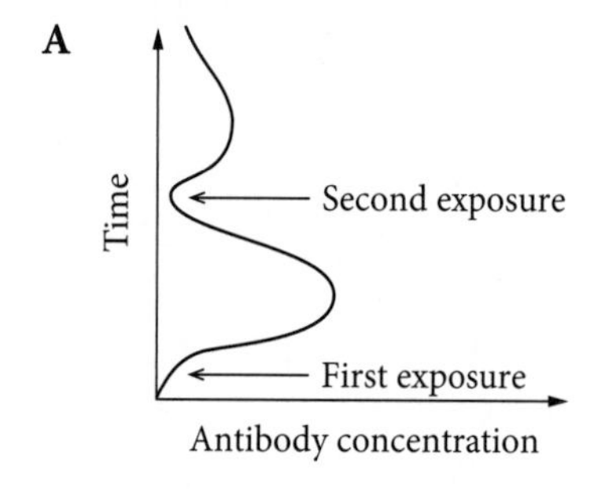

B

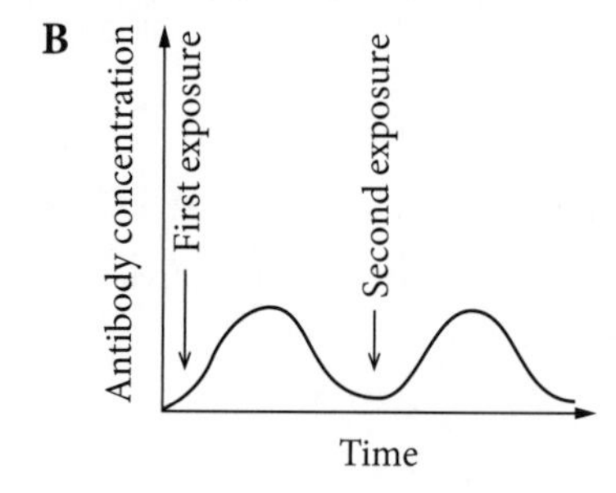

C

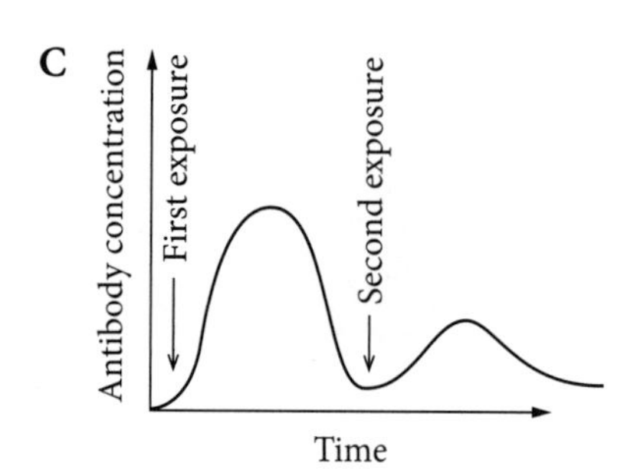

D

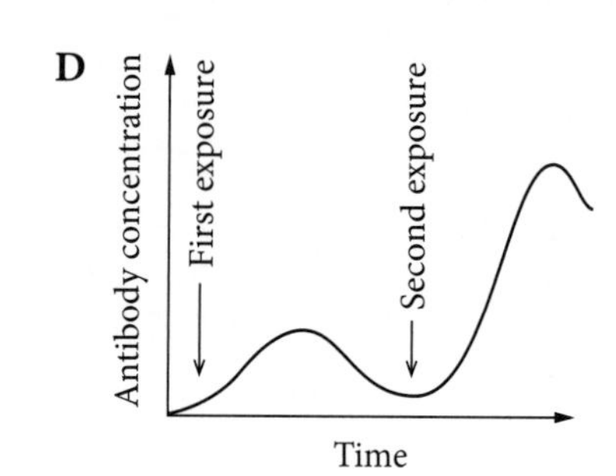

Question 12

What is the name of a molecule that is recognised by the immune system as being foreign?

A Antigen

B Antibody

C Antibiotic

D Anticodon

Question 13 ©NESA 2014 SI Q18

Two people were exposed to pathogen *P* on the same day. The graph shows the blood antibody levels for that pathogen over the following 28 days for each person.

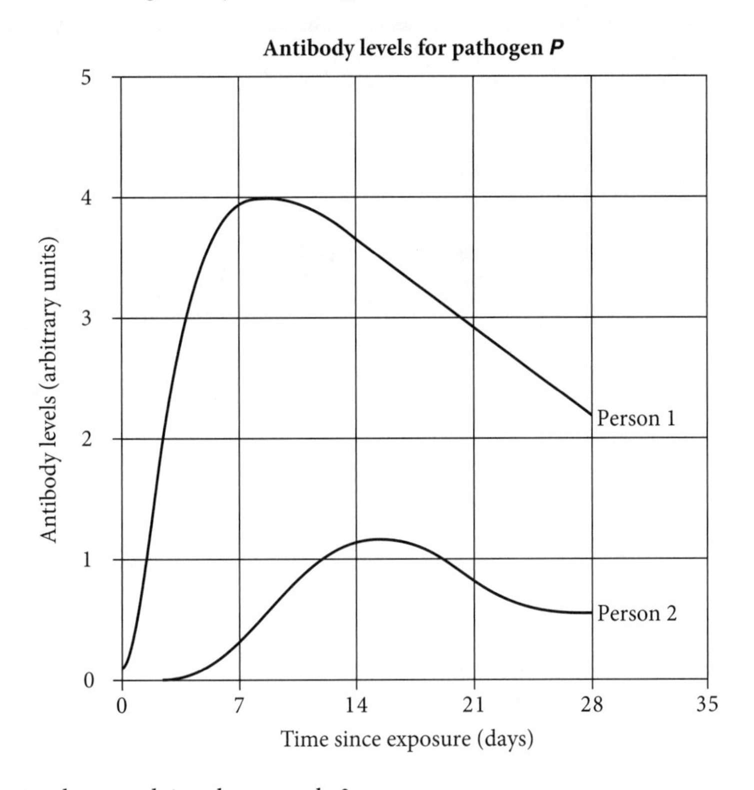

Which of the following best explains these results?

A Person 1 had not been previously exposed to pathogen *P* but had a recent organ transplant.

B Person 1 had been previously exposed to pathogen *P*.

C Person 2 had already been vaccinated against pathogen *P*.

D Person 2 had recent contact with a person infected with a similar pathogen.

Question 14

Which of the following correctly identifies the link between the innate and adaptive immune responses?

A Phagocytes present parts of engulfed pathogens on MHC class I antigens to B cells to initiate antibody production.

B Dendritic cells present antigens from engulfed pathogens on MHC class I molecules to helper T cells in the bloodstream.

C Neutrophils present antigens from engulfed pathogens on MHC class I molecules to killer T cells in the lymphatic system.

D Dendritic cells present antigens from engulfed pathogens on MHC class II molecules to helper T cells in the lymphatic system.

Prevention, treatment and control

Question 15 ©NESA 2019 ADDITIONAL SAMPLE EXAM MOD 7 SI Q9 (ADAPTED)

The map shown was drawn by Dr John Snow during the 1854 London cholera epidemic.

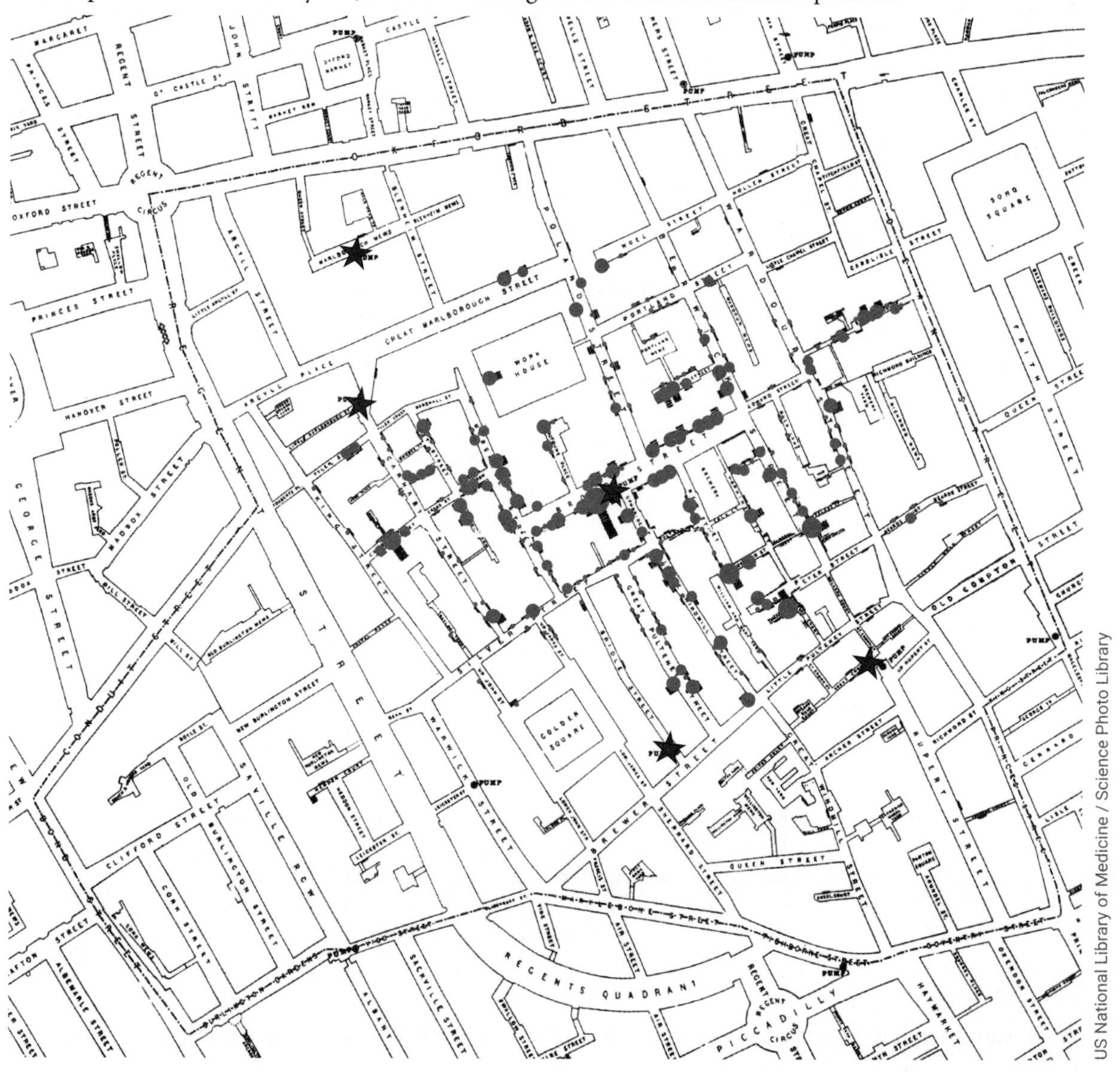

US National Library of Medicine / Science Photo Library

The circles indicate people who died from cholera and the stars indicate the location of water pumps in the Soho region of London. John Snow's conclusion that contaminated water was the cause of cholera was based on which of the following?

A Observational data correlated infected individuals with their water source.

B He was aware that *Vibrio cholerae* was transmitted via contaminated water.

C He isolated *Vibrio cholerae* after performing autopsies on deceased patients.

D He was aware that residents in the area had not been vaccinated against cholera.

Question 16

How does passive vaccination prevent disease?

A It generates the production of memory B cells.

B It generates the production of memory T cells.

C It generates the production of both memory B and T cells.

D It increases the number of antibodies against a pathogen.

Question 17

Foot-and-mouth disease is a viral disease that affects cloven-hoofed animals such as sheep and cattle. It is not endemic to Australia. What would be the most effective way to prevent the disease from entering Australia and spreading across the country?

A Monitor locally farmed animals weekly for signs of the disease.

B Ensure animals, or animal-based products, are quarantined until the contagious period has passed.

C Inspect animals and animal-based products when they enter Australia prior to transportation.

D Have state border checkpoints that inspect animals and animal-based products before they are allowed to cross.

Question 18

Which of the following options correctly distinguishes between incidence and prevalence?

	Incidence	Prevalence
A	The number of deaths caused by a disease	The number of new cases of a disease measured over a specific time
B	The number of deaths caused by a disease	The total number of cases in a population at a given time
C	The total number of cases in a population at a given time	The number of new cases of a disease measured over a specific time
D	The number of new cases of a disease measured over a specific time	The total number of cases in a population at a given time

Question 19 ©NESA 2019 SI Q7

Two types of bacteria were isolated from a patient's throat swab and grown in pure culture on separate agar plates. On each plate there were FOUR different antibiotic discs, *W*, *X*, *Y* and *Z*. The photograph shows the plates seven days later.

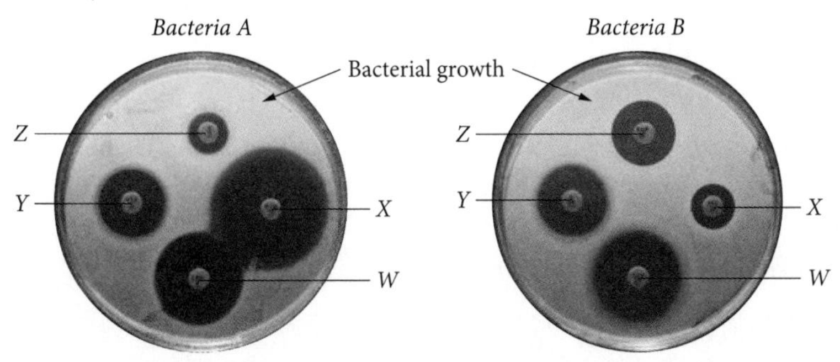

Which antibiotic should be used to treat the patient?

A *W*

B *X*

C *Y*

D *Z*

Question 20

A study of a disease starts with 7500 people. Of these, 150 have the disease in question. What is the prevalence of the disease per 1000 people?

A 0.02

B 20

C 75

D 200

Short-answer questions

Solutions start on page 263.

Causes of infectious disease

Question 21 (4 marks) ©NESA 2019 ADDITIONAL SAMPLE EXAM MOD 7 Q12 (ADAPTED)

Complete the following table to show the distinguishing characteristics of each pathogen and provide a disease caused by each.

Pathogen	Distinguishing characteristics of the pathogen	Disease caused by the pathogen
Virus		
Bacteria		
Fungi		
Protozoan		

Question 22 (4 marks)

In order to minimise microbial growth in food, a NSW food authority requires that food businesses keep all hot food at, or above, 60°C.

Write a valid methodology to test the effect of temperature on microbial growth from a specific food type.

Question 23 (4 marks)

HIV (human immunodeficiency virus) infects and destroys T lymphocytes. If untreated, HIV infection leads to AIDS, a syndrome that leads to the death of the patient from typically opportunistic infections.

Using your knowledge of the human immune system, explain why, over time, HIV infection leads to AIDS.

Question 24 (6 marks) ©NESA 2016 SII Q25

Rabies is a disease caused by a virus that affects mammals. In 1880 Louis Pasteur investigated dogs that were suffering from rabies in order to find the cause. He believed rabies was caused by a microorganism but could not culture it in broth nor observe it under the light microscope. However, he could cause the disease in healthy dogs by injecting them with saliva from infected dogs. He was able to repeat the disease cycle in this way.

a Why was Pasteur **not** able to observe the rabies virus? 2 marks

b Explain why Pasteur needed to identify and culture the microorganism in order to meet the scientific standards for establishing the cause of rabies. 4 marks

Question 25 (6 marks)

The majority of commercially grown bananas are seedless varieties that can only be propagated asexually. The Cavendish banana currently constitutes 50% of total world banana production and 99% of the world's export market. Using your knowledge of reproduction and agricultural production, assess the likely cause and effects of disease outbreak in Cavendish banana crops.

Question 26 (5 marks)

Construct a table like the one below to compare the adaptations of **two** pathogens that facilitate their entry into and transmission between hosts.

Pathogen	Adaptation for entry	Adaptation for transmission

Responses to pathogens

Question 27 (6 marks)

a Outline a specific physical **or** chemical barrier used by plants to defend against pathogens. 2 marks

b Explain a specific response of a named Australian plant to a named pathogen. 4 marks

Question 28 (4 marks)

If a pathogen gains entry to the body and the innate immune system is unable to clear the infection, the adaptive immune system is activated. Explain the cellular mechanism that causes this activation.

Question 29 (4 marks)

Identify **two** cell types involved in the innate immune response, and explain their role in chemical and physical changes that occur in the presence of pathogens.

Immunity

Question 30 (6 marks)

The graph shows the point of primary exposure to a pathogen and a later secondary exposure.

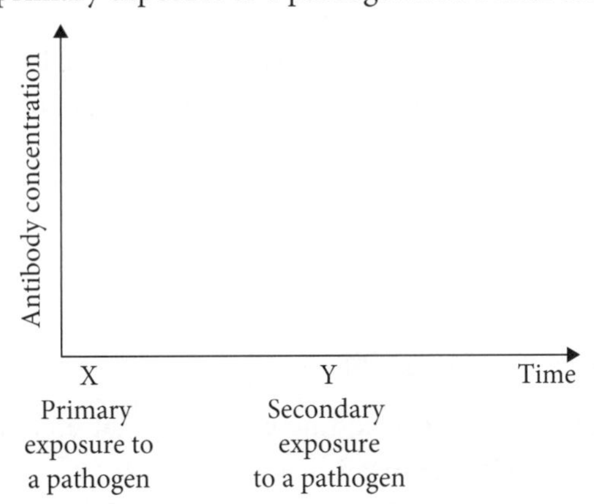

a Complete the graph to model the immune response after both primary and secondary exposure to the same pathogen. 2 marks

b Annotate your graph to explain its shape. 4 marks

Question 31 (3 marks)

The ABO blood groups represent different antigen molecules on the surface of red blood cells. Individuals who are either A or B blood group have only the one respective antigen molecule on their cell surface. Blood group O individuals have no antigen markers, while individuals with AB blood group have both antigen molecules present. Using your knowledge of the immune response, explain why a person with blood group O cannot receive a blood donation from any of the other blood types.

Question 32 (5 marks)

Draw a diagram that models the antibody-mediated and cell-mediated adaptive immune systems and the interaction between them.

Question 33 (4 marks)

Distinguish between passive and active vaccination.

Prevention, treatment and control

Question 34 (5 marks)

Assess the relative effectiveness of antiviral drugs and antibiotics to treat infectious diseases.

Question 35 (5 marks)

Explain how specific hygiene practices and genetic engineering can be employed to prevent the spread of infectious disease.

Question 36 (7 marks)

Evaluate the relative effectiveness of specific environmental management and quarantine methods used to control a named infectious disease outbreak in Australia.

Question 37 (7 marks)

a Define 'bush medicine'. 1 mark

b Outline reasons why pharmaceutical companies would be interested in researching bush medicines and isolating active ingredients. 2 marks

c With specific reference to smoke bush, explain the competing interests of pharmaceutical companies and Indigenous Australian communities, and assess the importance of recognising and protecting Indigenous Australian cultural and intellectual property. 4 marks

Question 38 (10 marks) ©NESA 2019 SII Q32

Dengue fever and malaria are examples of infectious diseases transmitted between humans by mosquitoes. Dengue fever is caused by a virus transmitted by mosquitoes of the genus *Aedes*. Malaria is caused by a single-celled organism transmitted by mosquitoes of the genus *Anopheles*.

The following data provide information about the global incidence of these two diseases over time.

Global malaria data for selected years from 1900 to 2010

Year	Global population ($\times 10^9$)	Number of countries with reported cases	Population at risk	
			($\times 10^9$)	(%)
1900	1.2	140	0.9	75
1946	2.4	130	1.6	67
1965	3.4	103	1.9	65
1975	4.1	91	2.1	51
1992	5.4	88	2.6	48
1994	5.6	87	2.6	46
2002	6.2	88	3.0	48
2010	6.8	88	3.4	50

S I Hay, C A Guerra, A J Tatem, A M Noor and R W Snow (2004). 'The global distribution and population at risk of malaria: past, present and future', *The Lancet Infectious Diseases*, 4(6): 327–336.

Distribution of reported cases of dengue fever in 1950 and 2010

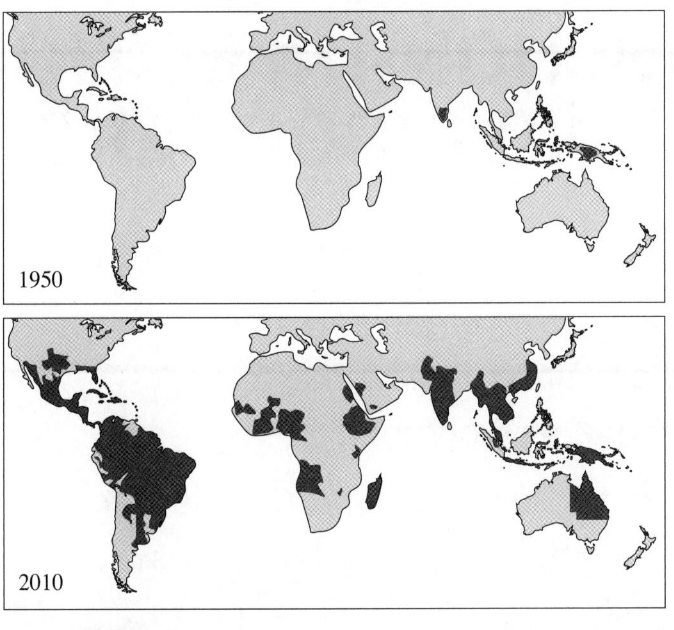

KEY ▬ Reported cases of the viral disease dengue fever

Max Roser and Hannah Ritchie (2018) – Dengue fever distribution maps in 1950 and 2010. Published online at OurWorldInData.org. Retrieved from https://ourworldindata.org/malaria Creative Commons BY license

a Based on the data provided, identify trends in the global disease burden for both malaria and dengue fever. 3 marks

b Analyse factors that could have contributed to the change in global distribution of both dengue fever and malaria over the last 100 years. Support your answer with reference to the data provided. 7 marks

CHAPTER 4
MODULE 8: NON-INFECTIOUS DISEASE AND DISORDERS

Chapter 4
Module 8: Non-infectious disease and disorders

Module summary

Module 8 explored homeostasis, the causes and effects of non-infectious diseases, epidemiology, and technologies used to assist people with a range of disorders. The role of negative feedback loops, adaptations, hormones and the neural system in maintaining a stable internal environment in plants and animals were addressed. A range of non-infectious human diseases were investigated, along with epidemiological methods and the effectiveness of prevention strategies. A key part of this module is the practising of analytical skills to interpret data from population-based studies. Finally, the structure and function of the ears, eyes and kidneys were addressed, providing the foundation to understand the application of technologies to treat and manage disorders affecting these organs.

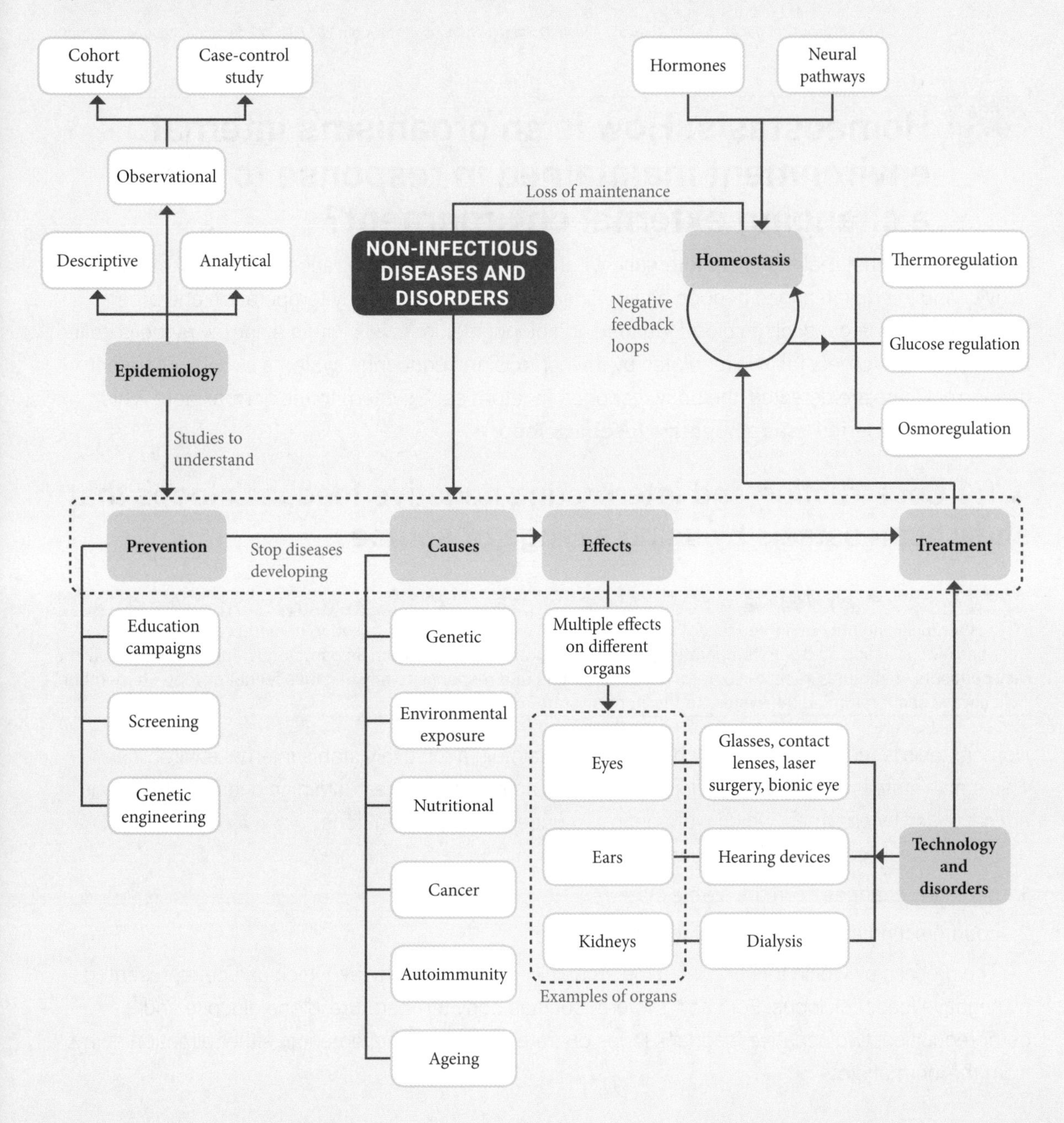

Outcomes

On completing this module, you should be able to:

- explain non-infectious disease and disorders and a range of technologies and methods used to assist, control, prevent and treat non-infectious disease.

Key science skills

In this module, you are required to demonstrate the following key science skills:

- analyse and evaluate primary and secondary data and information
- solve scientific problems using primary and secondary data, critical thinking skills and scientific processes
- communicate scientific understanding using suitable language and terminology for a specific audience or purpose.

4.1 Homeostasis: How is an organism's internal environment maintained in response to a changing external environment?

All organisms must maintain their internal environment within a narrow range for optimal metabolic activity and the maintenance of good health. The ability to maintain body temperature and other metabolic parameters such as blood **glucose** and blood sodium levels within a narrow range is called **homeostasis**. Homeostasis is regulated by the nervous and endocrine systems. When changes from the normal state are detected, the body responds to return each system to the normal state. This mechanism is referred to as a **negative feedback loop**.

4.1.1 Constructing and interpreting negative feedback loops that show homeostasis by using a range of sources

Note

This is an 'including but not limited to' dot point. In the HSC exam you may be asked to construct or interpret feedback loops associated with temperature, glucose or another example such as osmoregulation. You can apply the key concepts of stimulus, receptor, control centre, effectors and response to any negative feedback loop. Remember, it is always about returning the system to the 'normal' state.

Homeostasis is the process by which organisms maintain a relatively stable internal environment. This stable state is essential for efficient metabolic activity, as all life-sustaining chemical reactions are catalysed by enzymes, which work optimally under specific conditions.

Homeostasis consists of two stages:

1 detecting changes from the stable state
2 counteracting these changes from the stable state.

The process by which this occurs is the stimulus–response pathway, which can be represented by negative feedback loops. For each 'system', such as body temperature, blood glucose and osmoregulation, two negative feedback loops operate to counter movement in either direction away from the 'normal state'.

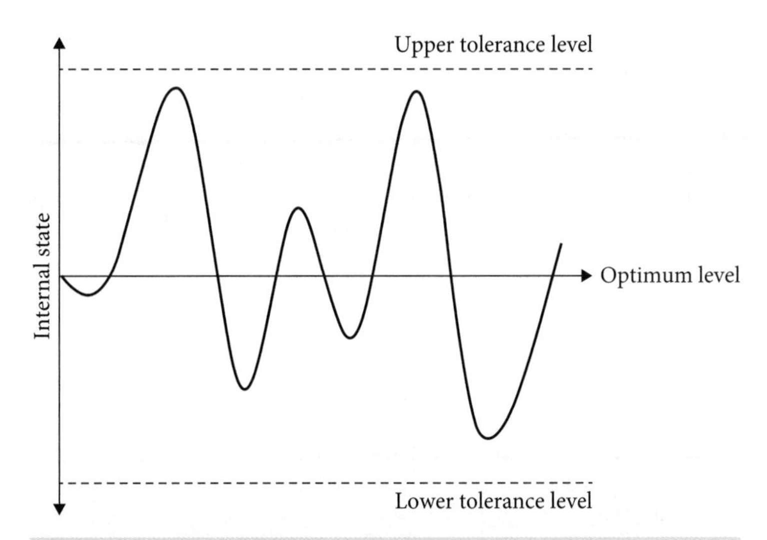

FIGURE 4.1 Graph demonstrating negative feedback mechanisms and fluctuation around the optimum level

Any change that induces a response is called a **stimulus**. This might be an increase in temperature or blood glucose level. The stimulus is detected by a **receptor**, such as a thermoreceptor or a chemoreceptor, which receives and transmits an impulse to a **control centre** (nervous system). The control centre interprets the message and sends a signal to the appropriate **effector** (e.g. muscles or glands), which brings about a **response** to restore the normal state.

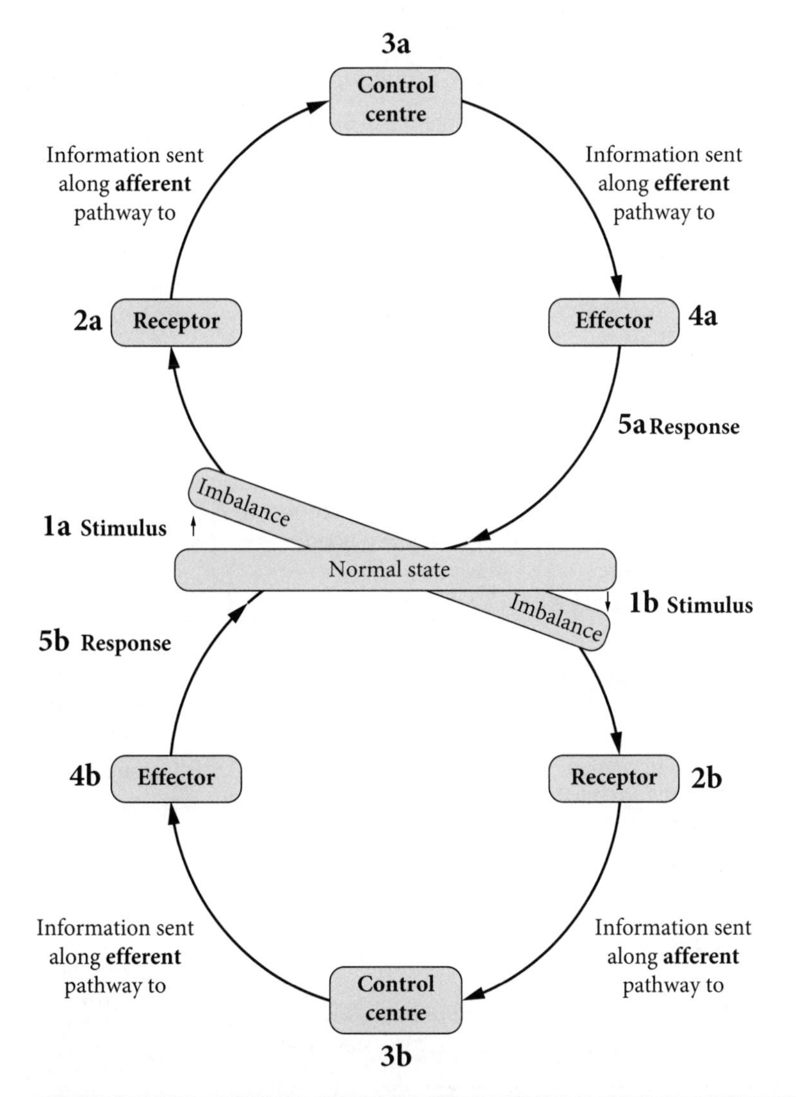

FIGURE 4.2 A typical negative feedback loop diagram

Note
Use Figure 4.2 with Table 4.1 to understand the stages in the different feedback loops.

TABLE 4.1 Features of three negative feedback loops

		Thermoregulation	Glucose regulation	Osmoregulation
Increase detected ↑	**5a** Response	• Body temperature decreases • Heat lost via radiation from skin • Perspiration secreted	• Blood glucose levels decrease • Liver cells activated to take up glucose and store as glycogen • Body cells take up more glucose	• Blood water levels decrease • Binding of ADH to collecting duct of nephrons reduced • Reabsorption of water reduced
	4a Effector	• Muscles cause dilation of blood vessels at skin surface • Sweat glands activated	• Insulin-secreting cells of pancreas stimulated to release insulin into blood	• Pituitary gland stimulated to produce less ADH
	3a Control centre	• Hypothalamus activates cooling mechanism	• Glucoregulatory neurons activate response mechanism	• Hypothalamus activates response mechanism
	2a Receptor	• Change detected by thermoreceptors in hypothalamus	• Increase in blood glucose detected by insulin-secreting (beta) cells of pancreas	• Change detected by osmoreceptors in hypothalamus
	1a Stimulus	• Increase in body temperature (e.g. exercise)	• Increase in blood glucose (e.g. after eating)	• Increase in blood water (e.g. over-hydration)
	Normal state	35.6–37.8°C	4.0–7.8 mmol L^{-1} glucose	135–145 mEq L^{-1} (sodium levels)
Decrease detected ↓	**1b** Stimulus	• Decrease in body temperature (e.g. cold environmental conditions)	• Decrease in blood glucose (e.g. fasting)	• Decrease in blood water (e.g. lack of hydration)
	2b Receptor	• Change detected by thermoreceptors in hypothalamus	• Decrease in blood glucose detected by glucagon-secreting (alpha) cells of pancreas	• Change detected by osmoreceptors in hypothalamus
	3b Control centre	• Hypothalamus activates heating mechanism	• Glucoregulatory neurons activate response mechanism	• Hypothalamus activates response mechanism
	4b Effector	• Muscles cause constriction of blood vessels at skin surface • Skeletal muscles activated	• Glucagon-secreting cells of pancreas stimulated to release glucagon into blood	• Pituitary gland stimulated to produce more ADH
	5b Response	• Body temperature increases • Blood diverted away from skin surface to reduce heat loss • Shivering	• Blood glucose levels increase • Liver cells break down glycogen and release glucose into the blood	• Blood water levels increase • Binding of ADH to collecting duct of nephrons increases • Increased reabsorption of water

Temperature

Thermoregulation is the ability of an organism to maintain its body temperature within a certain range independently of its external environment. For example, in humans, thermoregulation maintains the body temperature at 37°C to ensure the structural integrity of enzymes that catalyse cellular reactions.

> **Note**
> It is important to remember that homeostasis involves the detection of, and response to, changes in the *internal* environment.

Glucose

Glucose is an essential requirement for the production of adenosine triphosphate (ATP) during cellular respiration. ATP is the molecule that carries energy for most chemical reactions in a cell. Blood glucose levels in humans need to be maintained within a narrow range of 4.0–7.8 mmol L^{-1} in the fasting state (when your last snack or meal has been completely digested) for long-term health and survival.

Osmoregulation

Osmoregulation involves the maintenance of osmotic pressure (water and salt balance) in an organism. The normal blood sodium level in humans is 135–145 milliequivalents per litre (mEq L^{-1}). An abnormally low or high concentration of sodium in the blood can lead to health problems ranging from nausea, muscle weakness and confusion, to seizures, coma or even death.

Note

The osmoregulation negative feedback loops address the 'including but not limited to' component of the dot point. The structure and function of the kidney is covered in section 4.5.1. Look there to help you with some of the terms related to osmoregulation.

4.1.2 Investigating the various mechanisms used by organisms to maintain their internal environment within tolerance limits

Note

This is an 'including' dot point, so you may be asked a question about any of the mechanisms addressed in this section.

Trends and patterns in behavioural, structural and physiological adaptations in endotherms that assist in maintaining homeostasis

An **endotherm** is an organism that uses energy to regulate its internal body temperature. This is maintained at a relatively constant level, no matter what the external ambient temperature. Endotherms include mammals and birds. An **ectotherm** is an organism whose regulation of body temperature depends on the external temperature. Ectotherms include fish, reptiles, amphibians and invertebrates.

Endotherms have a range of **adaptations** that assist in maintaining homeostasis. These adaptations are categorised as behavioural, structural and physiological.

Behavioural adaptations

Behavioural adaptations are the conscious actions organisms take to warm up or cool down, to avoid **hypothermia** or **hyperthermia**.

TABLE 4.2 Thermoregulation and the behavioural adaptations required

Type of thermoregulation	Function	Behavioural adaptation
Hypothermia avoidance	To keep body heat	• Physical postures (basking, rolling into a ball) • Nest building • Grouping strategies (huddling, nest-sharing)
	To enhance body heat production	• Increased movement • Increased energy intake (i.e. more eating)
Hyperthermia avoidance	To dissipate body heat	• Habitat selection (places with shade and water) • Physical postures (stretching out body) • Panting
	To decrease body heat production	• Reduced movement • Decreased energy intake (i.e. less eating)

Animals will follow patterns of behaviour to regulate their body temperature. For example, koalas adopt different body postures in trees, depending on the heat of the day. This pattern of behaviour can be presented as a histogram, with explanations of the different body positions in relation to thermoregulation.

> **Note**
> This dot point is about investigating trends and patterns. This means you may be required to interpret data in diagrams, tables and graphs related to adaptations that maintain homeostasis.

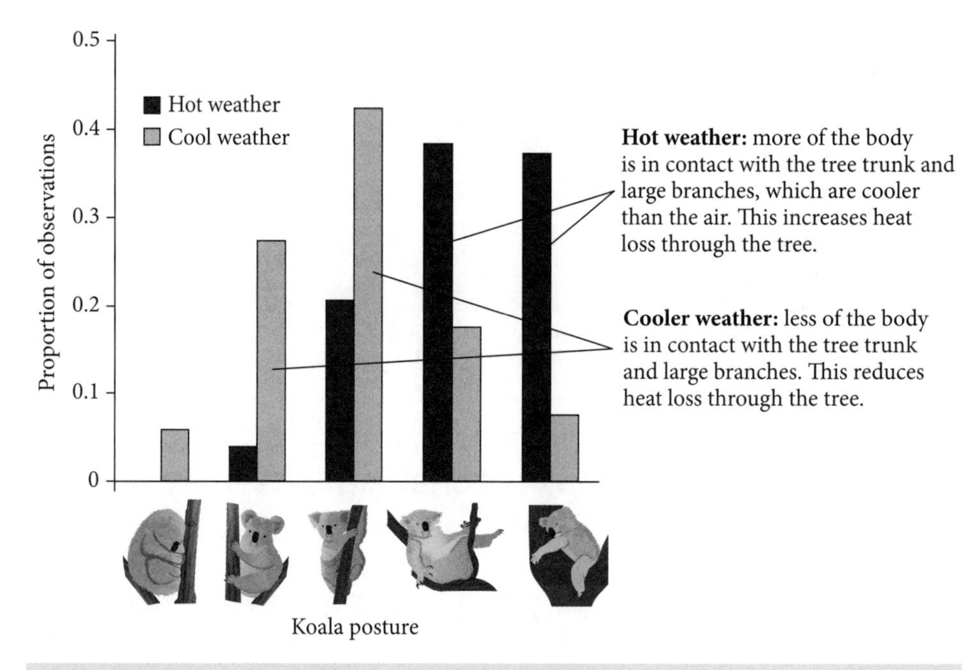

FIGURE 4.3 Patterns of behaviour that regulate body temperature in koalas

Structural adaptations

Structural adaptations are physical features of an organism that assist in maintaining homeostasis – examples are body shape, and the colour and thickness of fur. Allen's rule states that there is a trend in the length of limbs and appendages in relation to the climate. The surface area of an endotherm's body, relative to its body volume, affects the amount of heat lost to the surrounding air. Differences in body shape, limb length and ear size can be seen in hares living in habitats with a cold versus a hot climate.

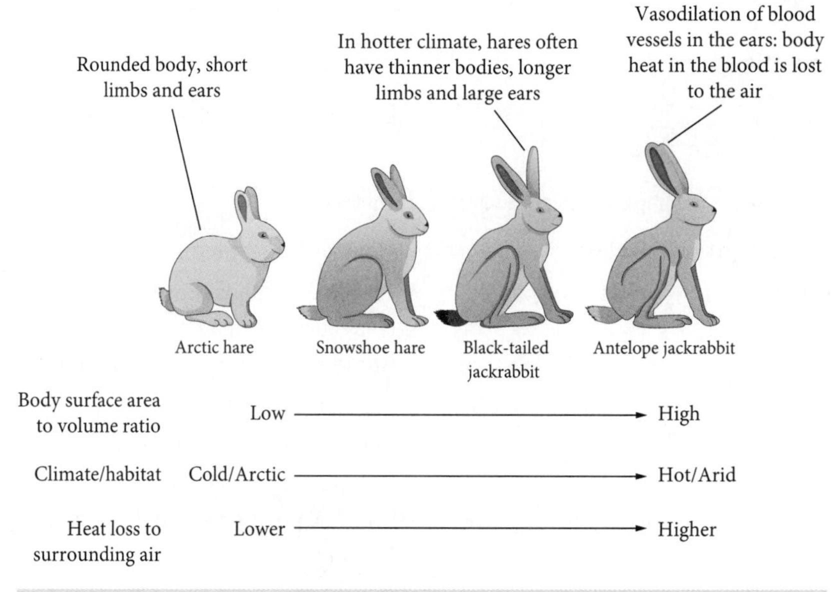

FIGURE 4.4 The trend in ear and limb length in hares in relation to the hares' habitat

Physiological adaptations

Physiological adaptations occur within an organism to regulate and maintain homeostasis. Examples include changes in heart rate, oxygen uptake and hormone levels. Antidiuretic hormone (ADH) regulates the amount of water in the body. When it is released from the pituitary gland, it prevents the kidneys from making too much urine and results in water retention. This will be used as an example to investigate a trend in a physiological adaptation in two Australian marsupials.

The spectacled hare wallaby (*Lagorchestes conspicillatus*) and Rothschild's rock wallaby (*Petrogale rothschildi*) both live in arid environments. The hare wallaby shelters in spinifex bushes during the heat of the day, when the air is dry and the temperature reaches over 40°C. The rock wallaby shelters in cooler caves with a temperature of about 30°C and higher humidity. The level of ADH in the blood of each species shows very different trends.

The hare wallaby uses ADH as a physiological adaptation to reduce urine production, to retain water in the body. The rock wallaby has a different physiological response. It relies on reduced blood flow to the kidneys instead of hormonal control. This, combined with its ability to seek out cool rock shelters in its habitat, allows this species to regulate its internal water levels.

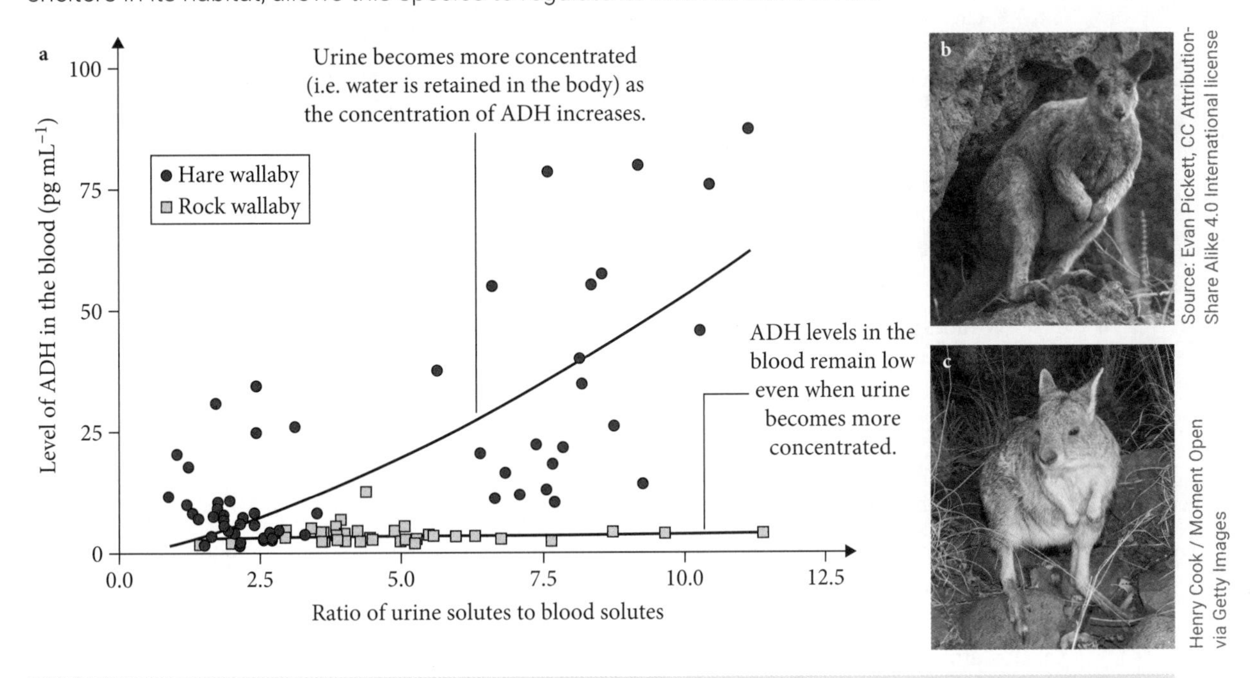

Source: Evan Pickett, CC Attribution-Share Alike 4.0 International license

Henry Cook / Moment Open via Getty Images

FIGURE 4.5 a Osmoregulation in two Australian marsupials: b Rothschild's rock wallaby and c spectacled hare wallaby

Internal coordination systems that allow homeostasis to be maintained, including hormones and neural pathways

Hormones in animals

Hormones are signalling molecules secreted by special groups of cells, called **endocrine glands,** to maintain homeostasis.

Hormones travel in the blood and in the fluid between cells. There are three types of hormones, categorised according to their chemical structure: **steroid hormones**, **peptide hormones** and **amine hormones**. The chemical structure of a hormone determines its solubility, the way it travels through the body and how it interacts with cells.

Note

For this dot point you need to know how hormones work, so focus on the 'mode of action' and solubility, and link this to how homeostasis is restored in the body.

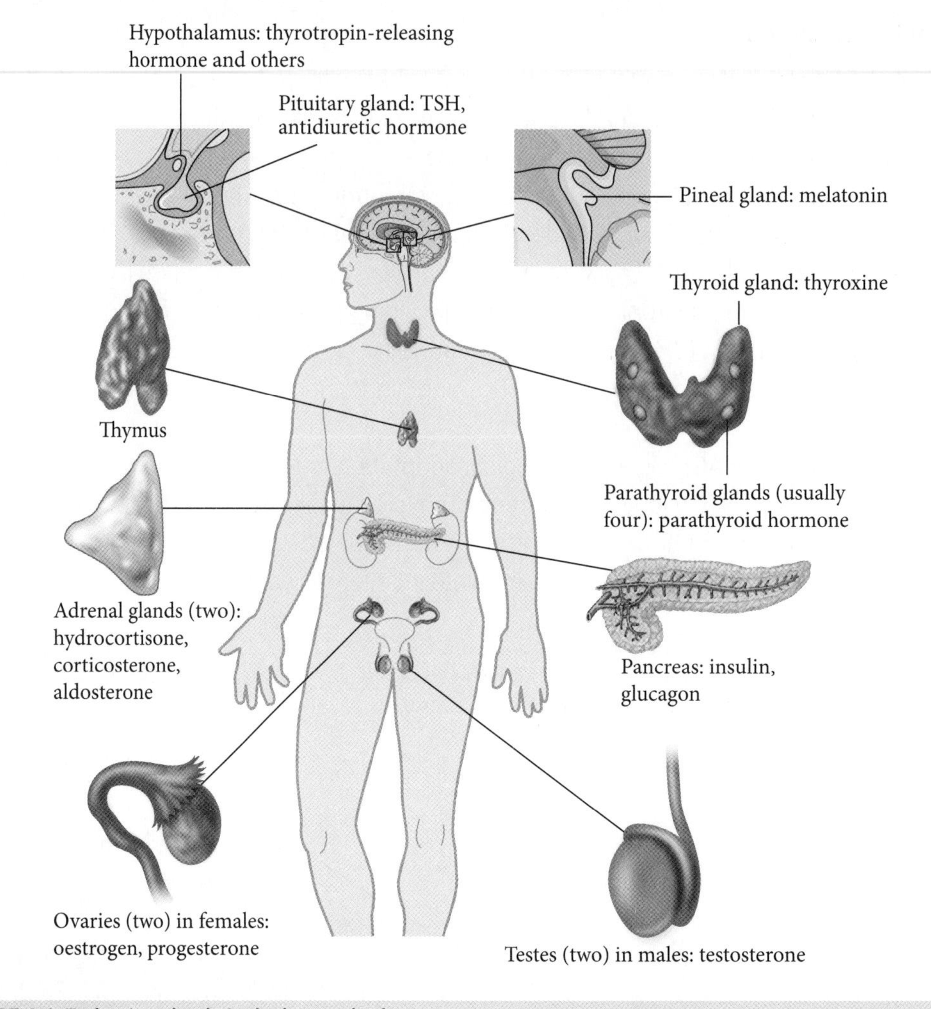

FIGURE 4.6 Endocrine glands in the human body

TABLE 4.3 The main hormone groups and their modes of action

Hormone group	Solubility	Chemical structure	Mode of action	Examples
Steroid	Fat soluble	Made from cholesterol	• Slow onset but long lasting • Travel in the blood attached to a water-soluble carrier protein • Freely diffuse across the plasma membrane • Bind to protein receptors in the cytoplasm and nucleus to activate gene transcription	Testosterone, oestrogen and progesterone
Peptide	Water soluble	Small chains of amino acids	• Fast acting and transient • Travel freely in the blood • Cannot move across the plasma membrane • Bind to receptors on the cell surface to activate **signal transduction** and gene transcription	Insulin and oxytocin
Amine	Fat or water soluble	Derivatives of the amino acid tyrosine	• Share properties with steroid and peptide hormones	Adrenaline

Hormones can signal to the same cell (**autocrine**), cells that are nearby (**paracrine**) or cells at distant sites in the body (**endocrine**). Each type of hormone affects cells at different distances and their effects can last for varying periods of time. Some hormones bind to a specific receptor on many different types of cells, while others only bind to a receptor on specific cells.

TABLE 4.4 Types of hormone signalling

Forms of hormone signalling	Description	Example
Autocrine	• Localised effect • Cell releases a hormone that is recognised by itself • Occurs during early development and infections	T-lymphocytes release a hormone in response to a viral infection, which stimulates the production of more lymphocytes to kill infected cells.
Paracrine Signalling cell Target cell	• Causes a quick response • Last a short time • Hormone is quickly degraded	Blood clotting. When a blood vessel is damaged, broken endothelial cells release von Willebrand factor (vWF), which stimulates the production of platelets. These blood cells clot the blood and 'patch up' the wound.
Endocrine Signalling cell Target cell Bloodstream	• Slow onset but long-lasting effects • Produced by cells in endocrine glands	Menstrual cycle. Luteinising hormone and follicle-stimulating hormone are produced in the pituitary gland and target the ovary to produce and release mature eggs.

Hormones in plants

Hormones are produced in different parts of a plant, such as the stem tip, buds or root tips. They travel through the **phloem** and **xylem** and pass between cells through **plasmodesmata**. There are five main groups of plant hormones: auxins, cytokinin, gibberellins, ethylene and abscisic acid.

TABLE 4.5 Plant hormones and their functions

Plant hormone	Function
Auxins	• Control the primary elongation of stems and inhibit the lateral growth of branches • Control the growth of stems towards light (**phototropism**)
Cytokinin	• Produced in the tips of roots and travels upward through the xylem • Promotes cell division in growth areas • Balance between auxins and cytokinins determines whether the plant produces roots or shoots
Gibberellins	• Trigger seed germination and growth of the stem region between internodes (places from which leaves grow)
Ethylene	• Gas produced by plants that stimulates ripening of fruit, loss of leaves (**abscission**) and flower death
Abscisic acid	• Synthesised in chloroplasts of leaves and is a signal of dehydration • Regulates water loss by controlling the opening and closing of **stomata** • Also promotes seed dormancy

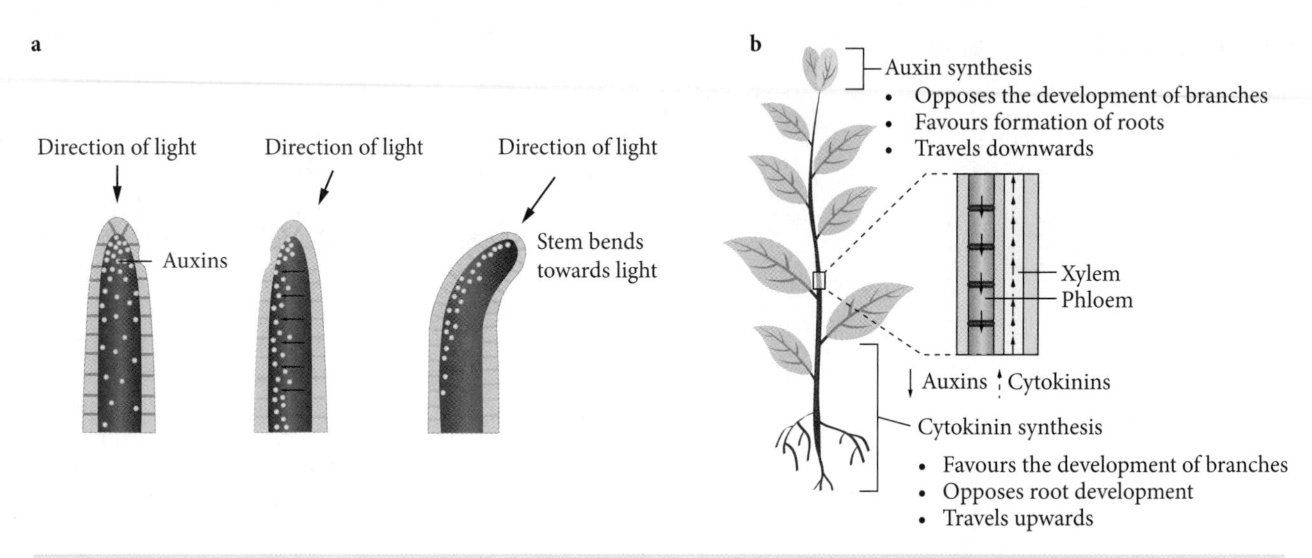

FIGURE 4.7 Effects of plant hormones: a auxins and phototropism, b auxin–cytokinin balance for root and shoot growth

Neural pathways

All animals have a nervous system. The vertebrate nervous system has two main parts: the **central nervous system** (CNS) and the **peripheral nervous system** (PNS).

The CNS is the brain and spinal cord. The brain is the processing centre of the body. The spinal cord is an extension of the brain that carries electrical and chemical signals between the brain and the PNS. The three main parts of the brain are the cerebrum, cerebellum and brainstem, and each controls different functions in the body.

Note

You should relate the structure and processes in the neural system to its role in maintaining homeostasis.

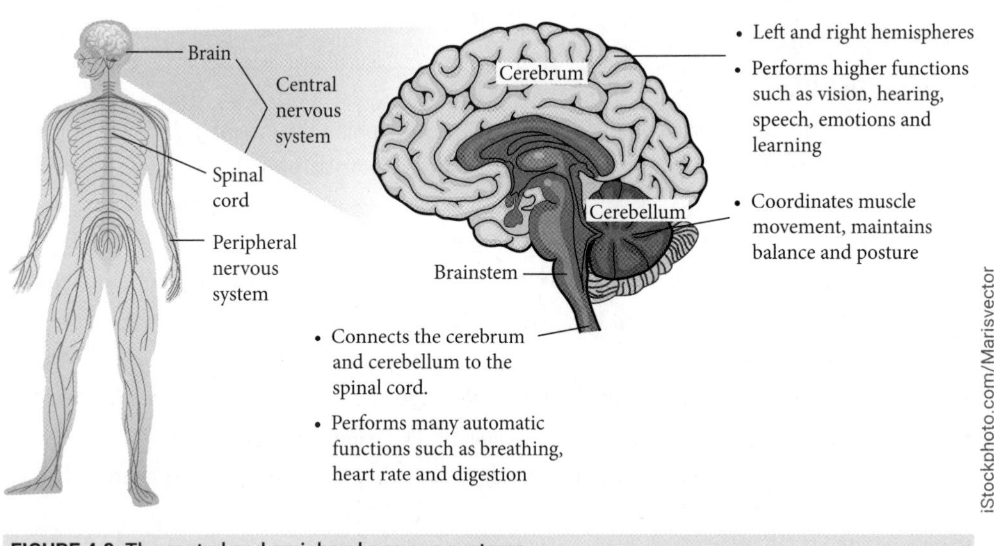

FIGURE 4.8 The central and peripheral nervous systems

The PNS consists of all the nerves that are not part of the CNS, that branch out to the organs, glands and muscles. The PNS is divided into the somatic and autonomic nervous systems, based on the parts of the body and the responses they control. Some actions, such as breathing, are done without thinking. Others, such as moving an arm, are done voluntarily.

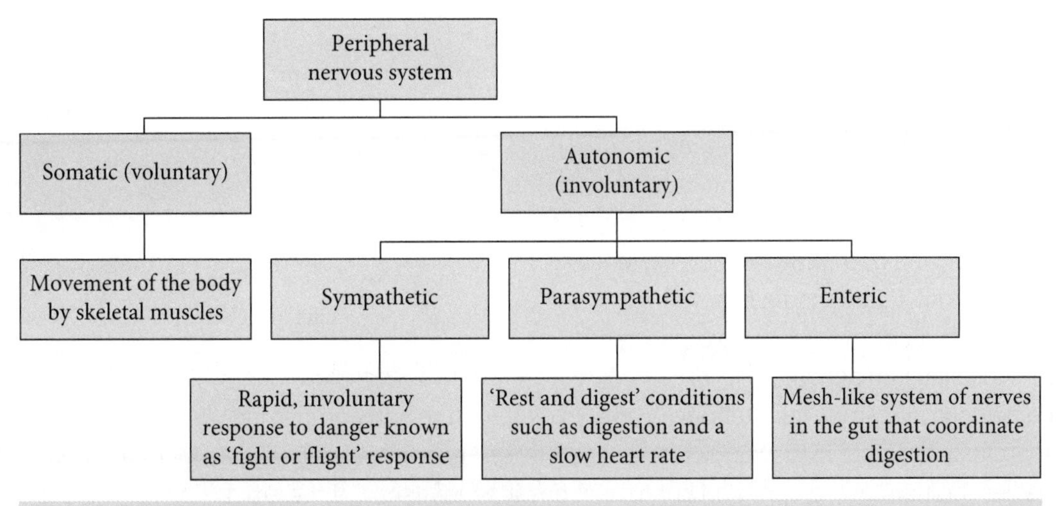

FIGURE 4.9 Categories of the peripheral nervous system

The basic building block of the nervous system is the **neuron**. A neuron is a specialised type of cell that has three main parts: **dendrites**, an **axon** and a **cell body** (soma). The dendrites are where a neuron receives a signal from another neuron. The axon is where a neuron sends a signal to a neighbouring neuron. The soma is where the cell nucleus is located. Transcription and translation occur here, and proteins are transported for use in dendrites and axons. Multiple axons bundle together to form a fibre, and fibres group together to form a **nerve**.

Signals travel through neurons as an **action potential**. The tips of neighbouring neurons do not touch each other; they are separated by a small gap called a **synapse**. When an action potential reaches the end of a neuron, it triggers the release of neurotransmitters. These chemical molecules travel across the synapse, bind to receptors on the next neuron, and the message continues along the nerve.

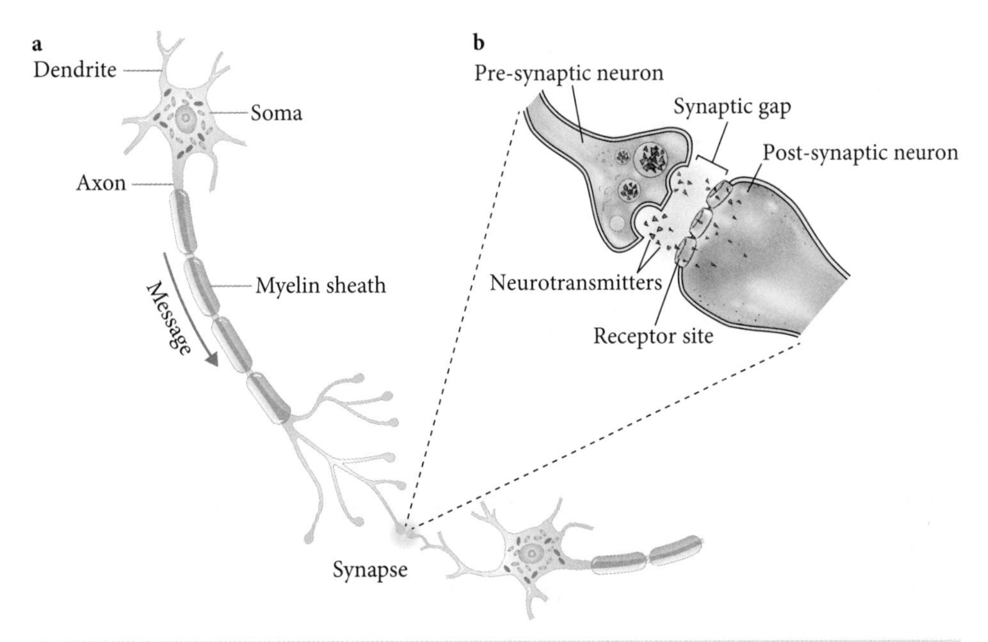

FIGURE 4.10 a The basic structure of a neuron. b Neurotransmitters crossing a synapse.

Motor neurons in the PNS transmit signals from the CNS to organs, glands and limbs, while **sensory neurons** provide feedback on the status of the action from the periphery of the body to the CNS.

Nerves are classified as **afferent** or **efferent** based on the direction of the information flow. Afferent nerves are made up of sensory neurons. They receive an input from the environment, such as light or sound, and transmit it to the CNS. Afferent nerves are also responsible for homeostasis in the body. Efferent nerves consist of motor neurons that transmit signals in the opposite direction from the CNS to organs and muscles to initiate an action.

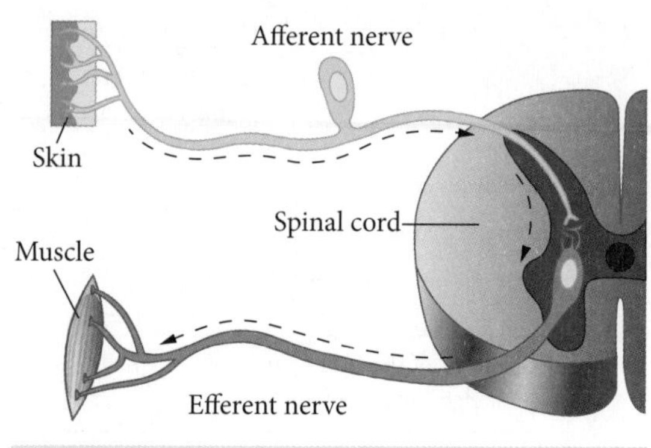

FIGURE 4.11 Afferent and efferent nerves

The nervous system of invertebrates varies in its structure and complexity. Even though some invertebrates have a brain, they do not have a centralised nervous system like vertebrates. This means that nerves throughout the body can control actions independently of the brain.

TABLE 4.6 The structure of the nervous system in invertebrates

Cnidarian (hydra)	Echinoderm (sea star)	Planarian (flatworm)	Arthropod (bee)	Mollusc (octopus)
Nerve net	Nerve ring Radial nerves	Eyespot Central ganglion (brain) Nerves cords Transverse nerve	Central ganglia (brain) Segmental ganglia	Brain Eye Nerve Ganglion Nerve
• Lack a true brain • They have a nerve 'net' of nerve cells dispersed across the body	• A nerve ring • Five radial nerves extend outward along the arms	• Have a small brain • Two nerve cords • Simple peripheral nervous system • Eye spots detect light	• Have a small brain • Ventral nerve cord • Ganglia (clusters of connected neurons along nerve cord)	• Most complex of the invertebrates • Neurons organised into specialised lobes • Eyes have a similar structure to human eyes

To summarise, the neural system is a control centre in negative feedback loops that maintain homeostasis. A change detected by a receptor is transmitted as an electrochemical signal along nerves to the brain, which results in a signal that is transmitted back along the nerves to cause a response.

Mechanisms in plants that allow water balance to be maintained

Plants need water for **photosynthesis**, **respiration**, and the transport of nutrients and hormones. Water is drawn upwards into the plant through the xylem by root pressure and capillary action. The passage of water from the roots to the organs, and eventual evaporation from the surface of leaves, is called the **transpiration stream**.

Stomata play a critical role in **transpiration** and water balance in plants. Stomata are openings (pores) on the underside of leaves. Each pore is surrounded by two **guard cells**, which can change the size of the pore's opening by expanding and contracting. The two main functions of stomata are gas exchange and regulation of water movement by transpiration.

Plants need to balance the intake of CO_2, needed for photosynthesis and growth, with the loss of water by transpiration. Stomata size varies depending on the time of day and environmental conditions. Their size is regulated by the active transport of potassium ions and osmosis, which changes the **turgor** of guard cells and the size of the pore opening.

Stomata open during the day when photosynthesis occurs, and close at night to reduce water loss. The closure of pores is also a physiological adaptation to conserve water during stress. When water is scarce or salinity is high, the hormone abscisic acid binds to proteins in guard cells, which results in the exit of water, shrinkage of the guard cells and closure of the pore. This disrupts photosynthesis and growth but will only change when the stress signal is reduced.

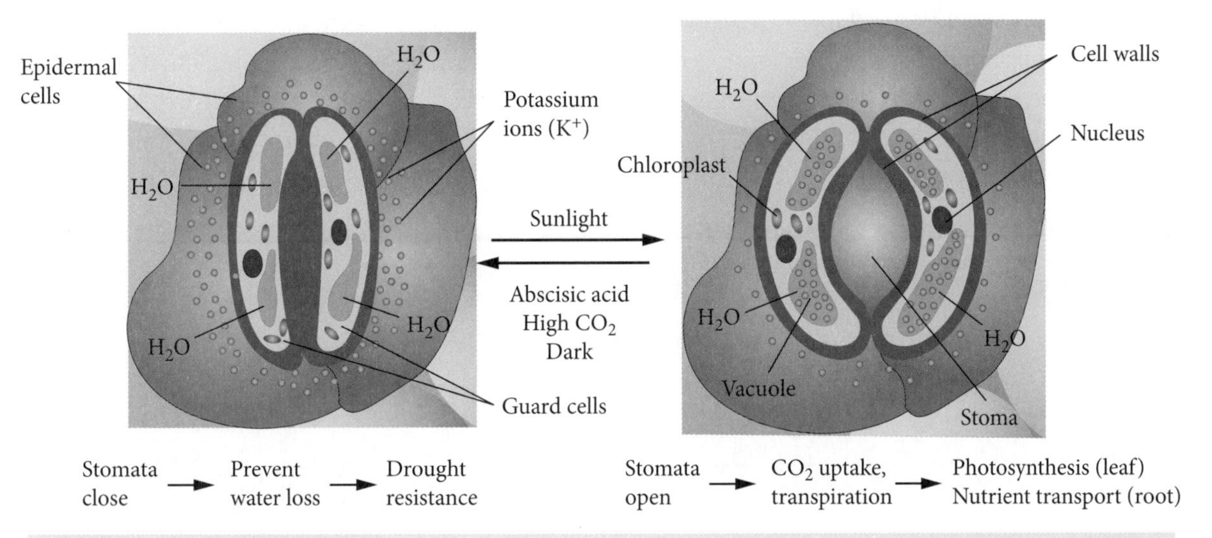

FIGURE 4.12 Water balance in plants

Plants also have a range of structural adaptations to minimise water stress. For example, plants in arid climates (e.g. cacti) have a reduced number of stomata, leaves with a waxy (hydrophobic) surface and a reduced surface area, to minimise water loss via transpiration. Succulent plants can store water in their leaves and stems, while white mangroves (*Laguncularia racemosa*) secrete excess salt through glands near the tip of each leaf stalk.

Note
Water balance in plants is another example of negative feedback maintaining homeostasis.

4.2 Causes and effects: Do non-infectious diseases cause more deaths than infectious diseases?

Non-infectious diseases and disorders are caused by complex interactions between genes and the environment. In higher-income countries they are becoming more common because of unhealthy diets, more sedentary lifestyles and longer lifespans. On the other hand, infectious diseases caused by pathogens are causing fewer deaths because of better sanitation, antibiotics, vaccines and quarantine methods. Scientists conduct research to identify the cause of non-infectious diseases and disorders, as well as strategies and drugs to prevent, diagnose and treat them.

4.2.1 Investigating the causes and effects of non-infectious diseases in humans

Note
This is an 'including but not limited to' dot point. One or two examples of each type of disease are provided to illustrate the concept of cause and effect. You should aim to understand one example of a genetic disease, a disease caused by an environmental exposure, a nutritional disease and cancer. This understanding can then be applied to other diseases.

Genetic diseases

Genetic diseases can be caused by germ-line and somatic mutations, as well as mutations in mitochondria. An autosomal recessive condition and a mitochondrial condition are provided as two examples of inherited **genetic diseases**.

Note
Cancer is classified as a separate disease category in the syllabus, but it is also a genetic disease, most commonly caused by somatic mutations.

Cystic fibrosis

Cystic fibrosis (CF) is an autosomal recessive disease *caused* by germ-line mutations in the cystic fibrosis transmembrane conductance regulator (*CFTR*) gene. This gene encodes a protein that maintains salt and water levels in various organs of the body. The disease is most commonly caused by the deletion of a single amino acid, called phenylalanine, at amino acid position 508 (F508del). The gene is transcribed but the polypeptide doesn't fold correctly, and it is degraded before it reaches the correct position in the cell membrane. The *effect* of *CFTR* mutations is the accumulation of thick mucus in the lungs, which leads to persistent coughing and difficulty breathing. The life expectancy of a person with CF is 40–50 years. Refer back to section 1.5.1 in Chapter 1 to recall how this mutation can be detected using agarose gel electrophoresis.

Note
Huntington disease (autosomal dominant) and haemophilia A (X-linked recessive) are two examples of other genetic diseases.

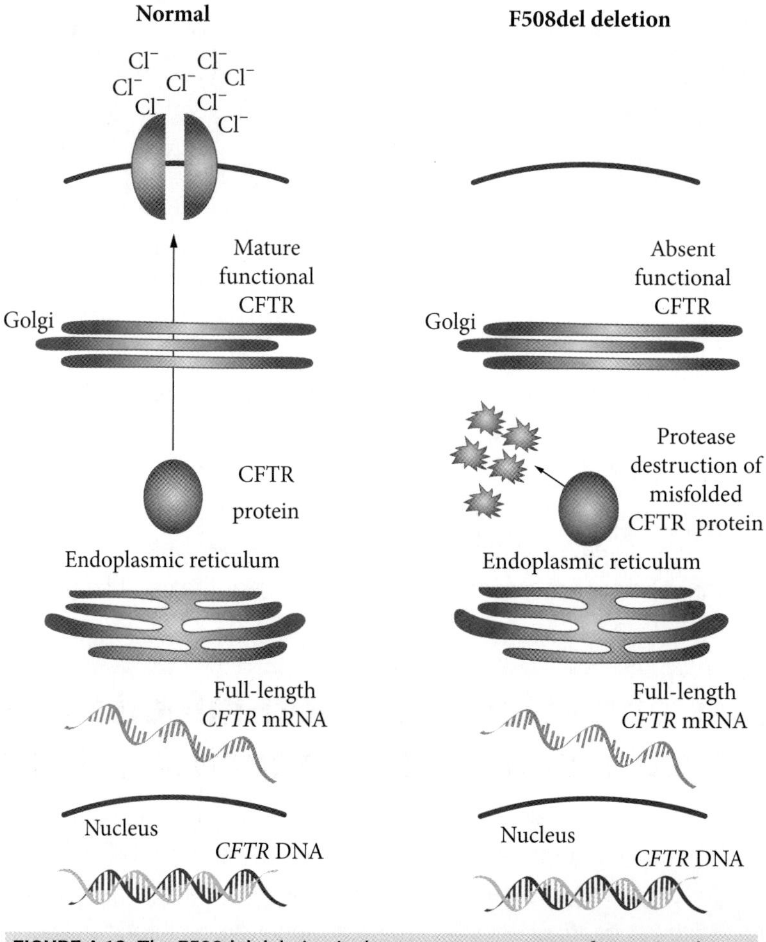

FIGURE 4.13 The F508del deletion is the most common type of mutation that causes cystic fibrosis.

Mitochondrial diseases

Each mitochondrion in a human cell has one small, circular chromosome. Genetic mutations in the mitochondrial genome can also cause diseases. A cell contains many mitochondria, so there will be a mixture of normal and mutant genomes in any one cell.

An egg contains thousands of mitochondria but the small number in sperm are degraded after fertilisation. Offspring therefore only inherit mitochondria, and any mutations, from the mother. Mitochondrial mutations generally cause muscular and neurological problems, because these cells need high amounts of energy (e.g. MELAS syndrome). The severity of a disease will depend on the proportion of mitochondria with the mutation.

Note
An understanding of mitochondrial mutations will help you to understand how gene therapy is used to prevent mitochondrial disease (section 4.4.1).

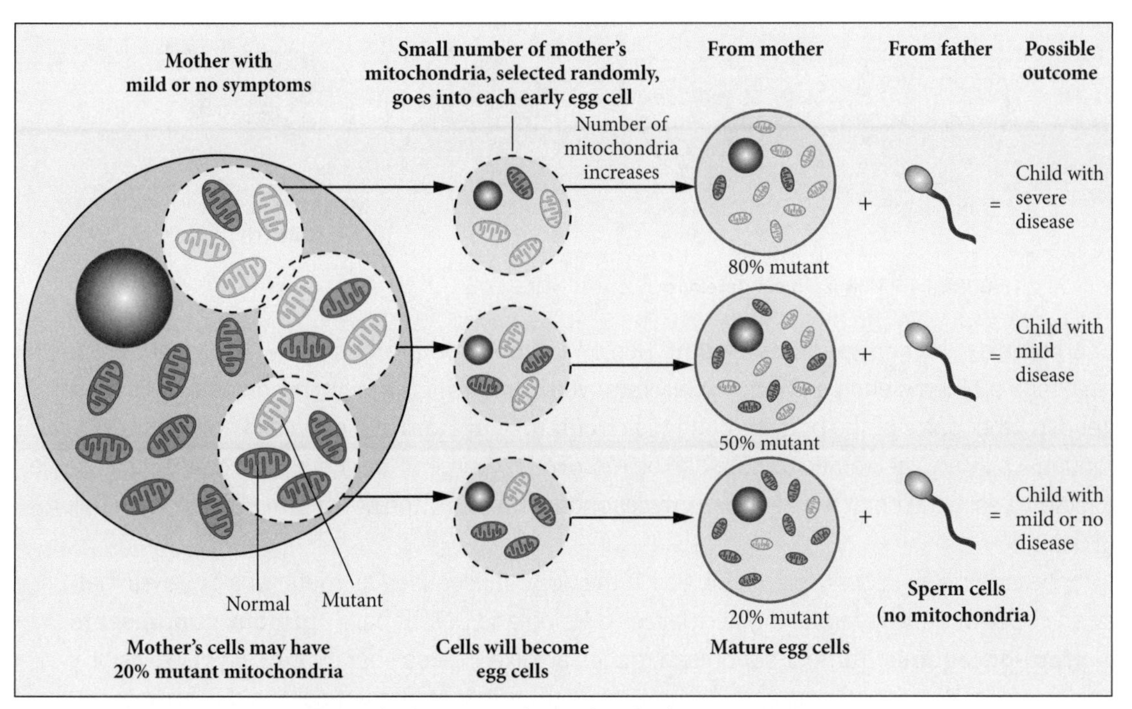

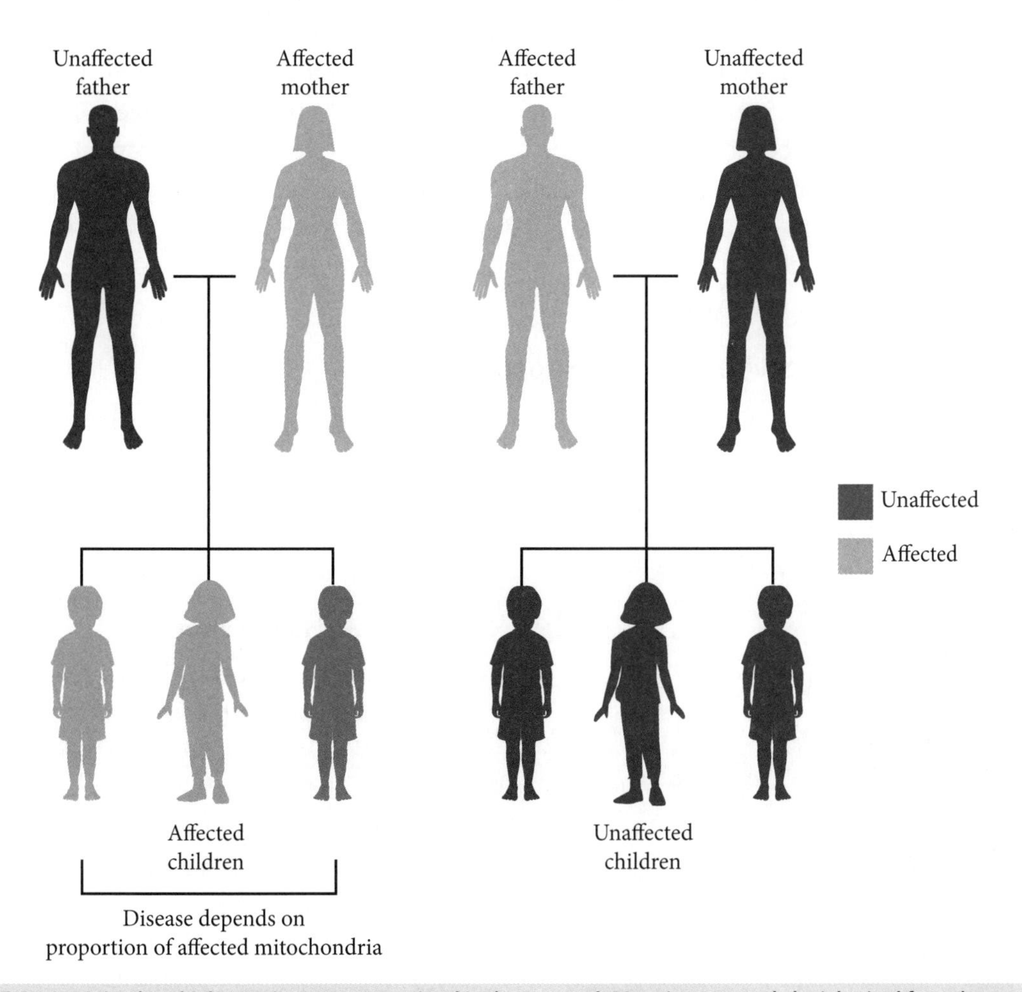

FIGURE 4.14 a Mitochondrial mutations are transmitted in the oocyte. b Mutations can only be inherited from the mother.

Cancer

Cancer is a disease caused by the uncontrolled growth of abnormal somatic cells. This results in the formation of a **tumour** or, in the case of blood cancers, the accumulation of abnormal cells in the circulatory system.

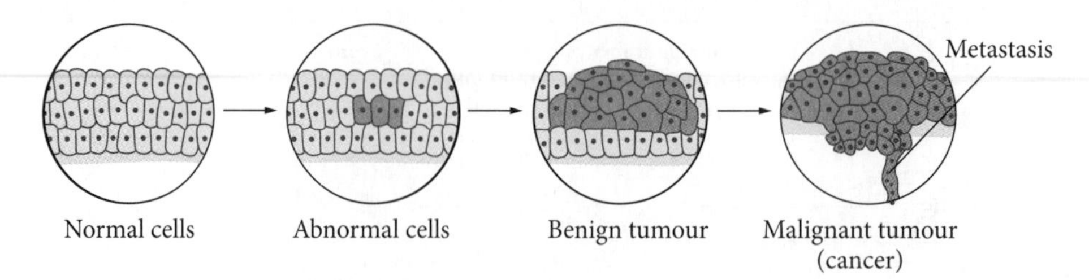

FIGURE 4.15 How a tumour develops

Tumours can be **benign** or **malignant**. Benign tumours remain localised, will not cause long-term health problems and can be surgically removed. Malignant tumours invade nearby tissues and spread to other parts of the body through the **circulatory system** in a process known as **metastasis**. Although a primary metastatic tumour can be removed by surgery, the metastatic cancer cells settle at distant sites in the body, where they form secondary tumours, the *effect* often being organ failure and death.

Cancer can begin in almost any cell type in the body. It is *caused* by mutations in genes that play an important role in regulating normal cell division and DNA repair – **tumour suppressors** and **proto-oncogenes**. Tumour suppressor genes are like 'brakes' that slow down cell division, repair errors in DNA or activate protein signalling pathways that instruct cells to die (**apoptosis**). Proto-oncogenes express protein receptor proteins, signalling proteins and transcription factors that promote cell division. They are called **oncogenes**, or 'accelerators', in their mutated form. A combination of mutations in both groups of genes can cause uncontrolled cell division.

Most cancers (90–95%) are caused by mutations in somatic cells (**sporadic cancer**). These cancers are more likely to occur as we grow older. The remaining 5–10% of cancers are **hereditary cancer** caused by germ-line mutations in tumour suppressor genes. People with a germ-line mutation are more likely to get cancer at a younger age. For example, a person who inherits a mutant *MLH1* gene will be diagnosed with Lynch syndrome and will be at an increased risk of bowel cancer. Similarly, inheritance of a *BRCA1* mutation causes familial breast cancer.

Diseases caused by environmental exposure

Diseases can be caused by exposure to the environment. This can include toxic substances in contaminated food or water, inhalation of chemicals or exposure to radiation. This section uses smoking as an example of an environmental exposure that causes lung cancer.

> **Note**
> It isn't always easy to categorise diseases. Melanoma is a type of cancer but it is caused by an environmental exposure (UV light).

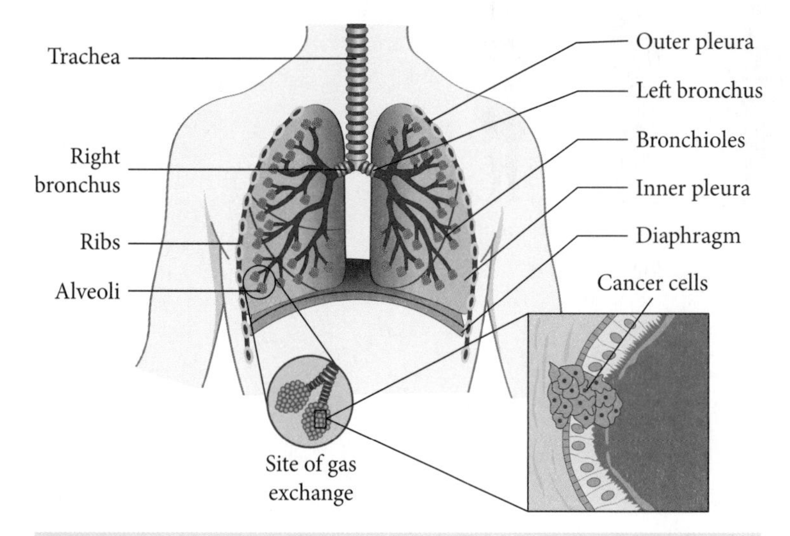

FIGURE 4.16 The structure of lungs and formation of lung adenocarcinoma

Smoking and lung cancer

The lungs are two sponge-like organs in our chest that deliver oxygen to the blood and remove carbon dioxide from the body. Lung cancers normally start in the cells that line the bronchi, bronchioles and alveoli.

Smoking tobacco causes 85% of lung cancers. The link between smoking and lung cancer was first demonstrated in epidemiological studies in the 1940s (see section 4.3.3). Animal experiments in the 1950s showed that mice developed cancer when the tar that forms from smoke inhalation was injected into their bodies. In the 1960s, molecular researchers discovered that chemicals in cigarettes bind to and mutate DNA.

So how does smoking *cause* lung cancer? There are about 7000 chemicals in cigarettes and 70 of these are **carcinogens** – benzo[*a*]pyrene (BP) is the most studied. It is produced when tobacco (the leaf from the plant *Nicotiana tabacum*) is burned and inhaled into the lungs. When it enters the body, enzymes metabolise BP into a compound that binds to guanine in DNA. This causes DNA to bend, resulting in G-to-T transversions (see section 2.1.2), a signature feature of lung cancer genomes caused by smoking. The *effect* of the mutations is the uncontrolled cell division in the lungs that metastasises to other parts of the body.

> **Note**
> Inhalation of asbestos is another type of environmental exposure that causes a type of lung cancer called **mesothelioma**. It is caused by fine asbestos fibres settling in the lungs. Immune cells try to ingest the fibres, but in the process they release reactive oxygen species that mutate the DNA in surrounding cells.

Diseases caused by autoimmunity

Our immune system normally recognises and attacks infections such as bacteria and viruses, but in some people it recognises cells of the body as 'non-self', attacking them and causing damage. When this occurs, it is called an **autoimmune disease**.

> **Note**
> Autoimmune diseases are examples of 'including but not limited to' non-infectious diseases that address this dot point.

Type 1 diabetes

Type 1 diabetes is a chronic autoimmune disorder that normally develops in childhood and is caused by the body's inability to produce **insulin** to regulate the level of glucose in the bloodstream.

When we digest food, carbohydrates are broken down into a simple sugar called glucose, which enters the bloodstream. This triggers the release of insulin from **beta cells** in the **islets of Langerhans** in the pancreas. Insulin binds to a receptor on the surface of cells, to activate a protein called GLUT4, which moves to the plasma membrane and shuttles glucose into the cell.

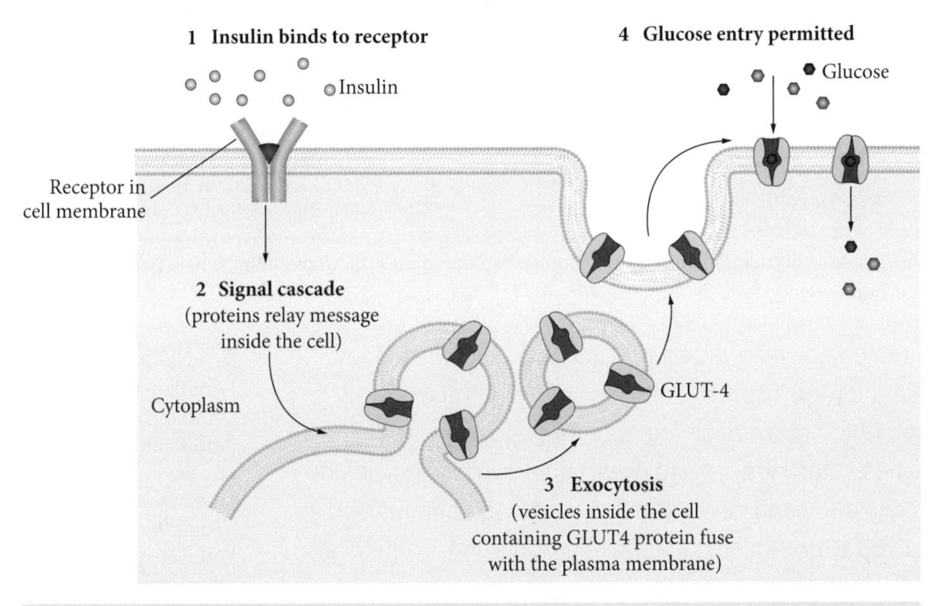

FIGURE 4.17 The mechanism of glucose uptake by cells

The amount of insulin secreted by beta cells is regulated by the level of glucose in the bloodstream; insulin increases after a meal when glucose levels are high, and declines after glucose leaves the bloodstream.

Type 1 diabetes is *caused* by the immune system attacking and destroying beta cells, which means the pancreas is unable to produce insulin. The exact reason for this autoimmune response is unknown, but it may have a genetic basis or be caused by exposure to viruses or other environmental factors.

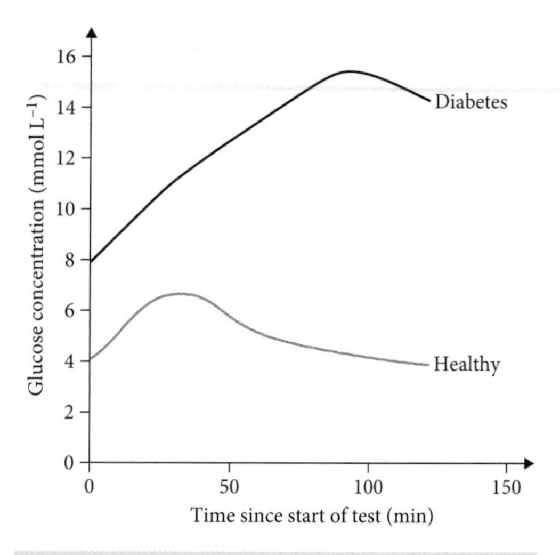

FIGURE 4.18 Blood glucose levels after a meal, in a diabetic person and a non-diabetic person

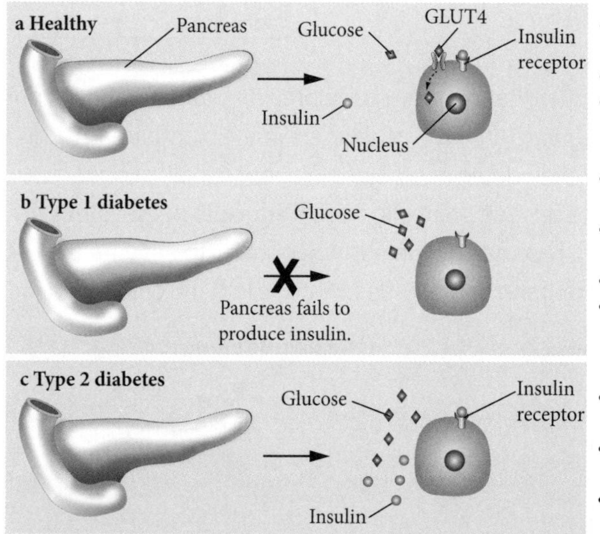

- Insulin produced by the pancreas in response to the increase in blood glucose.
- Insulin binds to insulin receptors on the cell surface.
- Glucose enters the cell through a membrane transporter protein (GLUT4).

- Immune system destroys beta cells in the pancreas that produce insulin.
- No insulin is produced in response to an increase in blood glucose.
- There is no insulin to bind to insulin receptors.
- Glucose cannot enter the cell because GLUT4 is not activated.

- Insulin produced by the pancreas in response to an increase in blood glucose.
- Insulin binds to insulin receptors on the cell surface.
- GLUT4 is not activated and glucose cannot efficiently enter the cell.

FIGURE 4.19 Insulin production and glucose uptake in a person a without diabetes, b with type 1 diabetes and c with type 2 diabetes

If left untreated, the *effects* of type 1 diabetes include kidney damage (diabetic **nephropathy**) and loss of vision (diabetic **retinopathy**), which occur because of damage to and narrowing of the capillaries in the kidneys and eyes. Other symptoms include poor blood circulation and nerve damage in the legs and feet (diabetic **neuropathy**), which can require amputation. There is also an increased risk of cardiovascular disease, stroke and sexual impotence. Type 1 diabetes can be treated with insulin injections (section 4.3.2).

Note

Another example of an autoimmune disease is rheumatoid arthritis. It is *caused* by the immune system attacking the synovium between the joints. The *effects* include joint pain, bruising of the skin, scarring in the lungs and an increased risk of cardiovascular disease.

Nutritional diseases

A nutritional disease is any type of non-infectious disease that is caused by deficiencies in the diet, absorption problems, overnutrition or eating disorders such as obesity and type 2 diabetes.

Obesity and type 2 diabetes

Obesity is an excessive accumulation of fat in the body. It is defined using a calculation called the **body mass index** (BMI). BMI is mass in kilograms divided by height in metres squared (kg/m^2). A BMI below 25 is classified as healthy, 25–30 as overweight, and greater than 30 as obese. Doctors also take other measurements such as waist circumference to help diagnose the condition.

Obesity is *caused* by the consumption of more calories than are burned by the body. It has become more common mainly as a result of the increased consumption of processed food, drinks high in sugar, and sedentary lifestyles. Due to increasing rates of obesity in younger people, more children, teenagers and young adults are being diagnosed with type 2 diabetes.

Type 2 diabetes accounts for 90–95% of all cases of diabetes. People with the disorder produce insulin but the body does not use it effectively, so glucose does not enter the cells, resulting in the accumulation of glucose in the bloodstream. The body may eventually stop producing insulin, and the *effects* of the disease are similar to those of type 1 diabetes. There is no cure for type 2 diabetes, but the symptoms can be managed through lifestyle choices, drugs and/or insulin therapy (see section 4.3.2).

Note

Rickets is another type of nutritional disease. It is *caused* by vitamin D deficiency. This results in decreased absorption of calcium by the digestive system and the *effect* is soft bones and skeletal abnormalities.

4.2.2 Collecting and representing data to show the incidence, prevalence and mortality rates of non-infectious diseases

Incidence, prevalence and mortality rates are three common measurements used in population-based studies. Data is usually presented as graphs or in tables to help visualise trends and patterns. Recall that **incidence** is the number of *new* cases of a disease in a population in a specified time period. It is presented as a *rate* to allow comparison of data sets between populations, time periods or locations. It is different from **prevalence**, which is the total number of disease cases in a population at a given time, regardless of when they first developed the characteristic. It is presented as a proportion, percentage or number of cases per 10 000 or 100 000 people. **Mortality rate**, sometimes called death rate, is the number of deaths in a population in a specified time period. The findings from epidemiological studies are used to inform public health strategies such as health campaigns and government investment in research programs.

Figures 4.20–4.24 are based on a collection of data that demonstrates how these measurements can be represented in population-based studies. The examples include mortality in Australia, obesity and diabetes (nutritional disease), smoking and lung cancer (environmental disease), and breast cancer (genetic disease).

Note

You should aim to understand the key terms, but it is more important to practise your skills in collecting and representing data in table and graph format.

Note

For this dot point you don't need to remember data for any particular disease. Examples of graphs and tables are provided to help you understand how data is collected and represented. Guidance is provided in the next section on epidemiology to help you *analyse* these data sets.

Mortality in Australia

Mortality data is useful to assess the general health of a population. Mortality can be represented as absolute numbers (e.g. if it is data for a single year) and is often categorised by gender or age to compare these groups in the population. Mortality is represented by a rate when the study measures the number of new deaths in a specified time period.

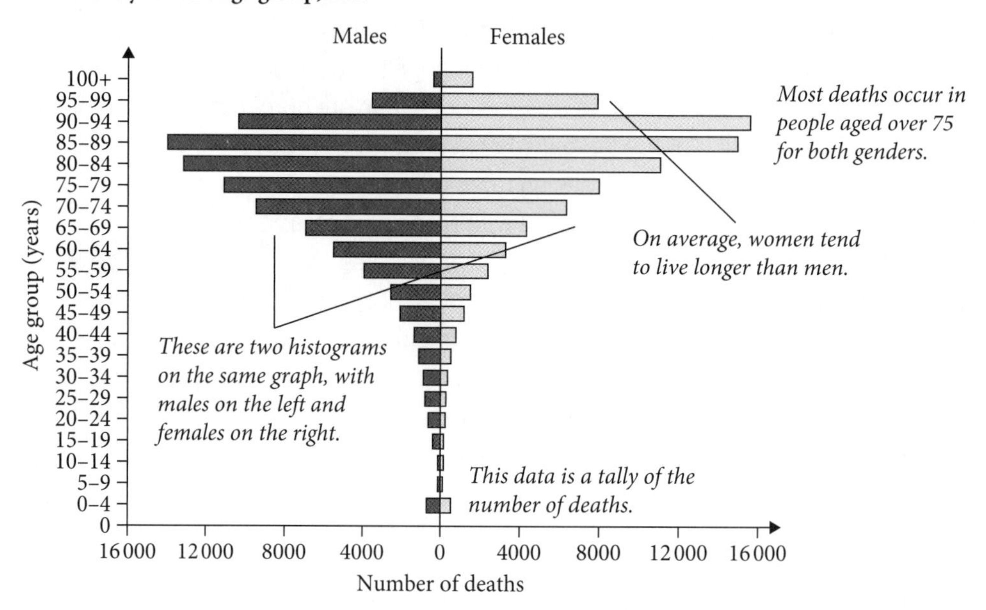

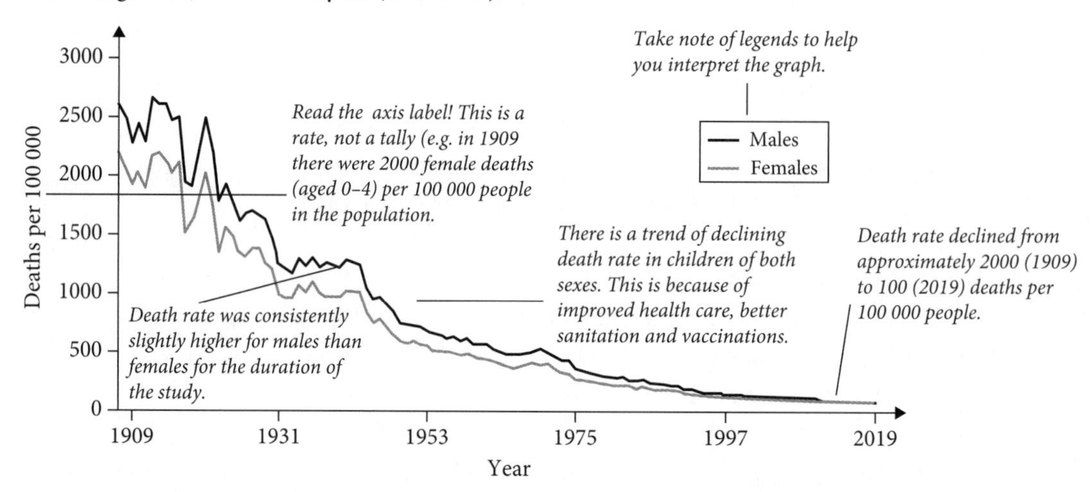

FIGURE 4.20 Examples of data representation for mortality in Australia: **a** number of deaths from all causes, by age and gender, and **b** overall mortality rate for children

Nutritional diseases – obesity and type 2 diabetes

Being overweight or obese is a risk factor for developing type 2 diabetes. Population-based data can be represented on maps as a way to observe correlations between a disease and geography, such as environmental conditions, wealth of a nation and standards of healthcare. Obesity is a global problem, but it is more common in some areas of the world than others.

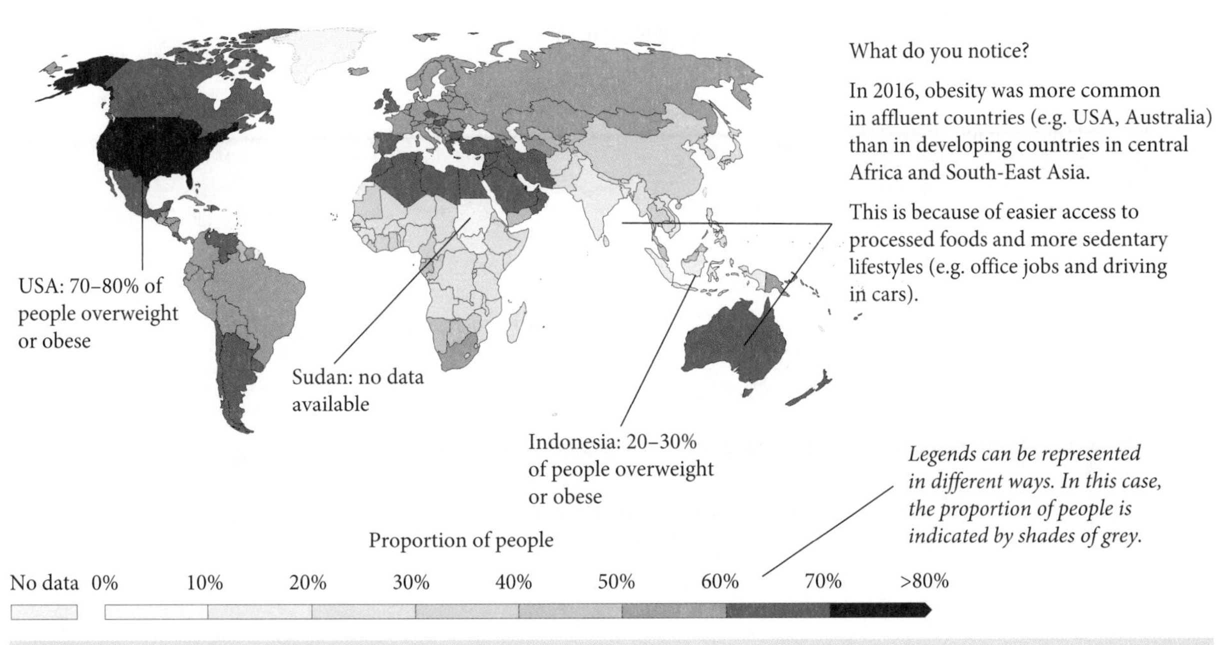

FIGURE 4.21 The global prevalence of being overweight or obese, 2016

Population-based data can also be represented in tables. By comparing data sets, it is possible to observe a correlation between a risk factor (BMI) and a disease (type 2 diabetes).

You might be asked to identify and describe a trend in data. Here you can see that the prevalence of diabetes increases with age in both sexes.

a Proportion of people with diabetes (2017–2018)

Age group (years)	Males (%)	Females (%)
0–14	0.2	0.2
15–24	1.1	0.4
25–34	1.6	0.3
35–44	1.8	2.6
45–54	5.2	4.0
55–64	12.0	8.5
65–74	18.7	12.4
75+	20.7	17.0

Obesity is more common in males than females in each age group, except people aged 35–44.

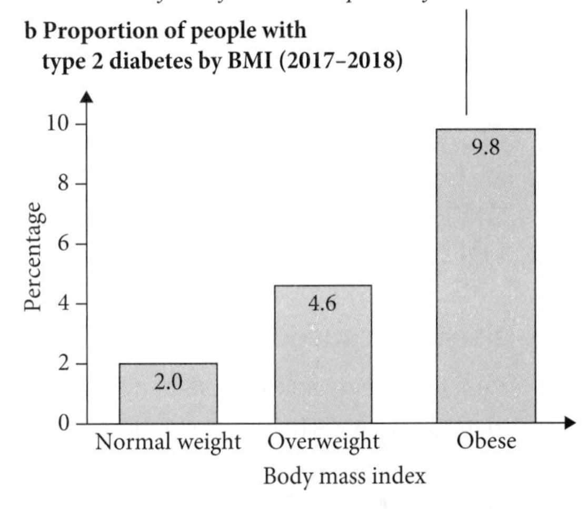

FIGURE 4.22 The prevalence of being overweight or obese, and diabetes, in Australia (2017–2018): a proportion of people with diabetes in 2017–2018, b proportion of people with diabetes, by BMI

Diseases caused by environmental exposure – smoking and lung cancer

Smoking tobacco is an environmental exposure that causes lung cancer. The incidence of a disease, such as lung cancer, is measured as a rate and the data is represented by a line graph. Incidence can be compared with other data, such as the prevalence of smoking, to assess whether an environmental exposure is a risk factor for a disease.

a Incidence of lung cancer (1982–2021)

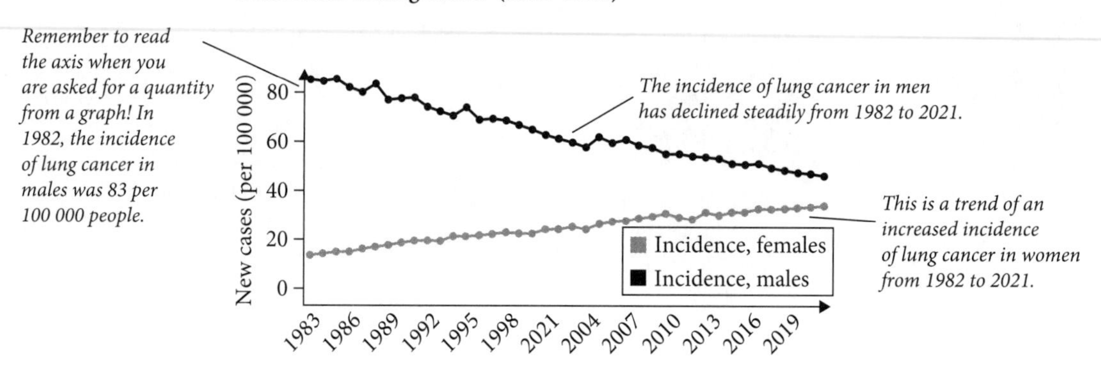

b Prevalence of current tobacco smoking among Australian adults (1945–2013)

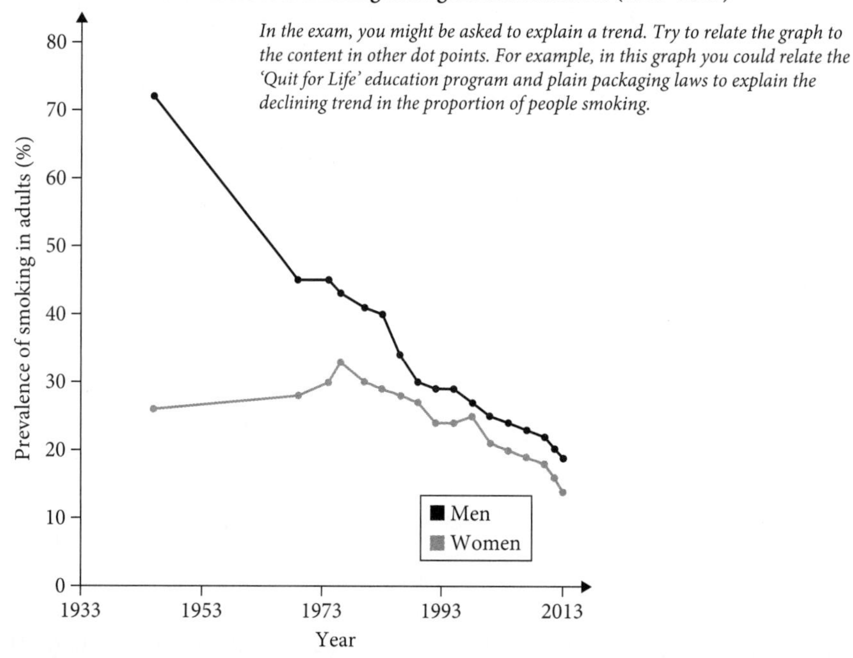

FIGURE 4.23 a The incidence of lung cancer in Australian men and women. b The prevalence of current tobacco smoking among Australian adults.

Diseases caused by genetics

Breast cancer is an example of a disease caused by genetic mutations in tumour suppressor genes and oncogenes. The graph in Figure 4.24 shows the incidence and mortality rate in Australian women.

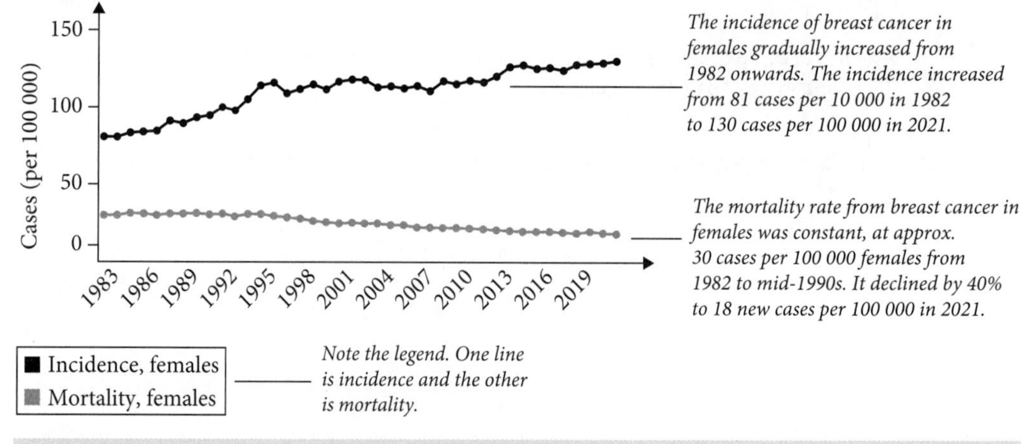

FIGURE 4.24 The incidence and mortality rate of breast cancer in Australian women

Note

If you would like to collect incidence, prevalence and mortality rate for other non-infectious diseases, go to the Australian Institute of Health and Welfare website.

4.3 Epidemiology: Why are epidemiological studies used?

Epidemiology is a field of research in which large populations of people are studied to provide insights into patterns, risk factors and exposures associated with diseases or events. In these studies, data is collected and compared between two groups of people, one with and one without the disease, treatment or exposure. Statistical methods are then used to determine whether there is a significant relationship between the factors being investigated. The conclusions from epidemiological studies are used to inform public health programs to control, prevent and treat diseases.

4.3.1 Analysing patterns of non-infectious diseases in populations, including their incidence and prevalence

This section analyses the patterns of three non-infectious diseases (diabetes, lung cancer and breast cancer) collected and represented in section 4.2.2.

Note

This is an 'including but not limited to' dot point.

In an 'analyse' question, try to interpret the table or graph before attempting the question. Read the title and the axis labels, look for any obvious trends and try to think of content from a dot point that relates to the data. A band 6 student would generally be expected to include the following in their response:

- a clear understanding of the key terms (e.g. mortality, incidence and prevalence)
- a description of trends in the data
- one or two pieces of data from the stimulus material
- a link between the data and content from a dot point, if asked to provide an explanation for a trend or pattern.

Nutritional diseases – obesity and type 2 diabetes (refer to Figure 4.21)

Step 1: Read the title. This immediately tells you that the data shows the proportion of people in each country who are overweight or obese. A proportion indicates that it is prevalence data.

Step 2: Read the legend (in this case, a scale) to understand what the different shades of grey mean. A darker shade means more people in that country are overweight or obese.

Step 3: Scan the map to identify any trends. Obesity is less common in central and southern Africa and South-East Asia, compared to the rest of the world.

Step 4: In an answer you might state proportions from specific countries to support your observation. You might be asked to relate this to a non-infectious disease, so you could draw on your knowledge of type 2 diabetes.

Note

Tips are provided in this study guide for all the graphs to help with the analysis.

Diseases caused by environmental exposure – smoking and lung cancer (refer to Figure 4.23)

Step 1: Read the titles. They tell you that it is incidence and prevalence data. You might have also gathered this from the units on the *y*-axis (new cases and a rate for incidence versus a percentage for prevalence).

Step 2: Read the axes and legends.

Step 3: Scan the graphs for increasing and decreasing trends.

Step 4: Interpret the trends based on your knowledge of the cause of cancer (Section 4.2.1) and prevention strategies (Section 4.4). The declining incidence of lung cancer in men correlates with a decline in the prevalence of smoking. Interestingly, the incidence of lung cancer in women has increased from 1982 until today. How can this be, when the prevalence of smoking in women has declined during the same time period? This is because there was a peak of smoking

9780170465267

by women in the early/mid-1970s and 1990s. Recall that lung cancer occurs many years after smoking. The increased incidence (i.e. new cases) in women is a 'lag' effect from the increased proportion of women who took up smoking before prevalence began declining.

Genetic diseases – breast cancer (refer to Figure 4.24)

There has been an increase in the incidence of breast cancer in Australian women, but at the same time there has been a slow decline in the mortality rate. How can this be? BreastScreen Australia is a collaboration between the federal, state and territory governments to reduce illness and death from breast cancer. Women aged over 40 can get a free mammogram every two years, and women aged 50–74 are actively invited to get a mammogram. The increased testing and better technology have meant that breast cancer is detected more often, but at an earlier stage. This means that cancer can be treated successfully, and the mortality rate has declined.

> **Note**
> Breast cancer is an 'including but not limited to' disease for this dot point.

4.3.2 Investigating the treatment/management, and possible future directions for further research, of a non-infectious disease using an example from one of the non-infectious diseases categories listed above

Type 2 diabetes is provided as an example to investigate treatment and management of, and future research into, a non-infectious disease.

> **Note**
> The treatment of cancer with surgery, chemotherapy, radiation therapy and targeted therapy is another non-infectious disease that could be investigated.

An early diagnosis of type 2 diabetes is essential to prevent long-term health complications. Blood glucose levels and the disease can be *managed* with simple lifestyle changes such as a healthy diet low in processed food, regular exercise and fewer sedentary periods. When this is unsuccessful, *treatment* and *management* plans can include medications, insulin therapy or weight-loss surgery.

People with type 2 diabetes need to regularly monitor their blood glucose levels during the day, especially before and after a meal or exercise. This can be done with a blood glucose meter, which measures the concentration of glucose in a small drop of blood, or electronic devices that continuously monitor the blood glucose level.

A doctor may prescribe drugs or recommend insulin therapy if blood glucose levels cannot be managed through diet and exercise. Each drug works by controlling either the level of blood glucose or insulin in the body. The mechanism and potential side effects of each drug are summarised in Table 4.7.

TABLE 4.7 The mode of action and side effects of diabetes medications

Drug	How the drug works	Side effects
Metformin	Lowers glucose production in the liver and makes the body more sensitive to insulin, so it is used more efficiently	Nausea, diarrhoea, abdominal pain
Sulfonylureas	Help the body to secrete more insulin	Weight gain and low blood sugar levels
Glinides	Stimulate the pancreas to secrete more insulin	
Insulin therapy	Increases the amount of insulin in the body	Risk of low blood sugar and high cholesterol

Weight-loss surgery may also be part of a *management* plan for diabetes. This type of surgery changes the shape of the digestive system to limit the amount of food that can be eaten. The side effects of this procedure include reduced uptake of nutrients and osteoporosis.

The best way to *manage* type 2 diabetes continues to be through lifestyle. This benefits individuals as it reduces the likelihood of medical intervention. It also benefits society because it reduces the cost of treating medical complications caused by the disease. However, technology and **genetic engineering** developed to treat type 1 diabetes may be used to *treat* and manage type 2 diabetes in the future. Possible advances include an artificial pancreas, replacing pancreatic beta cells through islet cell transplants, new drugs and improved genetic testing.

TABLE 4.8 Future management of diabetes

Future direction	Description
Artificial pancreas	A digital device that behaves in the same way as a pancreas by releasing insulin into the body in response to changes in blood glucose levels.
Islet cell transplantation	Healthy islet cells, which contain beta cells that produce insulin, may be transplanted into the pancreas of people who no longer produce insulin.
New drug development	There may be more drugs available in the future. Personalised medicine may mean individuals are treated with different combinations of drugs.
Genetic testing	Whole-genome sequencing and GWAS studies may identify SNPs or alleles that increase the risk of type 2 diabetes. Prevention strategies could start earlier in people carrying these genetic variants.

4.3.3 Evaluating the method used in an example of an epidemiological study

Epidemiological studies can be categorised as descriptive or analytical.

Note
You only need to evaluate one example of an epidemiological study, but two examples are provided here for comparison.

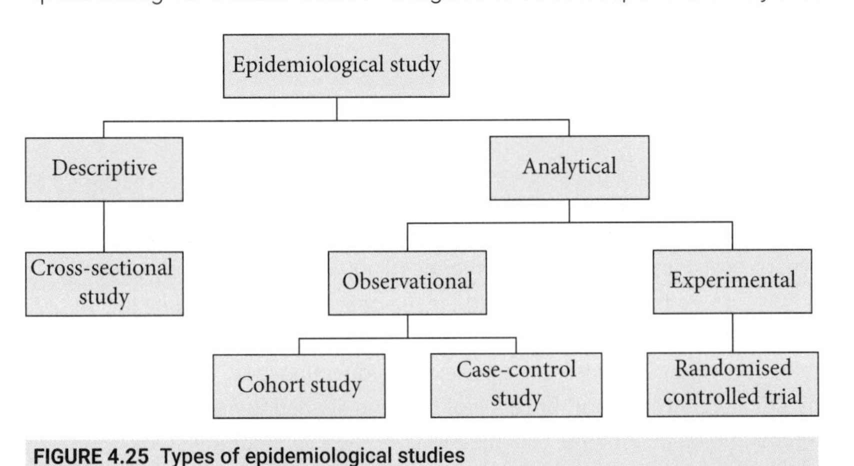

FIGURE 4.25 Types of epidemiological studies

A descriptive study uses existing data to understand patterns that can be attributed to a disease or event. It takes a 'snapshot' or 'cross-section' of the proportion of people with a disease or an outcome at a point in time (e.g. the global obesity data in Figure 4.21).

Analytical studies fall into two categories: observational and experimental.

Observational research aims to establish the cause of a non-infectious disease. This is done by analysing the association between an exposure, such as smoking, obesity, radiation or a chemical in the workplace, and an outcome, such as cancer or heart disease in a population. There are two main types of observational study: **cohort studies** and **case-control studies**.

Note
This is an 'evaluate' dot point. The advantages and disadvantages of each epidemiological study are therefore identified and a judgement is made about their value in identifying the cause of diseases.

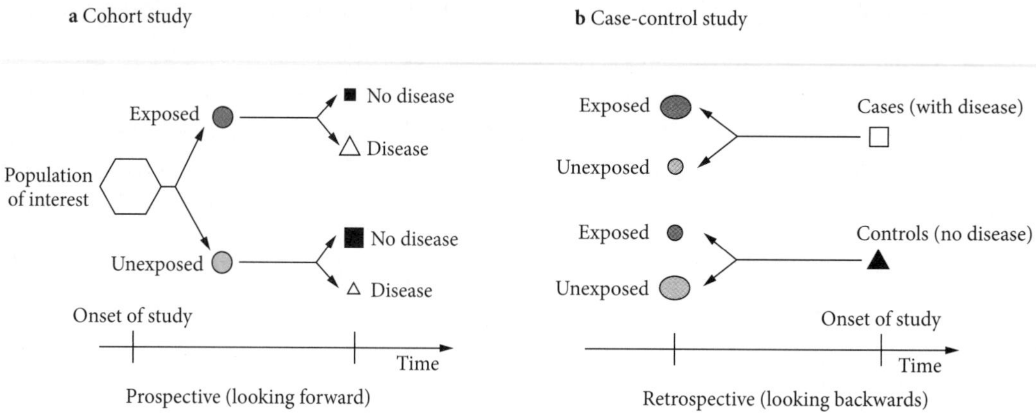

Note
The shapes represent different groups of people. The size of the shapes represents the proportion of people.

a Cohort study
Population of interest
Exposed
Unexposed
No disease
Disease
No disease
Disease
Onset of study
Time
Prospective (looking forward)

b Case-control study
Exposed
Unexposed
Cases (with disease)
Exposed
Unexposed
Controls (no disease)
Onset of study
Time
Retrospective (looking backwards)

FIGURE 4.26 The design of a a cohort study and b a case-control study

Observational study: cohort study

A cohort study is **prospective**. This means it follows a large group of people over a period of time and measures the frequency of disease in exposed and non-exposed individuals. A cohort study begins with the recruitment of a large group of people who do not have a disease and are representative of the population as a whole. Each person is interviewed at the beginning of the study to gather data such as their height and weight, diet, occupation, smoking and alcohol history. The individuals are then followed over time, ideally with regular interviews being conducted every few years, to gather information about their lifestyle and events during their life. Once data is gathered, a **relative risk** is calculated to indicate the likelihood that a particular exposure increases the risk of the non-infectious disease.

The following is a hypothetical example that illustrates how data from a cohort study could identify high BMI as a risk factor for cardiovascular disease.

EXAMPLE

In 1950, a research team began recruiting people to identify exposures that increased the risk of developing cardiovascular disease. It took 10 years, at a cost of $2 million, to recruit 10 000 people. The study began in 1960 and people were followed for 50 years. At the start of the study, researchers set criteria for inclusion, such as age over 18 years, and no pre-existing heart conditions that could influence the results. They also made sure that they invited a range of participants of different age, sex and education level, and who lived in different areas. The interviews had to be relatively short because of the time needed to interview all people, and to make sure participants completed responses accurately. Over the course of the study, 150 patients could not be located, and had to be excluded from the study. At the end of the study, 3900 people were identified as obese (BMI > 30) and 1200 of these people developed cardiovascular disease later in life. In the same time period, 5950 people maintained a healthy BMI (<25) and 500 of these people developed cardiovascular disease.

In a cohort study, a relative risk is calculated using the formula shown at right, to determine how much more likely a person exposed to a risk factor is to develop a disease.

$$\text{Relative risk} = \frac{\text{disease/exposure}}{\text{disease/no exposure}} = \frac{1200 / 3900}{500 / 5950} = \frac{0.31}{0.08} = 3.88$$

The study concluded that an obese person has a 3.88 increased risk of developing cardiovascular disease compared to a person with a healthy BMI.

In summary, a cohort study is the best way to identify risk factors that cause a disease with a minimal amount of bias, because it is prospective.

TABLE 4.9 Strengths and limitations of the cohort study

Strengths	Limitations
• Data was collected at regular intervals to get an accurate measurement of different risk factors such as obesity.	• Large numbers of people were required to identify the cases of cardiovascular disease. This study design may not have been suitable if they were investigating a very rare disease.
• The study followed a large population of participants for 50 years, which meant there was enough time for heart disease to be detected.	• It took a long time, and it was expensive to recruit people to the study.
• Every attempt was made to avoid bias in the study population by including people who were adults, did not have a pre-existing condition, and were representative of the population as a whole.	• It was difficult to get follow-up data for some patients who moved, or who had to be excluded from the study.

Observational study: case-control study

A case-control study is usually **retrospective**. This means the disease status of individuals in a population is known at the start of the study, and researchers 'work backwards' to measure the frequency and amount of exposure in people with (cases) and without the disease (controls).

A study begins by identifying cases and controls who come from the same population. This is important to ensure that other factors that might be associated with the outcome under study (known as **confounders**) are not introduced by having big differences between the two groups of people. Case-control studies also gather information using interviews or surveys. An **odds ratio** is calculated from the data to determine whether there is a significant association between an exposure and a disease. An odds ratio greater than 1 is evidence that the exposure causes the disease.

After World War I, there was a rapid increase in the number of people around the world who developed lung cancer. This appeared to correlate with an increase in tobacco consumption. At the same time there was also very bad air pollution from car exhaust fumes, coal fires, gas works and industrial plants. In 1950, Richard Doll and A. Bradford Hill published a landmark case-control study in London that identified smoking, rather than pollution, as the cause of lung cancer.

EXAMPLE

The study identified and interviewed patients from 20 London hospitals. A non-cancer control patient was interviewed at the same time, at the same hospital, of the same sex and within five years of age to the case. Interviews were conducted twice, many months apart, to confirm that the responses were consistent. The study took two years to complete (1948–1949). In total about 2000 cancer cases were recruited to the study. Some patients were excluded because of misdiagnosis, or because they were deemed too old to accurately recall information from earlier in their life. One important consideration was to determine how to measure the exposure (tobacco smoking). For example, the well-designed study was able to distinguish between someone who had smoked at any time in their life, smoked a pack a day, or had quit at some point in their life. This meant that separate analyses could be performed on the different categories of smokers to tease apart the effects of the amount and time of exposure.

An odds ratio of 14 was calculated in the simplest analysis, which compared people who had smoked at any age in their life to people who had never smoked. The study concluded that tobacco smoking was a significant risk factor for lung cancer.

	Smokers	Non-smokers
Lung cancer	647	2
No lung cancer	622	27

$$\text{Odds in exposed group} = \frac{\text{smokers with lung cancer}}{\text{smokers without lung cancer}} = \frac{647}{622} = 1.04$$

$$\text{Odds in not exposed group} = \frac{\text{non-smokers with lung cancer}}{\text{non-smokers without lung cancer}} = \frac{2}{27} = 0.074$$

$$\text{Odds ratio} = \frac{\text{odds in exposed group}}{\text{odds in non-exposed group}} = \frac{1.04}{0.074} = 14$$

FIGURE 4.27 A contingency table used to calculate the odds ratio

TABLE 4.10 Strengths and limitations of the case-control study

Strengths	Limitations
• The examination of the causes of a slow-developing disease such as lung cancer could be examined because of the large study population.	• Information on exposure and past history was primarily based on interview and may have been subject to recall bias.
• The study was relatively inexpensive because it was quick to conduct and required fewer subjects compared to a cohort study.	• The researchers had to trust the information provided to them by participants. It was difficult or impossible to validate information.
• The cases were matched as closely as possible to controls to reduce bias caused by age, sex and geographic location.	• Recruiting the control group from a hospital could introduce other confounding factors, such as other health issues that might contribute to lung cancer.
• The questionnaire was well planned and could distinguish between different types of smokers.	• Participants with lung cancer might provide more detailed or more accurate information than those without it, because they are motivated by having lung cancer and wanting to understand why.
• Careful analysis of medical records removed bias caused by misdiagnosis of lung cancer.	

In summary, case-control studies are well suited to investigating rare outcomes or outcomes that take a long time to develop, because the study begins with a large number of people who have presented with the disease.

Note

A randomised controlled trial is another type of epidemiological study that you could evaluate. It is an example of experimental epidemiology that introduces a treatment or targeted information to participants of a study and assesses its effectiveness in addressing a health outcome. Examples include testing a new drug to treat high blood pressure, or a targeted public health education program to prevent a disease.

Evaluating, using examples, the benefits of engaging in an epidemiological study

Individuals and society can benefit from engaging in an epidemiological study. These benefits are evaluated in Table 4.11 using the examples of a case-control study and a cohort study.

TABLE 4.11 Strengths and limitations of an epidemiological study

Study type	Benefits/strengths	Limitation
Case-control	• Individual satisfaction from contributing to knowledge about health for future generations. • Society benefits from the addition of knowledge about exposures and risk factors and their contribution to disease.	• Individuals are unlikely to benefit directly from participation. • Retrospective nature can make data more subject to bias from recall of exposures/ lifestyle choices and unmeasured differences with control population.
Cohort	• Individual satisfaction from contributing to knowledge about health for future generations. • Society benefits from prospective measurement of different exposures and correlating these with disease.	• Individuals are unlikely to benefit directly from participation. • Large cohort studies can be expensive, with many years of follow-up required.

In summary, engaging in epidemiological studies improves our knowledge of the causes of non-infectious diseases and helps to develop treatments or ways to reduce risk. Individuals also gain satisfaction because their contribution helps them to understand their disease and may benefit other people in the future.

4.4 Prevention: How can non-infectious diseases be prevented?

Non-infectious diseases are caused by genetic mutations, exposure to mutagens, and unhealthy lifestyles. These diseases can be prevented through education campaigns that change people's behaviour, and screening programs to detect diseases at an early and treatable stage. Genetic engineering prevents the progression of some non-infectious diseases, and in the future may be used to correct mutations to prevent genetic diseases.

4.4.1 Using secondary sources to evaluate the effectiveness of current disease prevention methods and develop strategies for the prevention of a non-infectious disease

Disease prevention refers to methods or strategies that are used to stop a disease from developing. This can include education campaigns to encourage lifestyle modification, population screening programs to detect disease at an early stage, or treatments for individuals (e.g. genetic engineering and drugs) that are administered before or soon after the symptoms of a disease occur.

Note
This is an 'including but not limited to' dot point. In the exam you may be asked a question on educational programs and campaigns, genetic engineering, or another method.

Note
This is an evaluate dot point. Remember to judge the benefits and limitations of each example.

Educational programs and campaigns

The SunSmart program is used here to evaluate the effectiveness of an educational campaign.

Note
'Quit For Life' is another campaign, aimed at reducing smoking and lung cancer.

SunSmart program

Skin cancer (**melanoma** and non-melanoma) is the most common cancer in Australia, but changes in behaviour that reduce our exposure to the sun also make it one of the most preventable. Over the past 40 years, the SunSmart program improved people's awareness of the link between skin cancer and sun exposure, using strategies such as television advertisements (e.g. Slip! Slop! Slap! Seek! Slide!) and a free app that provides daily information about UV light levels.

The SunSmart program has been one of Australia's most successful health education programs. Recent research estimates that it prevented more than 43 000 skin cancers and 1400 skin cancer deaths and saved the public healthcare system $92 million over its first two decades. This equated to a $2.22 return for every dollar spent.

Genetic engineering

Gene therapy can be used to treat or prevent non-infectious diseases. It replaces a defective gene with a functioning copy of the gene in the body. This can be done with viral vectors or CRISPR-Cas9 (recall from Module 6).

Note
It is recommended that you recall one example of genetic engineering for the exam. You may also be provided with stimulus information and be asked to evaluate the effectiveness of a technology.

In 2021, the first ever gene therapy was approved for use in Australia to treat a type of retinal dystrophy that is caused by an autosomal recessive mutation in the *RPE65* gene. *RPE65* is normally expressed in the retinal cells in the eye that detect light, but inherited mutations in the gene cause blindness at a young age. The gene therapy restores vision by introducing a functioning copy of *RPE65* into the retinal cells using a viral vector.

9780170465267

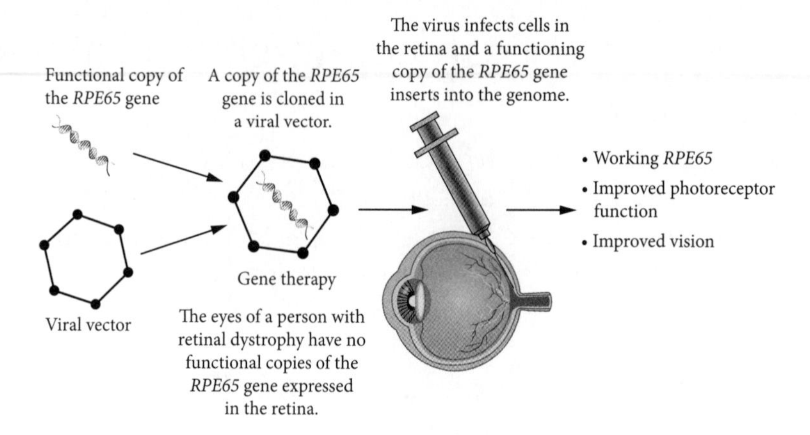

FIGURE 4.28 Gene therapy for patients with inherited retinal dystrophy

In 2021, CRISPR-Cas9 was used in the USA for the first time in an adult to treat sickle cell disease by editing genes involved in globin expression. In the future, CRISPR-Cas9 may be used to correct an inherited mutation in an embryo, such as the F508del mutation in the *CFTR* gene that causes cystic fibrosis. If this is done in a one-cell embryo, all cells in the body will have corrected copies of the gene. This would be an application that truly prevents disease (i.e. stops the disease from beginning).

Mitochondrial transfer is a technique that can prevent non-infectious diseases caused by mutations in mitochondrial genes (recall from Module 6). An example of such a disease is MELAS syndrome, which is caused by mutations in the *MT-TL1* gene, which affects the nervous system and muscles. A child born using mitochondrial transfer will have genomes from three parents – the nuclear DNA from the father's sperm, the nuclear DNA from the mother's egg, and the mitochondrial genome from the donor egg.

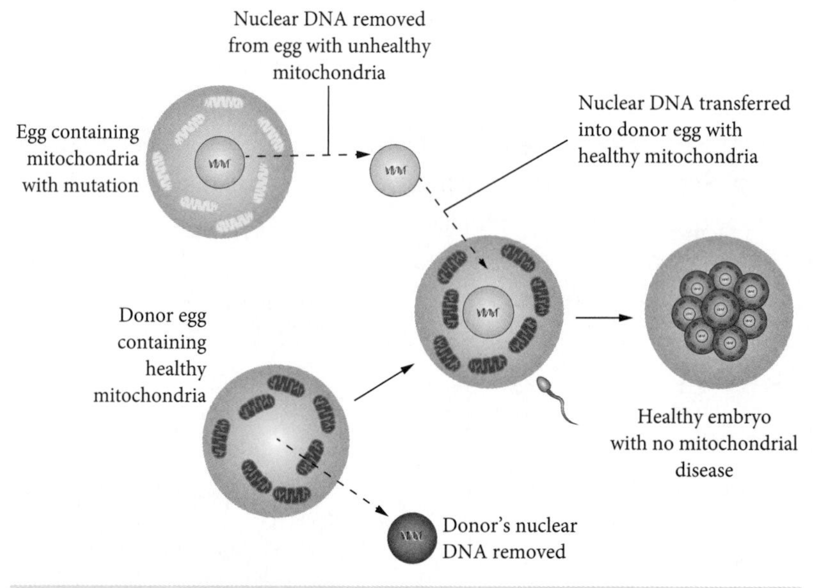

FIGURE 4.29 Mitochondrial donation to prevent non-infectious disease

All forms of genetic engineering are expensive and only prevent disease in a small number of people. For these reasons, you could argue that they are less effective at the population level than education campaigns and population screening at preventing non-infectious disease.

Population screening: National Bowel Cancer Screening Program

Screening programs use relatively cheap medical tests of at-risk members of the community to detect diseases at an early and treatable stage. For example, the National Bowel Cancer Screening Program aims to reduce the number of bowel cancer deaths by detecting it at an early stage with faecal occult blood tests.

> **Note**
> Population screening is an example of an 'including but not limited to' method to prevent disease for this dot point.

The incidence of bowel cancer increases after the age of 40 years and the tests are free to Australians aged 50–74 years. Bowel cancer begins as a slow-growing benign polyp that releases small amounts of blood, which mixes with faeces in the intestine. The blood can be very hard to see when you go to the toilet. Eligible people are sent a collection kit, so a faeces sample can be sent back to test for blood. If blood is detected, the person may have a colonoscopy, and the polyp can be removed before it turns into cancer. According to a 2017 study by Cancer Council Australia, screening for bowel cancer can reduce deaths from the disease by 15–25%.

Note

BreastScreen Australia is another example of population screening, in this case to prevent breast cancer.

4.5 Technologies and disorders: How can technologies be used to assist people who experience disorders?

The organs of the human body carry out specialised functions that help us to sense our environment. Ears are our organs of hearing and balance, eyes detect light and enable us to perceive the world around us, while our kidneys play a critical role in removing waste from our body. The normal function of organs can be affected by inherited genetic mutations, environmental exposures and accidents, and an ageing body. An understanding of the anatomy of organs and the function of their components has enabled scientists to develop technologies to alleviate the symptoms associated with disorders.

4.5.1 Explaining a range of causes of disorders by investigating the structures and functions of the relevant organs

Hearing loss

Sound is produced when an object vibrates, creating a **longitudinal wave**. The wave is propagated by vibrating particles that move back and forth in the same direction as the movement of the wave. The vibration is detected by specialised structures in the ear, converted into an electrical signal, and interpreted as a sound by the brain.

Note

Remember to practise using conjunctions, or linking words, to prepare for an 'explain' question. For example, exposure to loud sounds can damage the hair cells in the cochlea. *As a result*, sound is no longer transmitted correctly to the brain.

To understand how we hear, it is helpful to know some characteristics of longitudinal waves – wavelength, amplitude and frequency. Wavelength is the distance between two consecutive compressions, or **rarefactions**, and amplitude is the distance between the particles: the closer the particles, the higher the amplitude. Frequency is the number of compressions or rarefactions that pass a point in a fixed amount of time.

Amplitude is measured in decibels (dB) and determines the volume of a sound. Frequency determines the pitch of a sound; for example, an unpleasant squeal has a high frequency. It is measured as the number of vibrations per second, in units called hertz. Animals are able to detect different frequencies. The **auditory field** of humans is 20–20 000 Hz.

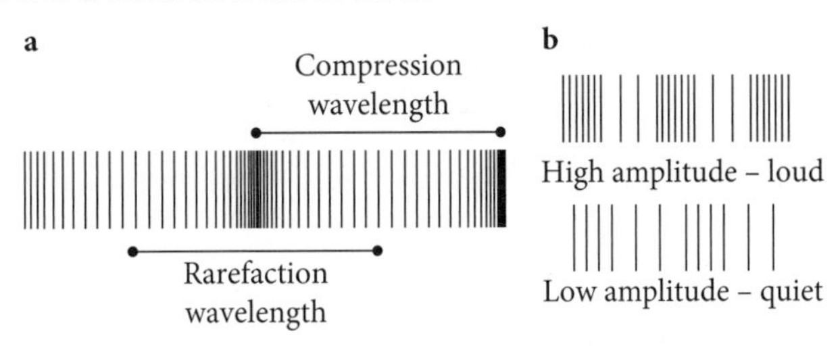

FIGURE 4.30 a Wavelength and b amplitude of sound waves

a

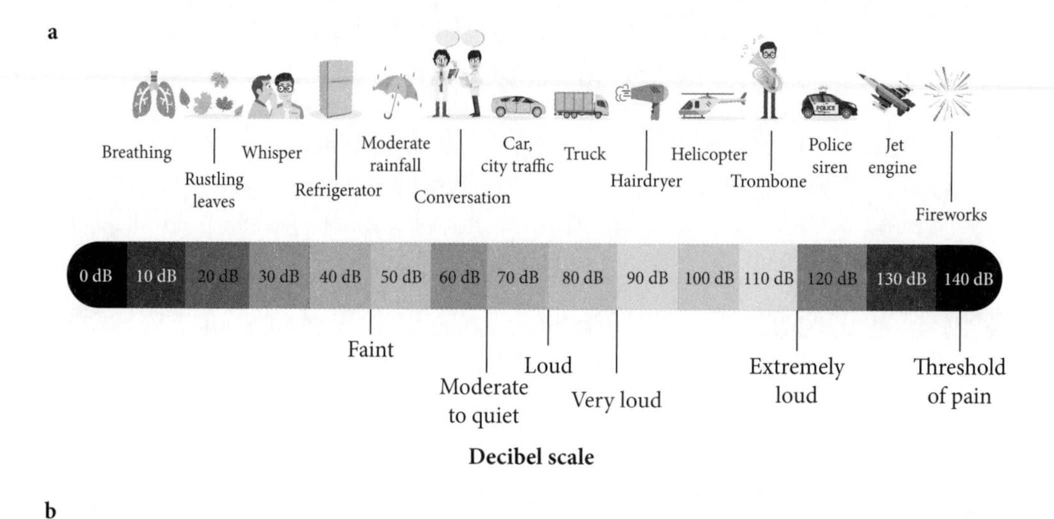

b

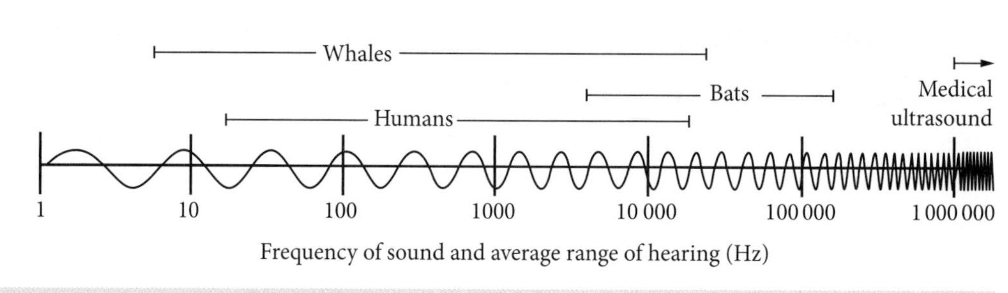

FIGURE 4.31 a Amplitude and b frequency of common sounds

Structure and function of the ear

Hearing enables us to communicate and navigate in response to audible stimuli. Ears are our organs for hearing, but they also help us keep our balance. The ear is divided into the **outer ear**, **middle ear** and **inner ear**.

The outer ear captures sound waves, and the middle ear transfers the vibrations to the inner ear, which converts them into nerve impulses to be interpreted by the brain.

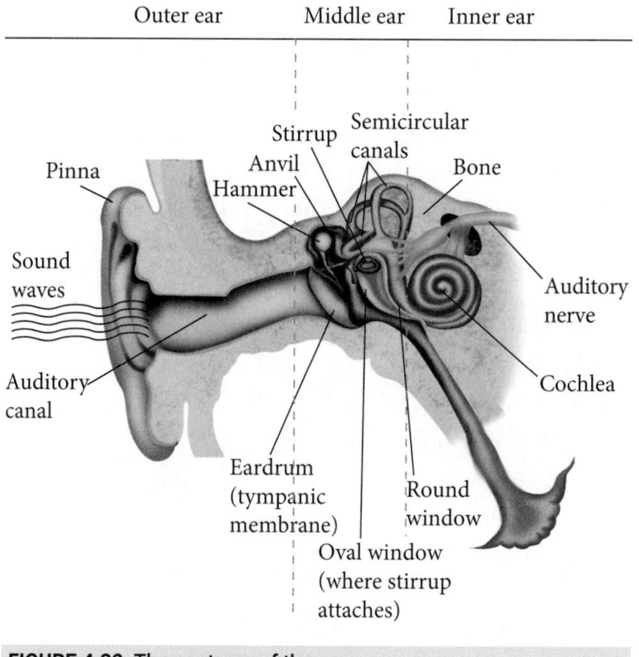

FIGURE 4.32 The anatomy of the ear

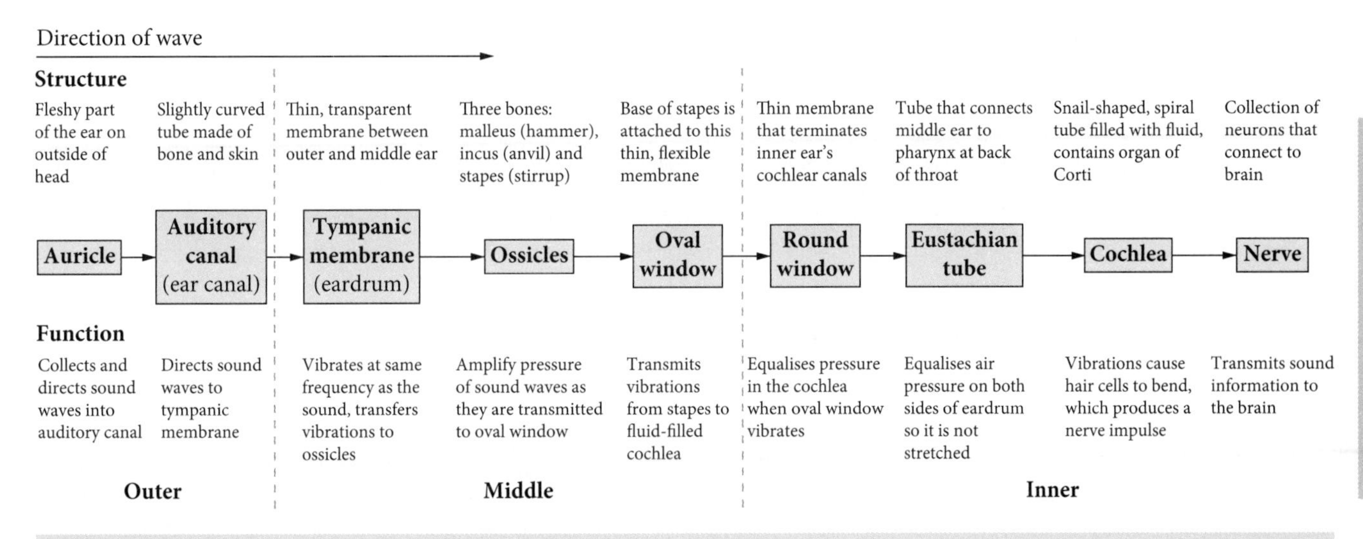

FIGURE 4.33 The structure and function of the parts of the ear

The **cochlea** is a hollow, fluid-filled, coiled structure in the inner ear that contains the **organ of Corti**, which is the organ of hearing.

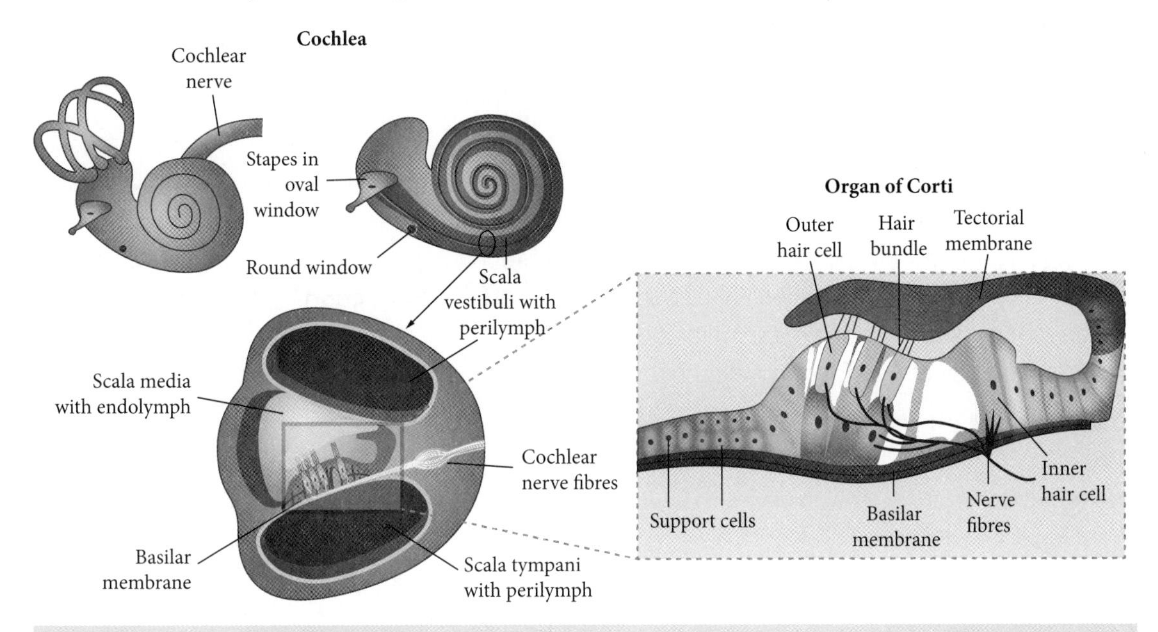

FIGURE 4.34 The structure of the cochlea and organ of Corti

When the oval window is compressed by the ossicles, waves travel through the fluid in the cochlea. This causes the basilar membrane, which consists of support cells and mechanosensory cells (**hair cells**), to flex. The hair bundles (cilia on the tips of the hair cells) vibrate and press against the tectorial membrane, resulting in the release of neurotransmitters. This auditory signal travels via the **vestibulocochlear nerve** to the brain.

Note
A cochlear implant is an option for people with no inner ear function.

Causes of hearing loss

Hearing loss can be **congenital**, which means hearing is absent at birth, or caused by events during our lives, such as exposure to loud sounds, disease and simply getting old.

Hearing loss can be mild or profound and is categorised based on which region of the ear is affected. **Conductive hearing loss** occurs when sound waves cannot reach the inner ear; it is caused by defects in the outer or middle ear. This can be caused by blockage from ear wax or benign tumours, infections in the middle and outer ear, or a perforated eardrum. This type of hearing loss is usually temporary and can be treated.

Sensorineural hearing loss occurs because of defects or damage to the cochlea in the inner ear or the auditory nerve and is often permanent. Possible causes include:

- congenital and acquired hearing loss from viral infections (e.g. German measles and chickenpox)
- side effects from toxic drugs (e.g. cisplatin used to treat cancer and quinine for malaria)
- syndromes (e.g. Treacher Collins syndrome) and germ-line mutations. Autosomal recessive mutations in the *GJB2* gene account for about 50% of hearing loss cases not associated with a syndrome. Loss of expression affects hair cells of the cochlea
- age-related hearing loss (or **presbycusis**), which is the gradual loss of hearing in both ears
- physical trauma, such as a fractured skull or loud noises.

Mixed hearing loss is when there is a combination of conductive and sensorineural defects in the ear.

Visual disorders

Eyes receive and convert light into nerve impulses that are interpreted by the brain as images of objects in the environment. The human eye perceives the colour or visible portion of the spectrum (recall the electromagnetic radiation spectrum from Module 6), while other animals, such as invertebrates and reptiles, can detect UV and infrared radiation. This section focuses on the structure and function of the human eye and conditions that cause loss of vision.

Before examining the eye, it is important to understand some properties of visible light waves. The amplitude is the height of the wave and determines the brightness of the light, while the wavelength determines the colour (e.g. red light has a longer wavelength than blue light). Longer wavelengths have a lower frequency than shorter wavelengths because fewer waves pass a point in a given time period, and vice versa.

It is also useful to know how light behaves when it encounters media of different densities, such as air and liquid. **Reflection** is when light 'bounces off' a surface. **Refraction** is when light changes direction as it passes from one medium to another. The angle of refraction is also dependent on the shape of an object's surface. For example, a **convex** lens and a **concave** lens refract light at different angles.

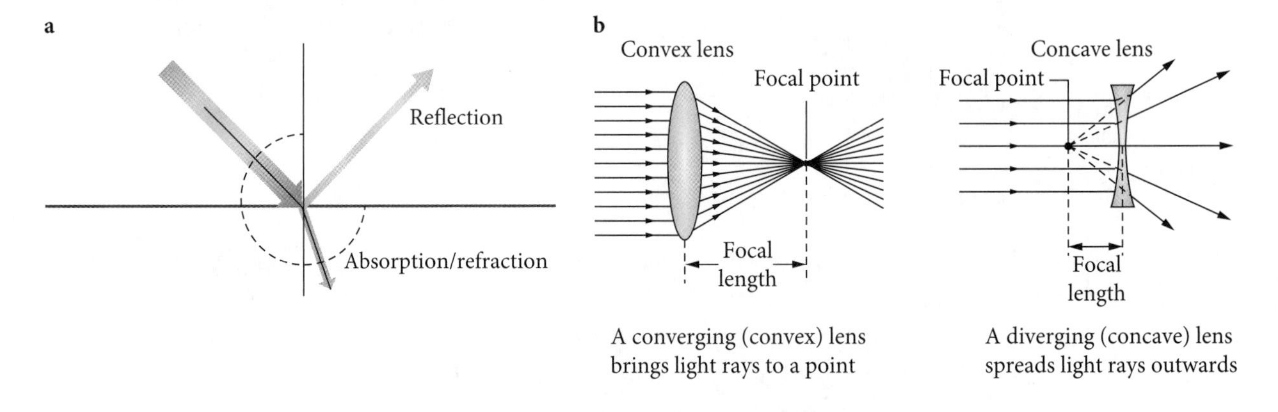

FIGURE 4.35 a Reflection and refraction of light. b The effect of surface shape on refraction.

Structure and function of the eye

Light enters the eye through the **pupil**, the black opening at the centre of the **iris**. Contraction and retraction of muscles in the iris control the amount of light that enters the eye through the pupil. Light then passes through the **cornea**, **lens** and fluid-filled eyeball, which refract light to a focal point on the **retina** at the back of the eye. The retina generates a nerve impulse that is transmitted via the **optic nerve** to the brain.

The retina consists of specialised sensory neuron cells, called **photoreceptors**, that detect colour and light and are localised in an area called the **macula**. There are two types of photoreceptors: **rod cells** and **cone cells**. Rod cells are very sensitive to light, but cannot differentiate between wavelengths of light, and function best in low light. Cone cells contain pigments (rhodopsins) for colour vision. Rhodospsins contain one of three opsins, sometimes called red (L-cone), blue (S-cone) and green (M-cone). The electrochemical message from a photoreceptor is transmitted through a bipolar cell, a ganglion cell, and then via the optic nerve to the brain.

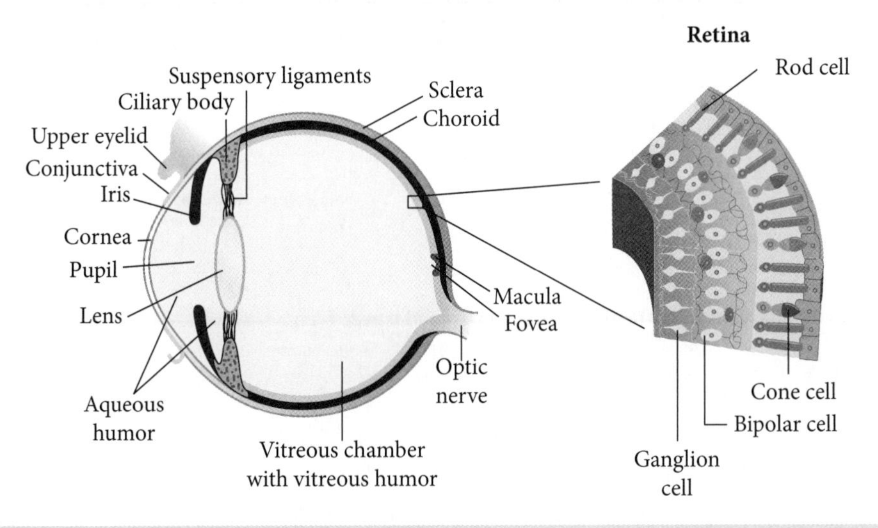

FIGURE 4.36 The structure of the human eye and the retina

Note

Remember that light refracts when it passes through media of different densities. The four refractive media in the eye are the cornea, lens, aqueous humour and vitreous humour.

TABLE 4.12 Functions of parts of the eye

Structure	Description	Function
Conjunctiva	Membrane that lines the sclera and inside of the eyelids	Protects and lubricates the eye with mucous and tears
Cornea	Transparent membrane made from collagen protein	Barrier protection for the eye and refracts light rays; contributes to the refraction of light as it passes through the eye
Sclera	Dense connective tissue around the whole eye	Provides support and maintains the shape of the eyeball
Choroid	Thin layer of tissue that is rich in blood vessels	Provides blood supply to the retina
Aqueous humour	Thin, transparent, water-like fluid	Maintains pressure, provides eye nutrition and refracts light
Iris	Coloured, muscular ring around the pupil	Controls the amount of light that enters the eye
Pupil	Opening at the centre of the iris	Allows entry of light into the eye
Lens	Transparent, biconvex, flexible protein disc	The main refractive structure of the eye that focuses light on the retina
Ciliary body	Ring of tissue containing the ciliary muscle	Holds the lens in position and changes its thickness
Vitreous humour	Transparent gel-like substance between the lens and the retina	Maintains the spherical shape of the eye; has some refractive ability
Retina	Light-sensitive layer of tissue at the back of the eye	Converts light to electrochemical signals
Macula	Pigmented area near the centre of the retina	Provides sharp, clear, straight-ahead vision
Optic nerve	Paired nerve that connects the eye to the brain	Transmits nerve impulses from the retina to the brain
Eye muscles	Sets of muscles attached from the eye to bone	Rotates the eyeball within the eye socket of the skull

Light enters the eye at different angles depending on the relative distance of objects from us; far-away objects reflect parallel light into the eye, while light from close-up objects enters the eye at an angle. The biconvex shape of the eye's lens and its ability to change shape to accurately focus light on the retina (**accommodation**) enables us to see objects at different distances. We also have binocular vision, which helps with depth perception.

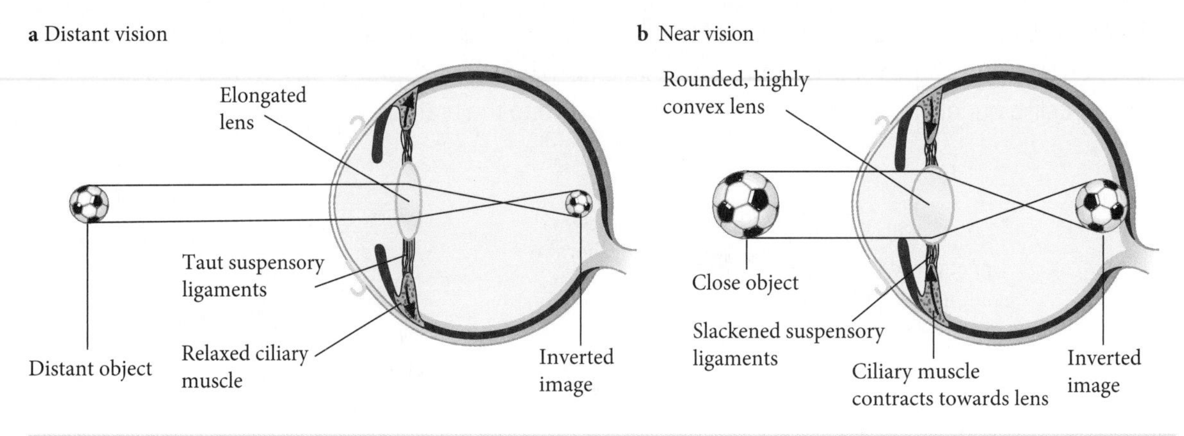

FIGURE 4.37 Accommodation of the lens for a distant vision and b near vision

Causes of visual disorders

Refractive errors are the most common type of vision disorder. They occur when light does not focus properly on the retina. The main types of refractive errors are **myopia** (short-sightedness), **hyperopia** (far-sightedness) and **astigmatism**. They tend to run in families and are diagnosed in childhood or adolescence.

Presbyopia is another refractive error that occurs when our eyes gradually lose the ability to focus on close-up objects, usually beginning in the early to mid-forties. Presbyopia is caused by a hardening of the lens, which becomes less flexible and therefore less able to accommodate light entering the eye.

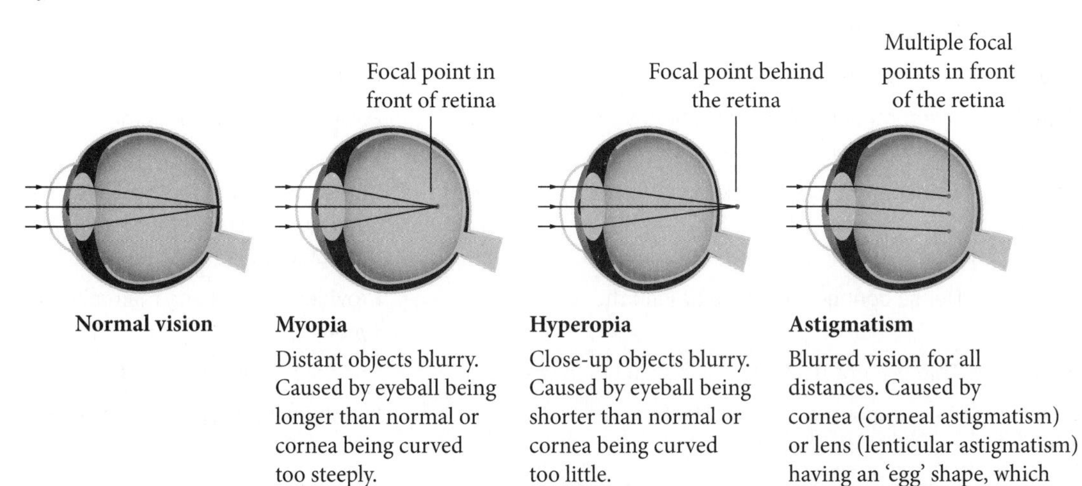

FIGURE 4.38 Types of refractive error visual disorders

TABLE 4.13 Other common visual disorders and their causes

Visual disorder	Cause
Cataracts	Blurry vision resulting from the breakdown of proteins and fibres in the lens
Glaucoma	Group of eye diseases seen in older people; results from progressive damage to the optic nerve caused by increased pressure in the eye
Macular degeneration	Deterioration of cells in the retina, resulting in loss of central vision
Retinopathy	Spotty, blurry vision, or loss of vision, caused by damage to the retina from abnormally growing or damaged blood vessels; a symptom of types 1 and 2 diabetes

Loss of kidney function

Structure and function of the kidney

The kidneys are two bean-shaped organs, each about the size of a clenched fist, located below the rib cage on either side of the spine. Kidneys are a part of the urinary system, along with the bladder and ureters. Their main function is to remove nitrogenous waste and excess fluid from the body in the form of urine. They also balance the level of water, salt ions and minerals in the body, and produce hormones (e.g. renin, erythropoietin and calcitriol) that regulate blood pressure, red blood cell production and the absorption of calcium in the intestine.

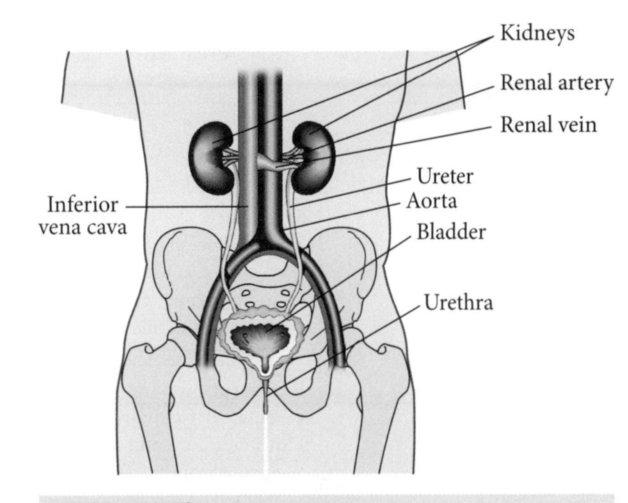

FIGURE 4.39 The urinary system

Blood enters the kidneys through the afferent arteriole, which branches into smaller blood vessels called capillaries, which enter millions of nephrons. **Nephrons** are the functional units of the kidney. Each nephron consists of a filtering structure called the **glomerulus** and a tubule. A glomerulus is a cluster of narrow blood vessels surrounded by a Bowman's capsule. The tubule consists of the proximal convoluted tubule, loop of Henle and the distal convoluted tubule, which leads into a collecting tubule.

Four main processes occur in the kidney – filtration, reabsorption, secretion and excretion. The product of these stages is urine, which passes through the collecting duct into the ureter and is released from the body through the urethra via the process of excretion. Urine consists of 95% water, 5% nitrogenous waste and ions such as sodium, potassium, hydrogen and calcium.

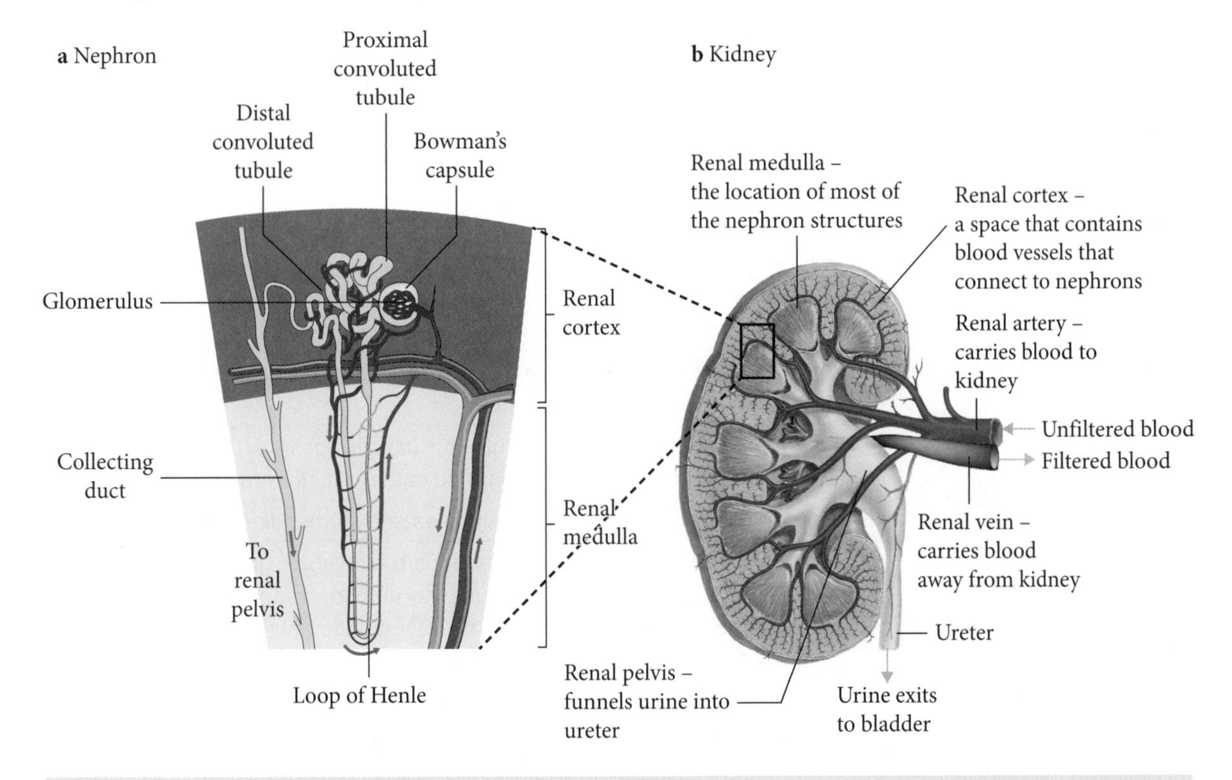

FIGURE 4.40 The structure of a a nephron and b a kidney

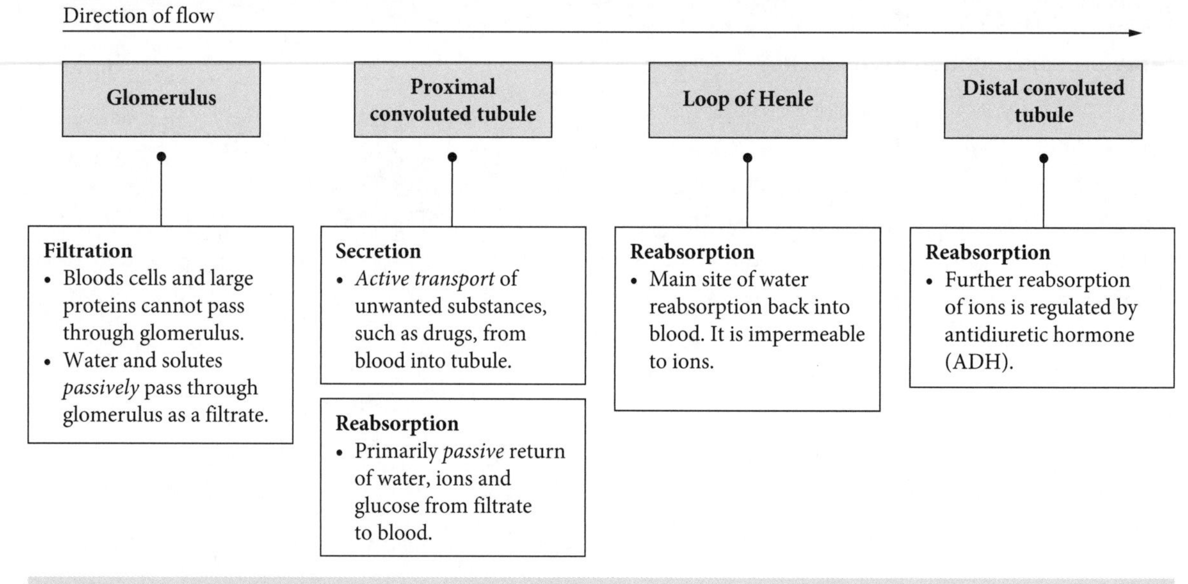

FIGURE 4.41 The main processes that occur in the kidney

Causes of kidney disorders

Kidney disorders can be divided into acute or chronic failure. Acute kidney failure is when the kidneys suddenly lose their ability to filter waste, which accumulates and changes the chemistry of the blood. This can be caused by:

- reduced blood flow to the kidneys – results from blood clots and high cholesterol (which block blood vessels leading to the kidney), severe dehydration, liver failure, infections, toxins and blood pressure medication
- urine blockages – the ureters can get blocked with kidney stones or tumours, which stop the flow of urine out of the body. Kidney stones are hard mineral deposits that build up in the kidney.

Chronic kidney failure is a gradual loss of function over many years. It results in the accumulation of excess fluid, electrolytes and wastes in the blood.

TABLE 4.14 Kidney disorders and their causes

Kidney disorder	Cause of disorder
Diabetic nephropathy	A symptom of types 1 and 2 diabetes. Abnormal filtering from the glomerulus into Bowman's capsule results in blood cells and large proteins leaking into the urine.
Hypertension (high blood pressure)	The walls of thin blood vessels (arterioles) in the kidney thicken and reduce blood flow to the kidney.
Glomerulonephritis	Inflammation of the glomeruli. This usually occurs after a bacterial infection and is the most common cause of kidney failure.
Interstitial nephritis	Inflammation of the kidney's tubules and surrounding structures
Polycystic kidney disease	An inherited genetic disorder that causes cysts to grow in the kidney. The most common type is caused by autosomal dominant mutations in the *PDK* genes. There is also a rarer autosomal recessive form caused by mutations in *PKHD1*.
Vesico-ureteral reflux	An abnormal ureter causes urine to flow back from the bladder into the ureters or kidneys. It can cause infections and scarring.
Pyelonephritis (recurrent kidney infection)	Recurring kidney infections caused by bacteria or viruses

4.5.2 Investigating technologies that are used to assist with the effects of a disorder

Note

This is an 'including but not limited to' syllabus dot point. This means you may be asked a question about technologies used to assist with hearing loss, visual disorders and loss of kidney function. The *bionic eye* and *intraocular lens* are provided as additional examples of technology you could use.

Hearing loss: cochlear implants, bone conduction implants, hearing aids

Cochlear implants can be used by people with profound sensorineural loss when hearing aids are not beneficial. Cochlear implants bypass the outer and middle ear and deliver the sound or vibrations directly to the auditory nerve. Normal hearing is not restored with a cochlear implant. A person needs therapy and practice to learn how to hear. The implant provides an increased awareness of sounds in the environment and improves the ability to understand through lip reading.

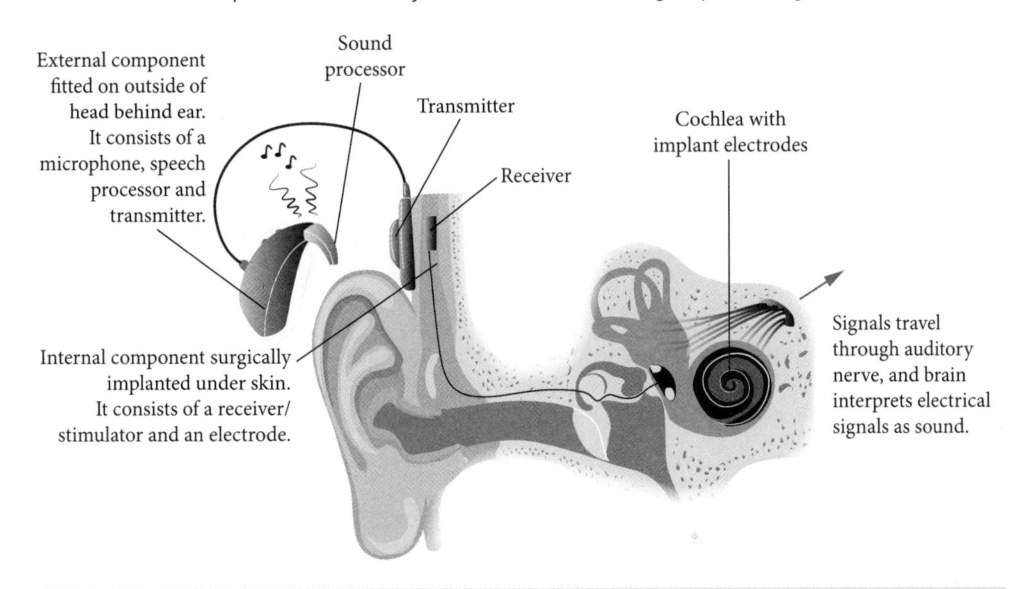

FIGURE 4.42 A cochlear implant

Bone conduction implants are used to treat conductive or mixed hearing loss or single-sided deafness. Similar to the cochlear implant, the system consists of an external sound processor and an implant.

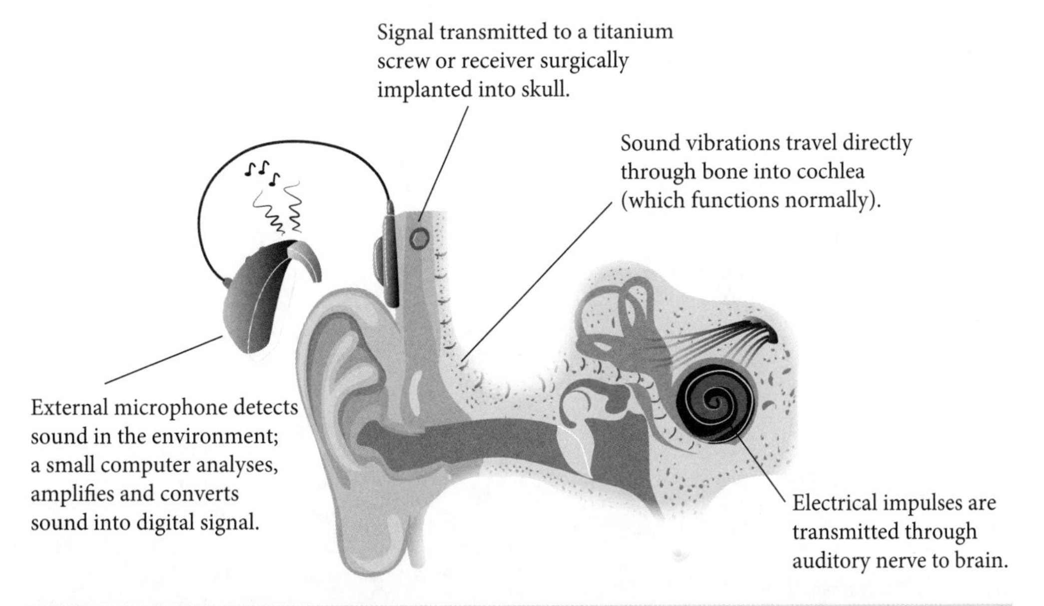

FIGURE 4.43 A bone conduction implant

Hearing aids are used to amplify sounds to improve hearing in people with sensorineural hearing loss. They are small, removable electronic devices that consist of a microphone, an amplifier and a speaker. The microphone detects sound waves and converts them to electrical signals, which are sent to an amplifier. The amplifier increases the power of the signals and then sends them to the ear through a speaker. Hearing aids only work if there are some surviving hair cells in the cochlea to detect sound vibrations. Hearing aids improve hearing and speech comprehension.

Visual disorders: spectacles, laser surgery

Glasses, contact lenses and laser surgery

Refractive errors can be treated with glasses or contact lenses. People who do not want to wear glasses or contact lenses can have the shape of the cornea permanently changed with corrective laser surgery. Each of these technologies corrects the angle of the refraction of light so that it focuses correctly on the retina.

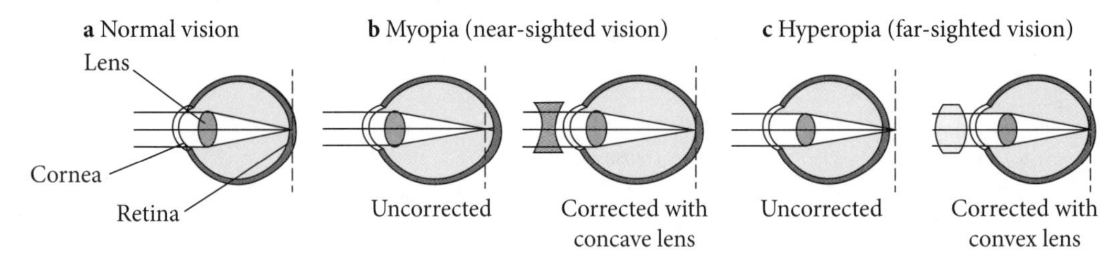

FIGURE 4.44 Correction of refractive error vision disorders with glasses and contact lenses: a normal vision, b myopia and c hyperopia

Additional technologies: intraocular lens implant and bionic eye

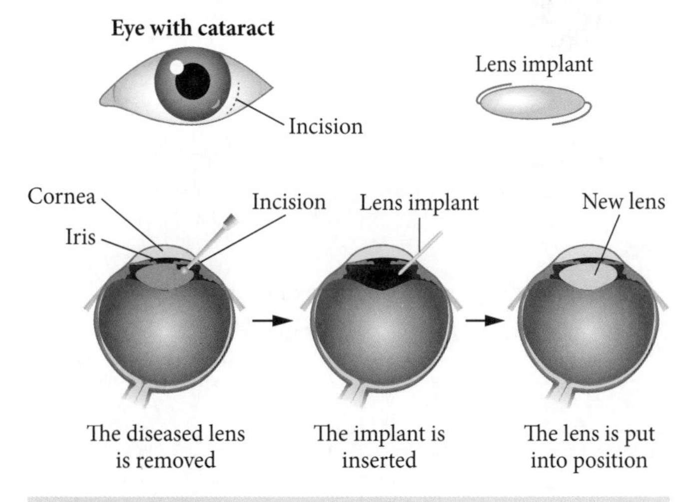

FIGURE 4.45 An intraocular lens implant to treat cataracts

An intraocular lens (IOL) is a lens placed inside the eye. These artificial lenses are made from plastic, and they work in the same way as a natural lens. Unlike contact lenses, an IOL is a permanent surgical implant. Microsurgery is used to remove a cloudy lens, usually caused by a cataract, and implant the IOL behind the iris. An IOL fits perfectly into the lens capsule and results in correct focus and clear vision.

The bionic eye is a future technology that is currently being trialled to restore sight in people with severe vision loss caused by degenerative diseases. Just like the cochlear implant, which bypasses the key sensory cells (hairs cells) in the ear, the bionic eye bypasses the photoreceptors in the retina.

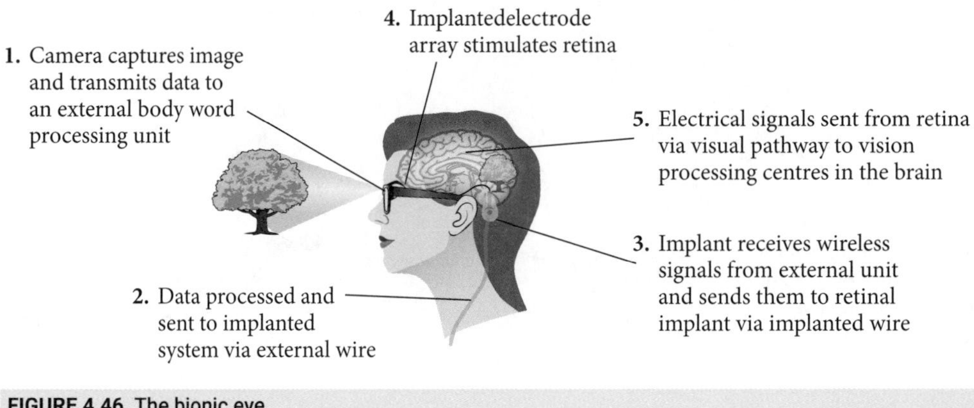

FIGURE 4.46 The bionic eye

Loss of kidney function

Each kidney performs half the kidney function in a body, but a person can survive with only one kidney. A single kidney can increase in size and carry out 75% of the normal function of both kidneys. If a person's kidney function decreases to about 10–15% of a normal kidney, the only treatments are **dialysis** or a kidney transplant.

Dialysis

Dialysis is a medical technique that replaces the normal filtering function of the kidneys. There are two types of dialysis: **haemodialysis** and **peritoneal dialysis**. In haemodialysis, the filtering of blood occurs outside the body. It is the most common type of dialysis and is ideal for people with reduced kidney function. It is done in a hospital, a dialysis centre or at home. A session takes about four to five hours and needs to be done three days per week.

In peritoneal dialysis, the filtering of the blood occurs inside the body. It allows continuous filtration, can be done at home or at work, and is not as disruptive to a person's daily activities as haemodialysis. It may not be suitable for people with obesity or abdominal scarring.

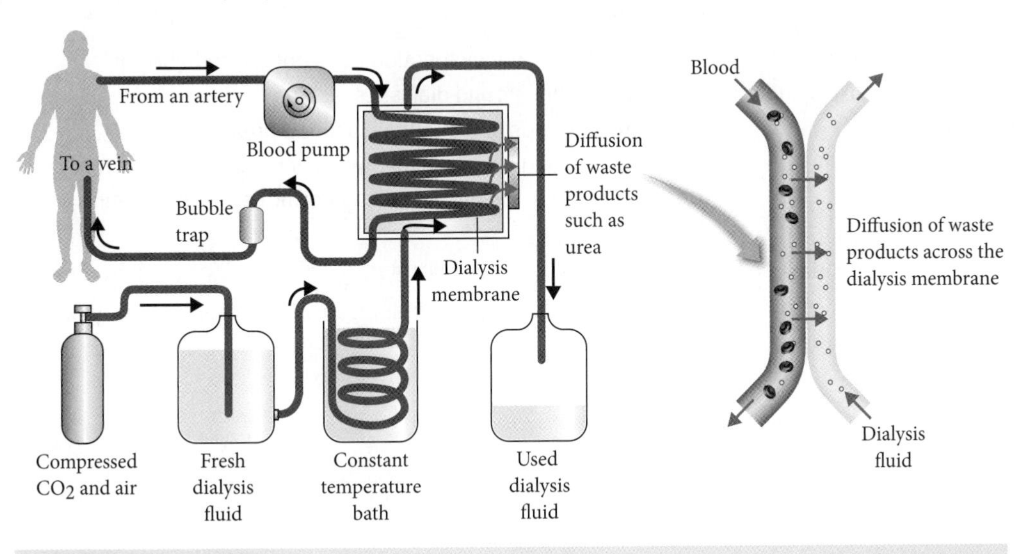

FIGURE 4.47 Haemodialysis

Kidney transplant

A kidney transplant can be done using a donated kidney from a living or deceased person. If a kidney comes from a living donor, the transplant can be done when the recipient's kidney is close to failing but before dialysis (a pre-emptive transplant). People can register as organ donors before dying. The recipient must take anti-rejection drugs for the rest of their life, to prevent the kidney being attacked by their immune system.

Kidney stones

Kidney stones can be treated by breaking them up with shock waves, and smaller stones are then passed naturally in the urine. They can also be surgically removed (nephrolithotomy) through a small incision in the back.

4.5.3 Evaluating the effectiveness of a technology that is used to manage and assist with the effects of a disorder

Technologies can vary in their efficacy in assisting with the effects of a disorder. For example, glasses are a relatively inexpensive technology that can restore normal vision to a person with myopia or hyperopia. Cochlear implants, on the other hand, are expensive and do not restore normal hearing to a person with profound hearing loss. They do, however, allow a person to perceive sound, which can enhance their ability to communicate and respond to audible stimuli in the environment.

Note
This is an 'evaluate' dot point. Constructing a table is an effective way of presenting the benefits and limitations of a technology. Remember to provide a value judgement in the form of a concluding sentence.

Dialysis to treat kidney failure is provided as a more detailed example of an evaluation of a technology.

TABLE 4.15 Benefits and limitations of dialysis

Benefits	Limitations
• Extends the person's lifetime • Filters out waste, toxins and excess water • Increases energy levels • Improves appetite	• It is not a cure – kidney function does not return to normal • Need to plan life around dialysis sessions • Can cause cramps and changes in blood pressure • Supplements and hormone therapy may be needed to correct deficiencies (e.g. vitamin D and the hormone erythropoietin)

In summary, haemodialysis improves the quality of life of a person with kidney failure. It does not, however, restore the normal function of the kidneys, and it is disruptive and requires daily planning around treatment times.

Glossary

abscission The natural shedding of leaves, flowers and fruit from a plant

accommodation The ability of the eye to focus on objects at different distances by changing the shape of the lens

action potential A change in polarity across the neuron membrane which travels down an axon as a nerve impulse

adaptation A characteristic or feature that aids in the survival and reproductive success of an organism

afferent Travelling inwards

amine hormones Hormones derived from amino acids

apoptosis Programmed cell death

astigmatism A refractive disorder that causes blurred vision at all distances

auditory field The range within which an organism can perceive sound

autocrine Describes a chemical signal released by a cell and acting on the same cell

autoimmune disease A disease that occurs when the immune system mistakenly attacks cells in its own body

axon The part of a neuron that carries nerve impulses away from the cell body

benign Describes a tumour or growth that does not spread to other parts of the body

beta cell A type of cell in the pancreas that synthesises insulin

body mass index A calculation used to estimate the total amount of fat in the body

bone conduction implant A surgically implanted hearing device that transmits sound vibrations through bone in the skull to the cochlea in the inner ear

cancer A disease in which some cells in a body grow uncontrollably and spread to other parts of the body

carcinogen A chemical mutagen that causes cancer

case-control study An observational study that aims to identify the cause of a disease by comparing a group of people with the condition (cases) to a very similar group of people who do not have the condition (controls)

cell body The circular part of a neuron that contains the cell nucleus; also called soma

central nervous system The brain and spinal cord

A+ DIGITAL FLASHCARDS
Revise this topic's key terms and concepts by scanning the QR code or typing the URL into your browser.

https://get.ga/aplus-hsc-bio-u34

circulatory system The system of connected vessels that transport blood around the body

cochlea A hollow, spiral-shaped bone in the inner ear that contains fine hairs that convert sound vibrations to nerve impulses

cochlear implant A surgically implanted electronic device in people with sensorineural hearing loss that stimulates the cochlear nerve to provide a sense of hearing

cohort study An observational study that follows a population through time to identify exposures that cause a disease or condition

concave Curving inwards

conductive hearing loss Hearing loss caused by a defect or blockage in the outer or middle ear

cone cell A type of photoreceptor in the retina that detects colour

confounder A variable that is associated with a disease but doesn't cause the disease

congenital Describes a condition that is present at birth

control centre A body structure (e.g. brain or pancreas) that determines the normal range of a variable such as temperature or blood glucose level

convex Curving outwards

cornea A transparent outer covering over the iris and pupil of the eye

dendrite A branch-like extension at the end of a neuron that receives electrochemical signals from another neuron

dialysis A technique that removes waste and excess fluid from the circulatory system of a person without functioning kidneys

ectotherm An organism that relies on the external environment to regulate its body temperature, e.g. reptile

effector A cell, tissue or organ that responds to signals from the control centre, providing a response to a stimulus in order to maintain homeostasis

efferent Travelling outwards

endocrine Refers to hormone signalling acting over a long distance

endocrine gland A gland that secretes hormones into the circulatory system

endotherm An organism that can regulate its body temperature independently of the external environment, e.g. mammal

epidemiology The study of patterns of disease factors that determine the presence or absence of diseases or conditions in populations

gene therapy Techniques that use genetic material to modify a gene or manipulate levels of gene expression to treat a disease

genetic disease A disease caused by a mutation in DNA

genetic engineering The use of technology to alter the DNA of an organism, usually for human benefit

glomerulus A cluster of blood vessels in the nephron through which blood is filtered

glucose A simple type of sugar

guard cell One of a pair of cells that surround each stoma opening in plants

haemodialysis A method of purifying the blood of people with kidney failure, using a dialysis machine

hair cell A mechano-sensory cell in the organ of Corti that converts sound vibrations into electrical signals

hearing aid An electronic device worn behind or in the ear that makes sounds louder, for use by people with a hearing disorder

hereditary cancer Cancer caused by the inheritance of a germ-line mutation in a tumour suppressor gene

homeostasis The ability of an organism to maintain a stable internal environment

hormone A signalling molecule that regulates homeostasis

hyperopia A common refractive eye disorder that results in blurred close-up vision but clear distance vision; also called far-sightedness

hyperthermia An abnormally high body temperature

hypothermia An abnormally low body temperature

incidence The number of new cases of a disease in a population in a specified time period

inner ear The innermost part of the vertebrate ear, containing the cochlea

insulin A hormone the regulates blood glucose levels

iris The coloured part of the eye; controls the amount of light that enters the pupil

islets of Langerhans Clusters of cells in the pancreas that synthesise and release hormones

lens A transparent, biconvex structure in the eye that refracts light onto the retina

longitudinal wave A type of wave in which the vibration of particles is in the same direction as the travelling wave

macula The part of the retina responsible for central vision

malignant Describes a tumour containing cells that grow uncontrollably and spread locally and/or to other parts of the body

melanoma A malignant cancer that forms in melanocytes in the skin

mesothelioma A rare type of lung cancer caused by the inhalation of asbestos fibres

metastasis The process by which cancer cells spread to other parts of the body

middle ear The region of the ear consisting of the eardrum and an air-filled chamber containing the ossicles

mortality rate The frequency of death in a defined population during a specified time period

motor neuron A neuron in the central nervous system that carries nerve impulses from the brain and spinal cord to muscles

myopia A refractive eye disorder that results in blurred distance vision

negative feedback loop A pathway that opposes a stimulus to cause a decrease in function

nephron The functional unit of the kidney, consisting of a glomerulus and tubule

nephropathy The deterioration of kidney function

nerve A bundle of nerve fibres that carries electrochemical signals to and from the brain and spinal cord

neuron A specialised cell that is the basic building block of the nervous system

neuropathy Damage to the peripheral nervous system

non-infectious disease A disease that is not caused by a pathogen and therefore cannot be transmitted between people

odds ratio A measure of association between a risk factor and a disease in a case-control study

oncogene A gene involved in cell growth that mutates and contributes to cancer

optic nerve A nerve that carries electrical impulses from the retina to the brain

organ of Corti An organ in the cochlea that generates nerve impulses in response to sound

osmoregulation Maintenance of the internal balance between water and dissolved materials regardless of environmental conditions

outer ear The fleshy part of the ear on the outside of the head

paracrine Refers to hormone signalling that acts on cells in a local environment in the body

peptide hormone A hormone made of peptides or proteins

peripheral nervous system All the nerves in the body except the brain and spinal cord

peritoneal dialysis A method of purifying the blood of people with kidney failure, in which the peritoneum is used as the filtering membrane

phloem Specialised tissue in plants that transports food made in the leaves to other parts of the plant

photoreceptor A specialised cell in the retina that detects light

photosynthesis The process by which plants use sunlight, water and carbon dioxide to create oxygen and glucose

phototropism The orientation of a plant or other organisms in response to light

plasmodesmata Channels that cross the cell walls between adjacent plant cells

presbycusis The gradual loss of hearing in both ears as a person grows older

presbyopia The gradual loss of the eye's ability to focus on close-up objects as a person grows older

prevalence The total number of disease cases in a population at a given time

prospective Expected to happen in the future

proto-oncogene A gene that controls normal cell division but causes cancer when it is mutated

pupil The hole in the centre of the iris through which light enters the eye

rarefaction A region in a longitudinal wave where the particles are furthest apart

receptor A molecule or cell in the body that senses a change in the environment

reflection When light bounces off an object

refraction A change in direction of a light wave caused by a change in speed as it moves between media

relative risk The ratio of the probability of an outcome in an exposed group to the probability of an outcome in an unexposed group

respiration The process by which cells obtain energy by breaking down glucose into ATP

response How the body reacts after a signal reaches an effector such as a muscle or gland

retina A light-sensitive layer of tissue at the back of the eye

retinopathy A disease of the retina in the eye

retrospective Examining events that happened in the past

rod cell A photoreceptor cell in the retina that is sensitive to low light

sensorineural hearing loss Hearing loss caused by defects in the inner ear

sensory neuron A neuron that detects stimuli in the environment and carries the signal to the central nervous system

signal transduction Transmission of a physical or chemical signal through a cell

sporadic cancer Cancer caused by genetic mutations in somatic cells

steroid hormone A chemical messenger in the body that is derived from cholesterol

stimulus A condition in the environment that is detected by a receptor

stoma A pore or opening on the underside of leaves that exchanges carbon dioxide and water with the atmosphere; plural *stomata*

synapse A small space between two neurons where neurotransmitters are released to transfer a nerve impulse to an adjacent neuron

thermoregulation The mechanism by which endotherms maintain a constant body temperature

transpiration Evaporation of water from the surface of plant leaves

transpiration stream An uninterrupted flow of water and solutes through the xylem from the roots to evaporation from the leaves

tumour A mass of tissue that forms from the abnormal growth and division of cells

tumour suppressor A gene that inhibits tumour formation by slowing down cell division, repairing DNA mutations and instructing cells when to die

turgor The water pressure inside a plant that keeps cells and tissues rigid

vestibulocochlear nerve A nerve that transmits auditory signals to the brain

xylem Specialised tissue in plants that transports water and minerals from the roots to the rest of the plant

Exam practice

Multiple-choice questions

Solutions start on page 268.

Homeostasis

Question 1

An endotherm is an organism that maintains a stable

A heart rate.

B body mass.

C internal temperature.

D external temperature.

Question 2

Which of these statements about a negative feedback loop is **false**?

A It causes instability in a system.

B It occurs in response to stimulus.

C It causes a decrease in a function.

D It is a reaction to a deviation from a set point.

Question 3 ©NESA 2020 SI Q1

In maintaining homeostasis, which of the following is a behavioural adaptation?

A Sweating to cool down

B Curling up in a ball to keep warm

C Speeding up or slowing down cell metabolism

D Skin going red as more blood flows to surface

Question 4

The opening and closing of stomata in plants depends on a water feedback loop. When water levels drop, the stomata

A close, evaporation from the leaf declines, and CO_2 uptake increases.

B close, evaporation from the leaf declines, and CO_2 uptake is limited.

C open, evaporation from the leaf surface declines, and CO_2 uptake is limited.

D open, evaporation from the leaf surface increases, and CO_2 uptake is limited.

Question 5 ©NESA 2019 SI Q12

The glucose tolerance test is used to investigate the control of glucose in the human body. Patients consume 75 g of glucose and their blood glucose is monitored.

Type 2 diabetes is a condition where the cells of the body do not respond adequately to insulin. Which graph could represent the results of glucose tolerance tests in a non-diabetic person and a person with untreated type 2 diabetes?

Non-diabetic — Type 2 diabetic

A

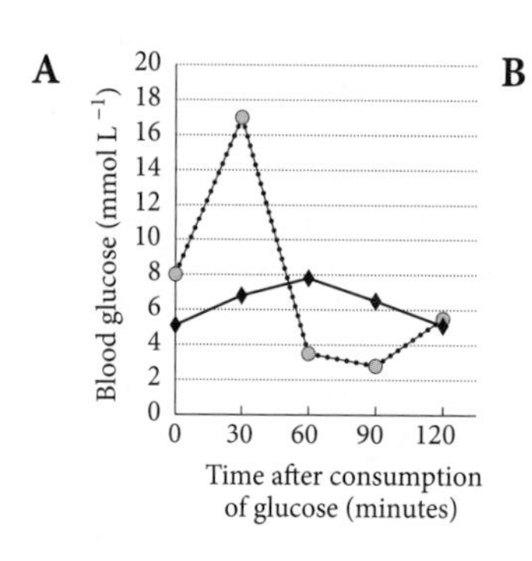

B

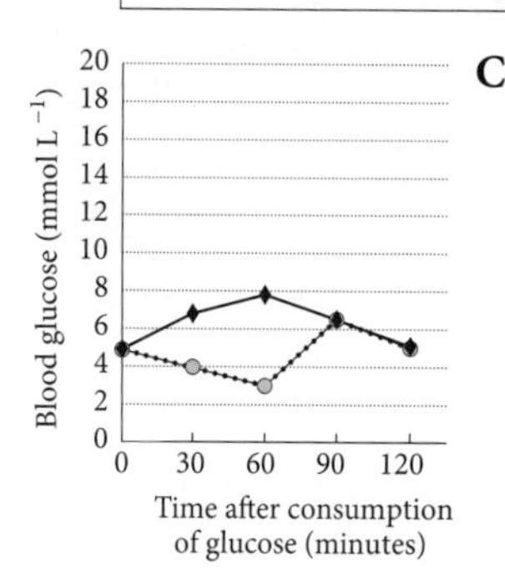

C

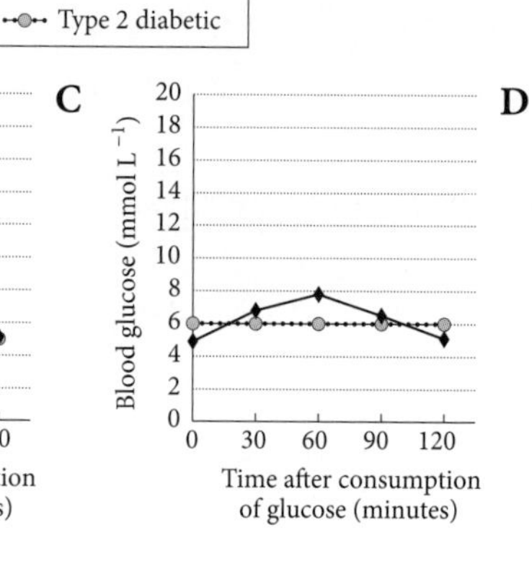

D

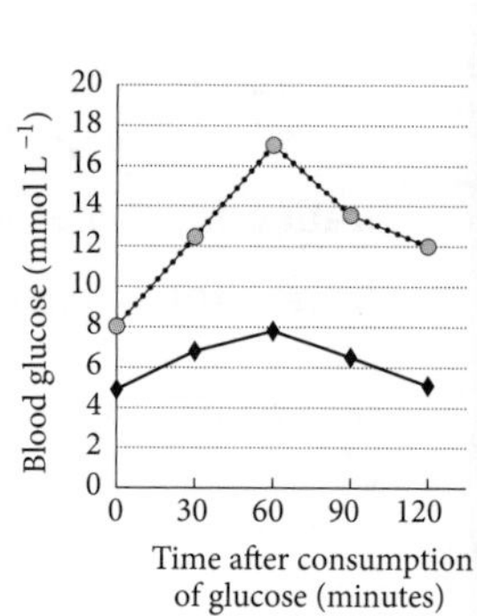

9780170465267

Causes and effects

Question 6

Genetic diseases can be caused by

A somatic mutations only.

B inherited mutations only.

C germ-line mutations only.

D germ-line and somatic mutations.

Question 7

Long-term exposure to ultraviolet light can result in metastatic skin cancer. This means the cancer cells

A develop into benign tumours.

B spread but stay localised in the skin.

C are destroyed by the adaptive immune system.

D move from the skin to other sites in the body and form secondary tumours.

Question 8

Which of the following is a type of nutritional disease?

A Scurvy, because it is caused by a severe deficiency in vitamin C

B Type 1 diabetes, because blood glucose levels cannot be regulated

C Mesothelioma, because it is caused by inhalation of asbestos fibres

D Skin cancer, because it is caused by absorption of too much ultraviolet light

Question 9

The graph shows the global percentage of people with diseases or conditions caused by injury, infectious and non-infectious diseases. Which conclusion is correct?

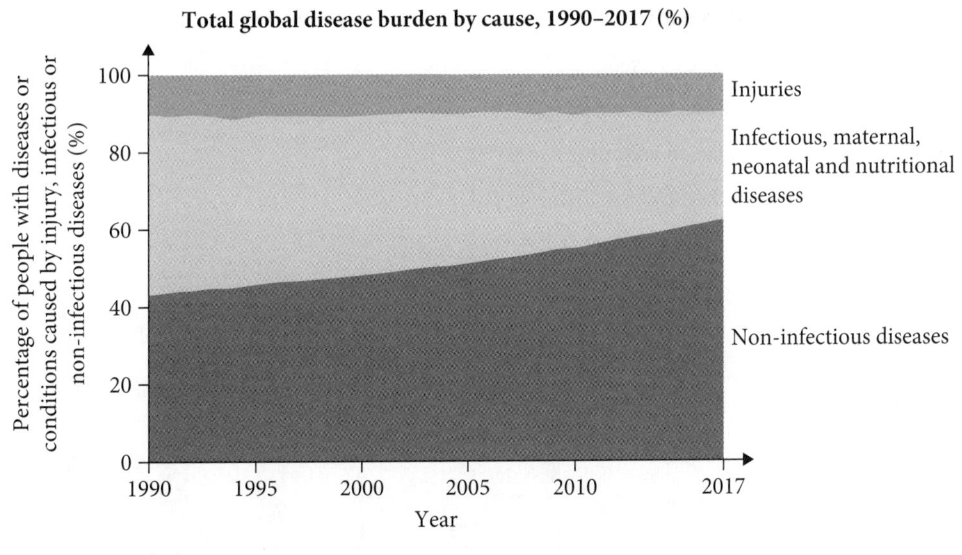

A Injuries are responsible for the majority of the disease burden.

B The percentage of infectious disease cases has increased from 1990 to 2017.

C Non-infectious diseases have increased because of more exposure to pathogens.

D Non-infectious diseases made up the greatest proportion of disease burden in 2017.

Question 10

Which of the following statements is true about a non-infectious disease?

A It can be transmitted between hosts.

B It depends on exposure to a pathogen.

C It can be inherited and affect a person all their life.

D It can only be treated by genetic engineering technology.

Epidemiology

Question 11

The graph shows the proportion of people in Australia with the five most common diseases, conditions and disorders, together with data on fatality. Which statement about the graph is correct?

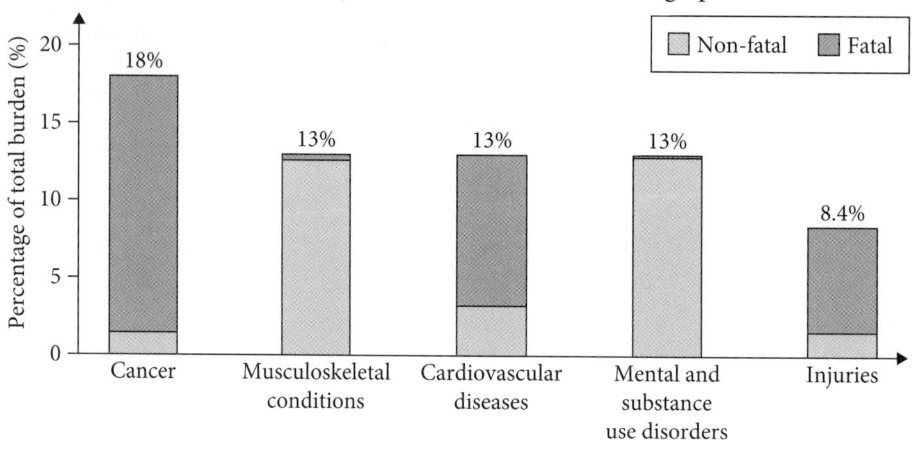

Based on Australian Institute of Health and Welfare material.

A Injuries are the least common and most are fatal.

B Mental and substance use disorders are almost always fatal.

C Cancer is the most common and the majority of cases are not fatal.

D Cardiovascular disease is the third-highest contributor to total burden.

Question 12

An epidemiological study recruited a study population of 5000 people with cardiovascular disease and compared exposures over their lifetime with 5000 control cases. The researchers found that cardiovascular disease was more common in obese people. What can be concluded from this study?

A Obesity causes cardiovascular disease.

B Obesity does not cause cardiovascular disease.

C Obesity is associated with a more sedentary lifestyle.

D Obesity is a risk factor associated with cardiovascular disease.

Question 13

The graph shows the prevalence of dementia in Australia in 2021. Which of the following is a correct interpretation of the graph?

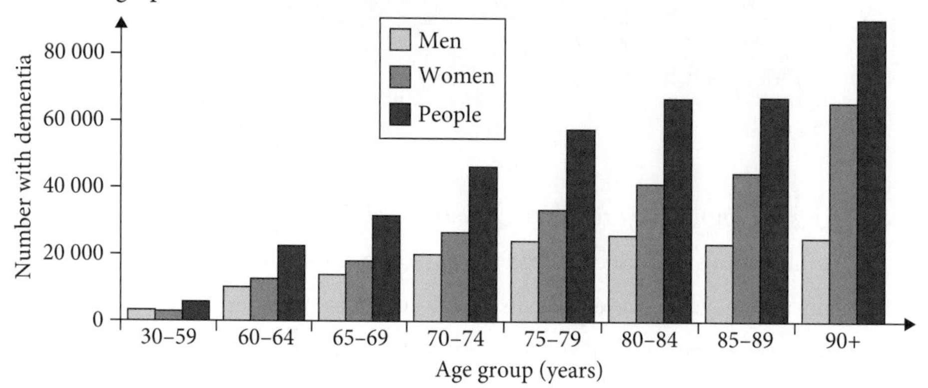

Based on Australian Institute of Health and Welfare material.

A Men do not develop dementia.

B More men than women develop dementia.

C An increase in age is the cause of dementia.

D More women than men have dementia, and the risk increases with age.

Question 14

A population-based study recruited 12 000 people and followed them through time with regular interviews in an attempt to identify risk factors that caused a disease. This study is

A retrospective. It studies risk factors that occur after the start of the study.

B prospective. It identifies risk factors that occur after the start of the study.

C prospective. It identifies risk factors that occur before the start of the study.

D retrospective. It identifies risk factors that occur before and after the start of the study.

Prevention

Question 15

In 2012, Australia became the first country in the world to require cigarette manufacturers to have cigarette packets with plain packaging. This law is an example of

A diagnosis because it identifies people who like to smoke.

B prevention because it aims to discourage people from smoking.

C treatment because it cures people with diseases caused by smoking.

D screening because the government could identify people who bought these cigarettes.

Technologies and disorders

Question 16

A blockage in the auditory canal of the ear that prevents sound waves reaching the inner ear is called

A cochlear hearing loss.

B auditory hearing loss.

C conductive hearing loss.

D sensorineural hearing loss.

Question 17 ©NESA 2019 SI Q6 (ADAPTED)

How does the cochlear implant assist people with severe hearing loss?

A It amplifies sound.

B It stimulates the pinna.

C It stimulates the eardrum.

D It directly stimulates the auditory nerve.

Question 18 ©NESA 2020 SI Q11

The diagram shows a model of the human eye. Which of the following correctly identifies a labelled part and its function?

	Label	Name	Function
A	I	Cornea	Refract light
B	I	Retina	Transmit light
C	II	Retina	Focus light
D	II	Cornea	Absorb light

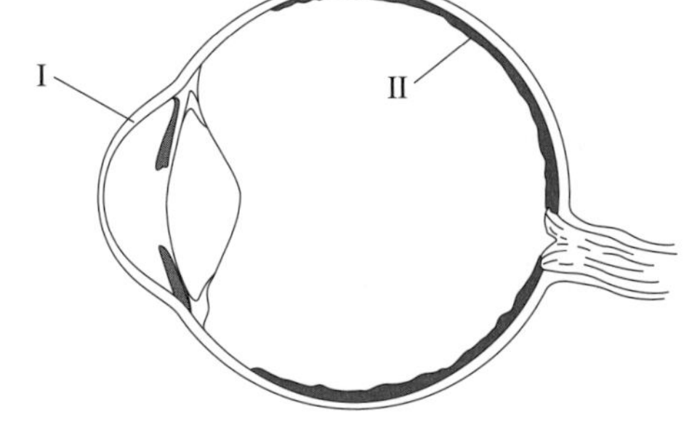

Question 19

Hyperopia (far-sightedness) is a visual disorder that makes close-up objects blurry. It occurs because

A the retina is damaged.

B light is refracted to a focal point behind the retina.

C light is refracted to a focal point in front of the retina.

D photoreceptors fail to send nerve impulses to the brain.

Question 20

The function of a nephron in a kidney is to

A release urine directly into the urethra.

B filter urine before it enters the bladder.

C concentrate nutrients and return them to the blood.

D produce urine in the process of removing waste from the blood.

Short-answer questions

Solutions start on page 271.

Question 21 (4 marks)

a Apart from glucose and temperature, name **one** example of a negative feedback loop that shows homeostasis. 1 mark

b Explain why this negative feedback loop is essential to the survival of an organism. 3 marks

Question 22 (3 marks) ©NESA 2019 SII Q21

The diagram shows a flow chart of the reaction of a human body to an increase in temperature. Fill in the three blank steps on the flow chart.

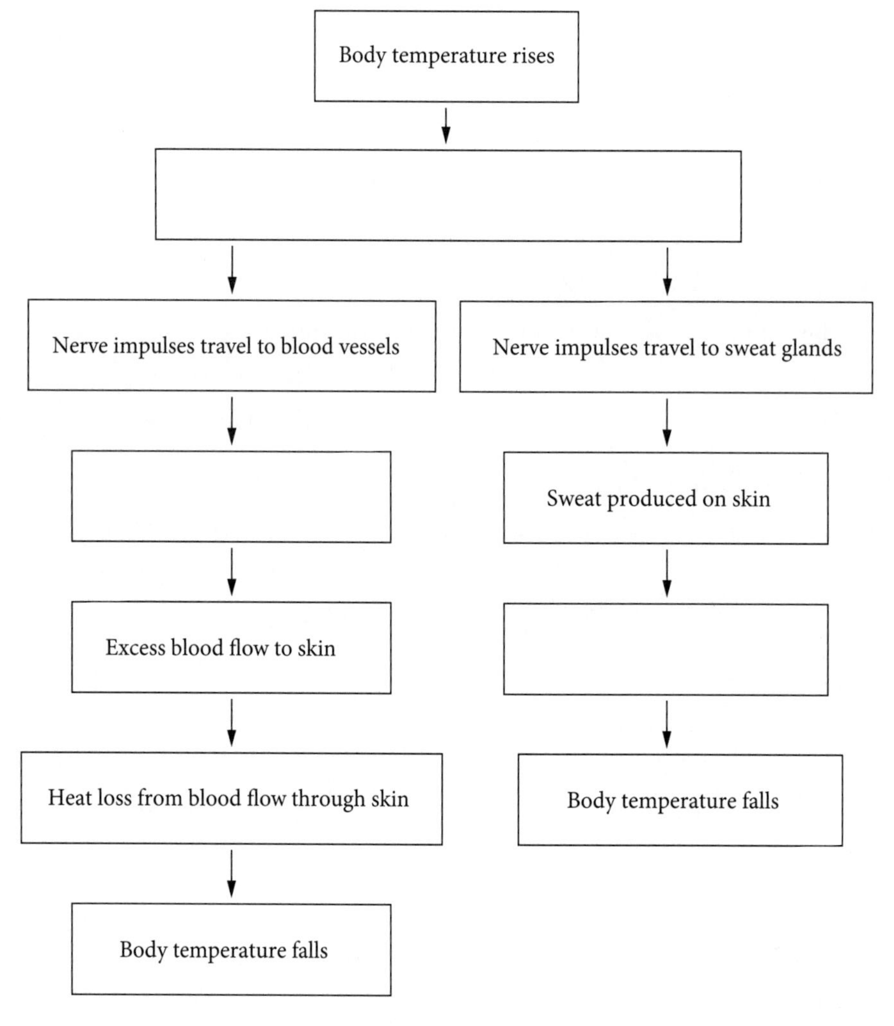

Question 23 (3 marks) ©NESA 2019 SII Q29

Describe **one** mechanism by which plants maintain internal water homeostasis.

Question 24 (6 marks)

Copy and complete the table by providing an example, cause and effect of three types of non-infectious disease.

Non-infectious diseases			
Type	**Example**	**Cause**	**Effect**

Question 25 (3 marks)

Choose **one** disease caused by an environmental exposure.

a What is the name of the disease? 1 mark

b What causes the disease? 2 marks

Question 26 (3 marks)

The A1C test can be used alone or in combination with a blood glucose test to diagnose and monitor people with diabetes. The A1C test calculates the average level of blood glucose over three months by measuring the amount of glucose attached to haemoglobin in red blood cells. The table shows the values used to diagnose someone with diabetes, and the graph shows A1C results and blood glucose levels for a person over four days.

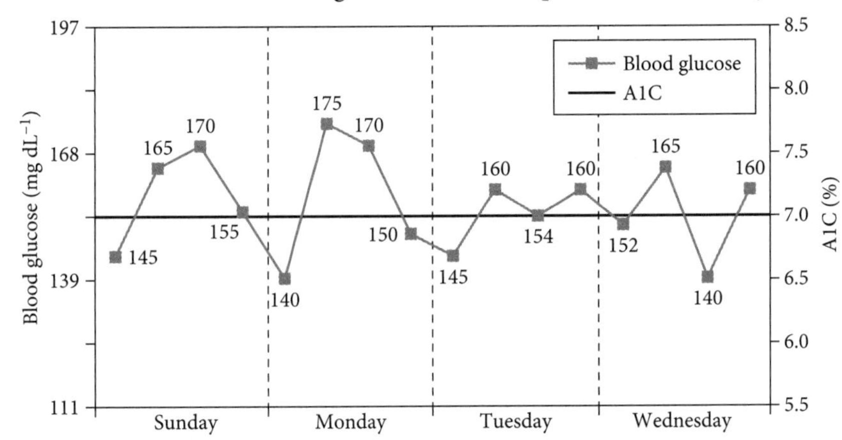

Diagnosis	A1C level (%)
Normal	Below 5.7
Pre-diabetes	5.7–6.4
Diabetes	6.5 or above

Source: National Institute of Diabetes and Digestive and Kidney Diseases (NIDDK)

a Would the person be diagnosed as normal, pre-diabetic or diabetic based on the A1C test? 1 mark

b Explain the cause of the pattern of blood glucose (mg dL^{-1}) over the four days. 2 marks

Question 27 (2 marks)

Explain the difference between incidence and prevalence in relation to non-infectious diseases.

Question 28 (5 marks)

a Describe the treatment and management strategies for **one** example of a non-infectious disease. 3 marks

b Describe **one** possible future direction of research that could be used to treat this disease. 2 marks

Question 29 (3 marks)

Draw a flow chart to illustrate the process of **one** type of epidemiological study.

Question 30 (2 marks)

Evaluate the value of participating in an epidemiological society by providing **one** benefit and **one** limitation for an individual and for society.

	Benefit	**Limitation**
Individual		
Society		

Question 31 (5 marks) ©NESA 2019 SII Q23

Explain how educational programs can be effective in reducing the incidence of non-infectious diseases. Support your answer with examples.

Question 32 (5 marks)

a What structures are indicated by A, B and C in the diagram of the ear? 1 mark

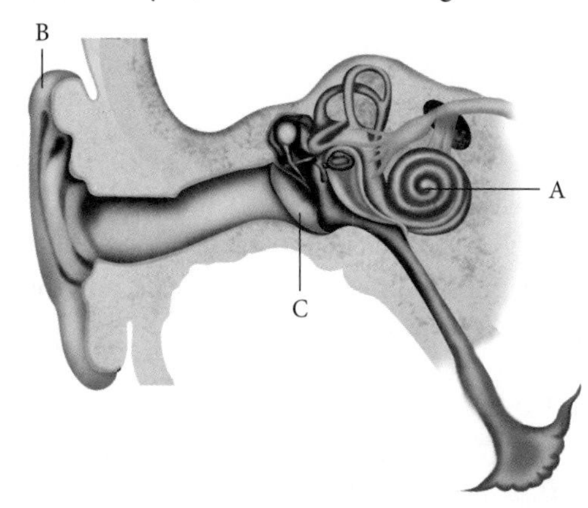

b Name a hearing device that is suited to a person with profound hearing loss caused by a damaged inner ear. 1 mark

c Explain how this hearing device transmits sound to be interpreted in the brain. 3 marks

Question 33 (5 marks) ©NESA 2020 SII Q24 (ADAPTED)

An indicator of kidney function is the volume of filtrate formed at the glomerulus in 1 minute (GFR).

GFR of healthy adult	>100 mL min^{-1}
GFR needing dialysis	<15 mL min^{-1}

A patient's kidney function was monitored, and the following data recorded.

Year	2011	2012	2013	2014	2015	2016	2017	2018	2019
GFR (mL min^{-1})	81	76	77	77	79	65	60	45	35

a Draw a graph to determine the year that the patient is predicted to require dialysis. 2 marks

b Explain how dialysis compensates for the loss of function of the kidneys. 3 marks

Question 34 (4 marks)

a In which organ of the body are nephrons located? 1 mark

b Describe a technology that can be used to treat a person with chronic failure of these organs. 3 marks

Question 35 (3 marks)

a Name a vision disorder. 1 mark

b What structure of the eye is affected in this disorder? 1 mark

c Describe a technology that can be used to assist with the effects of this disorder. 1 mark

CHAPTER 5
THE SCIENTIFIC METHOD

Chapter 5
The scientific method

The scientific method refers to the series of steps taken to carry out an investigation. A good understanding of the Working Scientifically skills outcomes and how to apply them as they relate to the scientific method is essential, as you will be assessed on them in the HSC exam, with the content providing the context for the questions.

Working Scientifically: skills outcomes

The Biology syllabus recognises seven key skills required for Working Scientifically:

- WS1 Questioning and predicting
- WS2 Planning investigations
- WS3 Conducting investigations
- WS4 Processing data and information
- WS5 Analysing data and information
- WS6 Problem solving
- WS7 Communicating

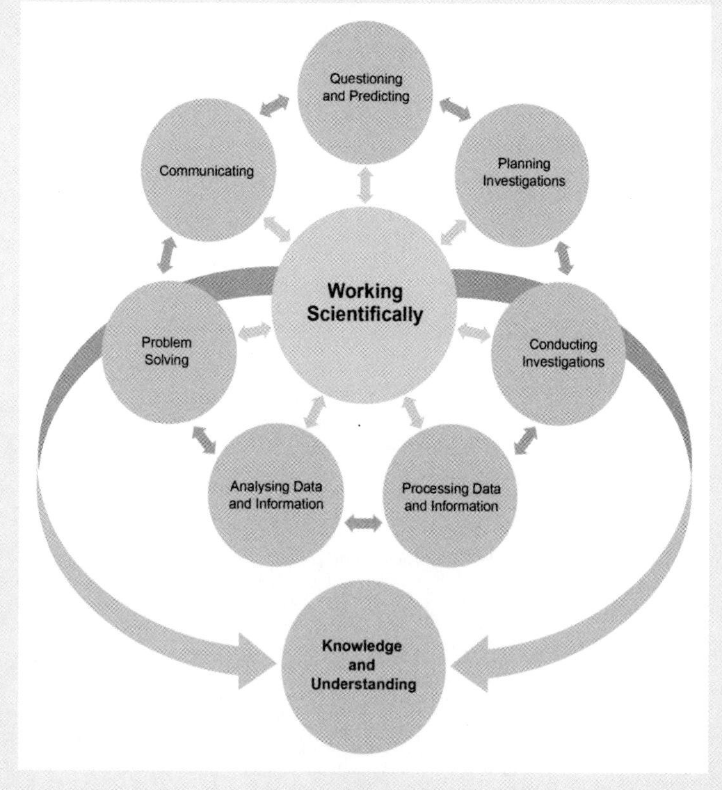

These outcomes are interdependent, essential to the scientific method and intrinsically linked to the accumulation of knowledge in each of the four modules studied.

Your school-based assessment will have had a 60% weighting towards these skills. Your mandatory depth study will have had a weighting of 20–40% and was required to assess the Working Scientifically skills outcomes:

- WS1 Questioning and predicting
- WS7 Communicating
- a minimum of two additional Working Scientifically skills outcomes
- at least one knowledge and understanding outcome.

The questions you can expect in the HSC exam typically take the form of:

- writing a valid procedure to test a given observation
- assessing the quality of a given procedure in terms of validity
- constructing and extracting information from appropriate tables and graphs.

Variables

In most experiments, a single **variable** is changed to determine its effect on another variable. All other influencing factors are kept constant. This is known as a **controlled experiment**. The variable you choose to investigate is known as the **independent variable**. The **dependent variable** is the measurable or observable value that changes in response to changes in the independent variable. All other variables, the **controlled variables**, are kept constant (or changed in constant ways) during an **investigation** to avoid them influencing the outcome of the experiment.

A good understanding of variables is essential because they form a common thread through all sections of the scientific method. This is explored in this chapter, in the context of an experiment investigating the effect of antibiotic concentration on the inhibition of bacterial growth, specifically *Escherichia coli* (*E. coli* K-12 strain). The independent variable is antibiotic concentration (μg mL^{-1}), and the dependent variable is inhibition of bacterial growth (mm).

It is important to note that, while in most cases there will be a single independent variable, it is possible to have more. For example, if comparing the effect of antibiotic concentration (μg mL^{-1}) on the inhibition of growth in different species of bacteria, both antibiotic concentration (μg mL^{-1}) and bacterial species are independent variables.

You must also be confident in distinguishing between the controlled variables and an **experimental control**. In many investigations, two experiments are run in parallel: the control group and the experimental group. They are identical in every respect, with one exception. The control group is not exposed to the independent variable. The control group serves as a baseline against which changes in the independent variable can be compared to determine their true influence on the dependent variable.

The experiment shown in Figure 5.1 is an example. As mentioned, controlled variables are kept constant, so they do not influence the outcome of the experiment. In contrast, the experimental control does not include the independent variable, in this case antibiotics. You can be confident that any results observed are due to the antibiotics and no other extraneous factors.

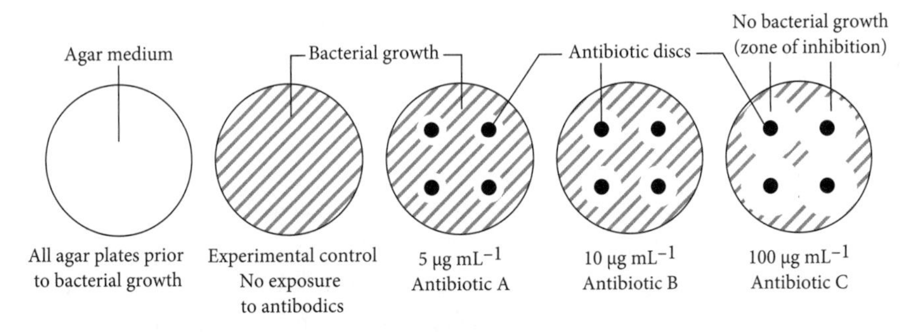

FIGURE 5.1 An example of an experiment

Questioning and predicting

Investigations start with an **observation** where information is acquired using one of your senses or using scientific equipment to collect and record data. For example, you may have observed that scientists are concerned about the increasing incidence of antibiotic-resistant strains of bacteria.

From this observation you would next pose a specific *question* that can be investigated. This requires an explicit link between the independent and dependent variables. For example, what concentration of antibiotics is most effective at preventing the growth of *Escherichia coli* (*E. coli* K-12 strain)?

The next step is to develop a **hypothesis**. All stage 6 science syllabuses define a hypothesis as 'a tentative explanation for an observed phenomenon expressed as a precise and unambiguous statement that can be supported or refuted by investigation'. You should therefore be thinking, 'If ..., then ... because ...', where you explicitly use scientific reasoning to predict how the changing independent variable will affect the dependent variable. For example, *if* antibiotic concentration increases, *then* bacterial growth will be impaired more significantly *because* the higher doses will kill more bacteria, preventing their reproduction and spread.

Planning investigations

In the HSC this outcome is often assessed by requiring students to write a valid **procedure (method)** based on the stimulus provided, or to assess the quality of a given procedure in terms of validity and reliability. When starting the planning process, it is therefore important that you are aware of the following terms, are able to distinguish between them, and address each of them independently in your experimental procedure.

TABLE 5.1 Validity, reliability, accuracy and precision

Validity	**NESA says** 'An extent to which tests measure what was intended, an extent to which data, inferences and actions produced from tests and other processes are accurate.' **What this means for you** In short, your procedure section should test, and only test, the aim of the experiment. To do this, the procedure should have all controlled variables in place to ensure that any changes in the dependent variable are the result of changing the independent variable.
Reliability	**NESA says** 'An extent to which repeated observations and or measurements taken under identical circumstances will yield similar results.' **What this means for you** Where possible, an experiment should include repetition to demonstrate consistency of results. Commit to the number of repeats appropriate to the experiment – a minimum of three. Alternatively, a large sample size with one trial can allow for a sufficient sample size to be obtained, outliers can be removed, and an average obtained. To be deemed reliable, the results obtained should also be **reproducible** by an independent researcher. If your data is **repeatable** and reproducible, it is deemed to be reliable.
Accuracy	Accuracy refers to how close a measurement is to the true or accepted value. The best way to ensure **accuracy** in an experiment is to use the best available equipment correctly, to avoid both **systematic** and **random errors**. Reducing errors increases accuracy.
Precision	Precision is the ability to obtain the same measurement when the experiment is repeated or reproduced. It is important to understand that it is possible to have high **precision** and low accuracy at the same time, and vice versa. For example, a poorly calibrated machine can produce the same result time after time, but due to its incorrect calibration, the result is not accurate.

During the planning stage, you should consider and assess possible risks specific to your experiment and put in place the necessary safety measures. For example, even though the *E. coli* K-12 strain of bacteria is deemed safe for use in schools, care must be taken to avoid ingestion or entry into any wounds. By wearing the necessary personal protection equipment (PPE) this risk can be minimised.

As part of the planning stage, you will also select the necessary equipment and materials (and their quantities) and technologies required for you to perform your investigation. The independent and dependent variables, and experiment control where necessary, must be chosen and a valid procedure written, including all appropriate controlled variables, to allow the reliable collection of data.

It is recommended that the procedure be explicit in terms of:

- the quantities of equipment and materials used in the experiment
- how the independent variable is changing (e.g. add antibiotic discs at 0, 5, 10 or 100 μg mL^{-1} to each plate)
- how the dependent variable is being measured (e.g. measure the zone of inhibition using a ruler with millimetre graduations)
- what controlled variables are in place (e.g. 30 mL agar added to each plate and exposed to the antibiotics for 48 h).

Conducting investigations

The nature of this outcome makes it less likely to be assessed in a formal written exam, as the focus is on you completing a first-hand investigation. There are, however, aspects of the outcome that may make an appearance.

For example, as part of conducting an investigation you are required to select and extract information from a wide range of reliable **secondary sources** and acknowledge them using an accepted referencing style.

To achieve this, you must know how to apply the terms **reliability** and **validity** as they relate to secondary sources. You cannot assume that an individual secondary source is reliable without having other sources to compare it with. For your sources to be deemed reliable, there must be consistency of information between them.

The validity of a secondary source is determined based on the author's credentials, whether the article has been peer reviewed, the reputability of the publisher or journal, and the purpose of the article. You must be confident that the author has no inherent bias in writing the article.

For example, the following article is a valid source, because it was written by experts in a specific field and they have published their results in a peer-reviewed journal.

Chai, A., Lam, H., Kockx, M., Gelissen, I. (2021). 'Apolipoprotein E isoform-dependent effects on the processing of Alzheimer's amyloid-β', *Biochimica et Biophysica Acta – Molecular and Cell Biology of Lipids*, 1866 (9).

Processing data and information

The processing and analysing data and information outcomes are two of the most widely assessed Working Scientifically skills outcomes. You will be required to determine whether the data and information collected are **qualitative data** or **quantitative data** (continuous or discrete), which will in turn inform your decision on how to best represent them.

TABLE 5.2 Types of data: quantitative and qualitative

	Quantitative data		Qualitative data
	Discrete	Continuous	
Definition	The result of *counting*; can only take certain numerical values	The result of *measuring*; can take any numerical value	Descriptive; obtained from observations but not measured
Data type	Numerical	Numerical	Words, letters
Example	Number of malaria cases per year in different countries	Measuring changing levels of carbon dioxide in the atmosphere over time	Type of blood group

Tables and graphs are visual representations used to organise data and information collected during an investigation. Tables are used to provide easy access to specific data points, while graphs are used to illustrate trends and make comparisons and predictions to inform future investigations.

When you are setting up a table, in most cases the independent variable is placed in the left column, the dependent variable with the different trials is in the next column, and the calculated average is on the far right. The units should always be placed in the title row of the column and not repeated with each data point entry.

Independent variable with units — Dependent variable with units — Graph averages

Antibiotic concentration ($\mu g\ mL^{-1}$)	Zone of inhibition (mm)				
	Trial 1	Trial 2	Trial 4	Trial 4	Average
0	0	0	0	0	0
5	1.5	1.5	2.0	1.0	1.5
10	4.5	4.5	4.0	5.0	4.5
100	37.0	41.5	39.0	38.5	39

FIGURE 5.2 A results table with independent variable and dependent variable

A graph should also be constructed in a specific manner to demonstrate the effect the independent variable has on the dependent variable.

- The title should be clear and descriptive, highlighting the independent and dependent variables.
- The independent variable is placed on the *x*-axis and the dependent variable on the *y*-axis.
- Both axes should include labels and units of measurement.
- The axis scales should be incremental and proportional to the data.
- Only plot the data provided. For example, do not assume that all line graphs start at the origin.
- When plotting the data, use the average calculation unless told otherwise.

But what type of graph should you choose?

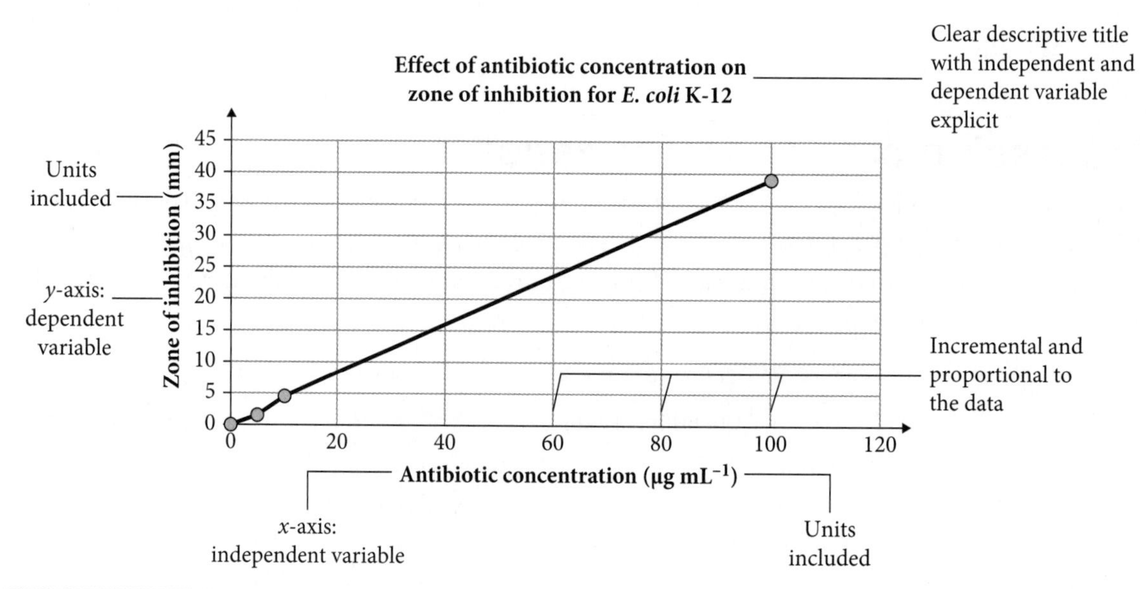

FIGURE 5.3 A graph of experimental results

Table 5.3 lists the main types of graphs used in presenting scientific data, with examples of when they should be used.

TABLE 5.3 Types of graphs and when to use them

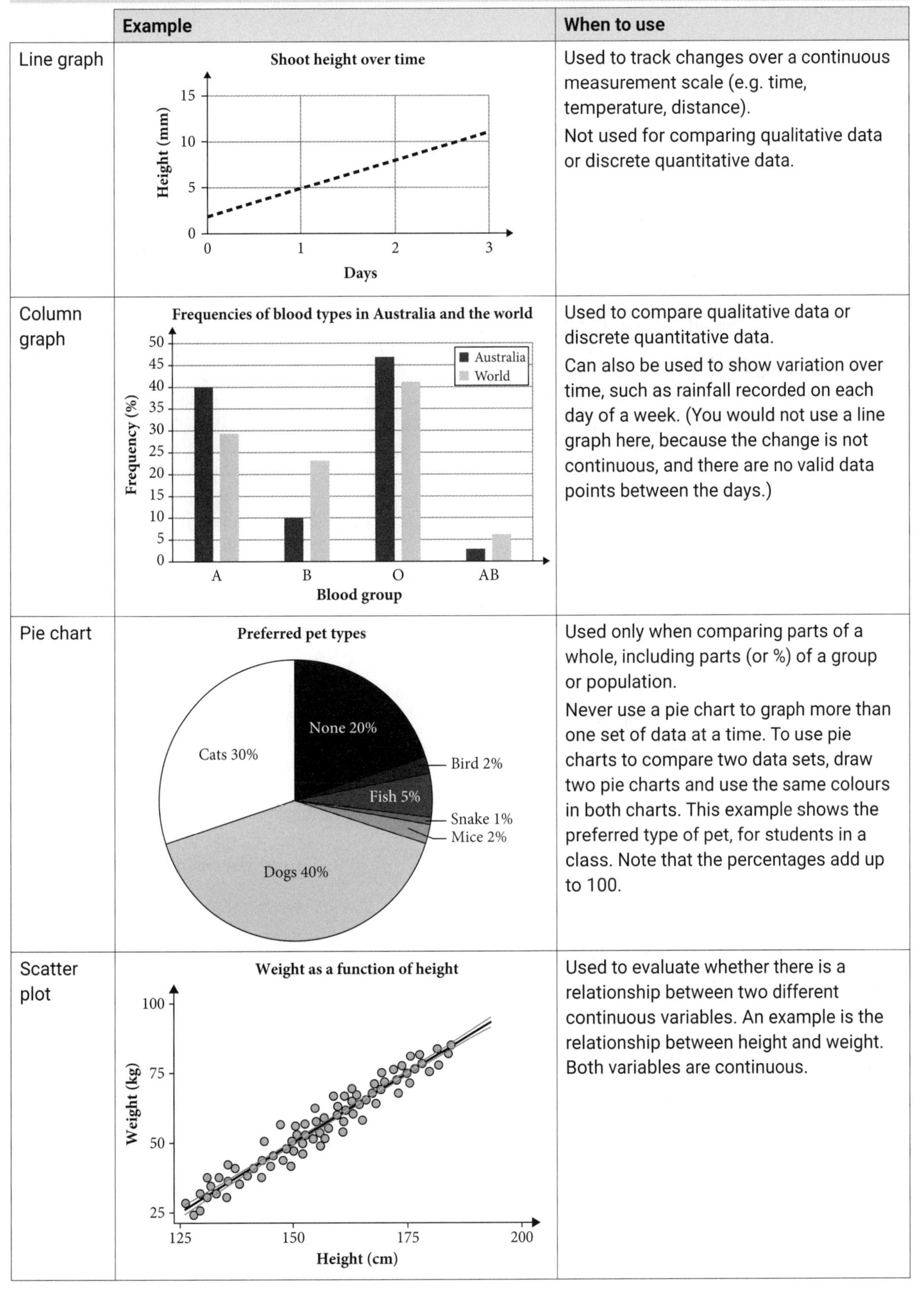

	Example	When to use
Line graph	Shoot height over time	Used to track changes over a continuous measurement scale (e.g. time, temperature, distance). Not used for comparing qualitative data or discrete quantitative data.
Column graph	Frequencies of blood types in Australia and the world	Used to compare qualitative data or discrete quantitative data. Can also be used to show variation over time, such as rainfall recorded on each day of a week. (You would not use a line graph here, because the change is not continuous, and there are no valid data points between the days.)
Pie chart	Preferred pet types	Used only when comparing parts of a whole, including parts (or %) of a group or population. Never use a pie chart to graph more than one set of data at a time. To use pie charts to compare two data sets, draw two pie charts and use the same colours in both charts. This example shows the preferred type of pet, for students in a class. Note that the percentages add up to 100.
Scatter plot	Weight as a function of height	Used to evaluate whether there is a relationship between two different continuous variables. An example is the relationship between height and weight. Both variables are continuous.

Analysing data and information

Students are often required to analyse data sets to derive trends, patterns and relationships between the independent and dependent variables. An important component of the analysis is to explore any errors or limitations that may have occurred in the experiment, so a judgement can be made about the validity, reliability and accuracy of the data collected. This will in turn inform the design of future studies.

Sources of error

Even in the best designed and conducted experiments, errors cannot be eliminated entirely, and this must be considered as part of your analysis. There are two main types of error that you should consider: systematic and random. Both can influence the quality of the data collected, and understanding how to reduce these errors will improve the accuracy and precision of your results.

TABLE 5.4 Types of experimental error and their effects

Type of error	Description	Example
Systematic	• Affect the accuracy of results. • Occur because of faulty or consistently incorrect use of equipment. • One-sided – consistently produce a result that differs by the same amount from the true value. • Accuracy of measurements cannot be improved by repeating the experiment. • Relatively hard to detect via statistical analysis. However, once detected and corrected, accuracy will improve.	An incubator with an incorrectly calibrated thermostat will introduce systematic error where the temperature recorded is not a true representation of the actual temperature. Until rectified, the systematic error will remain.
Random	• Affect the precision of a measurement. • Occur because of unpredictable variations in the measurement process. • Two-sided – lead to results that fluctuate above and below the true value. • Accuracy of measurements can be improved by repeating the experiment. • Relatively easy to detect via statistical analysis. Once detected and corrected, precision will improve.	Problems associated with estimating the values between graduations on equipment. This could include using a 1 m ruler to measure the zone of inhibition in the agar plates, or using a 1 L beaker to measure out 30 mL of agar solution, instead of using a 50 mL graduated cylinder.

Analysis of secondary data

Students often struggle with analysing data and information presented in tables and graphs. Most often, the answer provided lacks detail, identifying a trend but failing to provide an in-depth analysis. An 'analysis' question requires you to identify different components of the data and information, and to expand upon any relationships between them and the possible implications of the relationship.

For example, the graph in Figure 5.4 shows global temperature fluctuations, and atmospheric levels of carbon dioxide (CO_2) over time. If asked to analyse this graph, many students will stop short at identifying the trend that both temperature fluctuations and atmospheric levels of CO_2 increase over time.

To obtain higher marks, it is essential to provide details about this trend. The graph in Figure 5.4 shows significant fluctuations in temperature anomalies; for example, in 1976 there was a −0.15°C anomaly, lower than the previous 10 years (*1 on graph), and in the 2-year period between 1998 and 2000 there was a significant drop in temperature fluctuation (*2 on graph). These drops are despite the continued increase in atmospheric CO_2 levels. Overall, however, there is a clear trend that both temperature fluctuations and atmospheric levels of CO_2 are increasing over time. At a minimum there is a correlation between increasing levels of atmospheric CO_2 and increasing global temperature, which would need to be investigated further to determine causation.

9780170465267

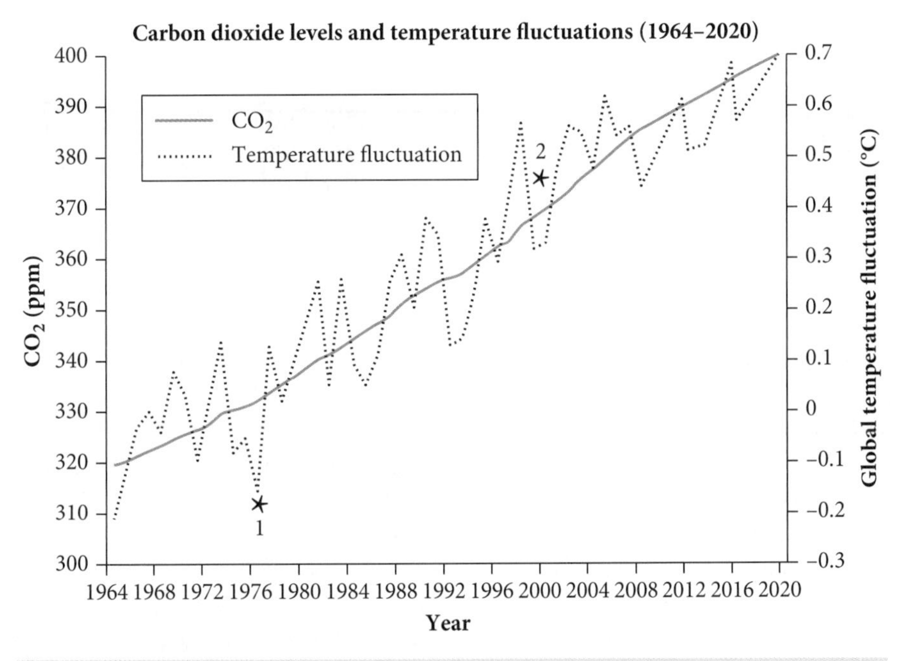

FIGURE 5.4 A graph of temperature fluctuations and CO_2 levels over time

Problem solving

NESA defines a model as any representation that describes, simplifies, clarifies or provides an explanation of the workings, structure or relationships within an object, system or idea. This includes mathematical models (tables, graphs and formulae), physical models and computer simulation models. You should expect to be asked questions where you must engage with a stimulus that requires you to use critical thinking to make predictions and solve problems associated with the model.

For example, in reference to the above graph, you could extrapolate the data to make a prediction about temperature anomalies and atmospheric levels of CO_2 in the year 2040. You could be given the Hardy–Weinberg equations (see Module 6) and an appropriate data set that you must manipulate to derive information. Based on the results obtained, you must use your knowledge of genetics to evaluate the model's strengths and limitations.

Communicating

Scientists communicate their research using three main modes of delivery: in written form via peer-reviewed journals, as oral communications (e.g. at a research conference) and via posters. All modes follow the scientific method and present key aspects of the research. Whatever mode of delivery you use, you must communicate your knowledge succinctly, in a logical manner and always using the correct scientific language. Another important consideration is that you ensure the mode and nature of the communication are appropriate to the audience.

Strong communication skills are essential, as they are a necessary requirement for addressing all other course outcomes. For example, questioning and predicting, planning an investigation, processing and analysing data and information all require the correct use of scientific terminology and nomenclature and an ability to write concisely. Similarly, for questions relating to the knowledge and understanding outcomes, better-performing students typically have well-structured responses, include the correct use of scientific language, and use diagrams where necessary to support their response. Exam questions, particularly longer responses, often have marks allocated for concise writing and correct use of scientific terminology.

Glossary

accuracy How close a result is to the true value

controlled experiment An experiment where usually one variable, the independent variable, is manipulated

controlled variable The variable that is kept constant during an investigation to avoid it influencing the outcome of the experiment

dependent variable The variable that is measured and whose value depends on the independent variable, i.e. it responds to the independent variable

experimental control An experimental group that does not have the independent variable; serves as a baseline against which changes in the independent variable can be compared to determine the true influence on the dependent variable

hypothesis A tentative explanation for an observed phenomenon expressed as a precise and unambiguous statement that can be supported or refuted by investigation

independent variable The variable that is changed or manipulated by the scientist to determine its effect on the dependent variable

investigation The scientific process of answering a question, exploring an idea or solving a problem, which requires activities such as planning a course of action, collecting data, interpreting data, reaching a conclusion and communicating these activities; can include primary and/or secondary-sourced data or information

observation The acquisition of knowledge or data using one of your senses or through the use of technology

precision How repeatable a given result is when the same procedure is followed

procedure (method) The steps taken to carry out a scientific investigation

qualitative data Descriptive data obtained from observations but not measured

https://get.ga/aplus-hsc-bio-u34

A+ DIGITAL FLASHCARDS

Revise this topic's key terms and concepts by scanning the QR code or typing the URL into your browser.

quantitative data Numerical data obtained by measuring (continuous data) or counting (discrete data)

random error An unpredictable variation in measurement that affects precision and can be improved by taking multiple measurements and calculating an average

reliability The extent to which repeated observations and/or measurements taken under identical circumstances will yield similar results

repeatable Describes an investigation that can be conducted again by the same investigator under the same conditions to generate similar results

reproducible Describes an investigation that can be conducted by a different investigator under different conditions to generate similar results

secondary source Data and/or information sourced from other people, including written information, reports, graphs, tables, diagrams and images

systematic error A predictable variation in data that affects accuracy and can only be improved by correcting the error (e.g. correctly calibrating equipment)

validity The extent to which tests measure what was intended; the extent to which data, inferences and actions produced from tests and other processes are accurate; addressed by ensuring all variables other than the independent variable are controlled

variable Something that can change or be changed, as distinct from a constant, which does not change

Exam practice

Multiple-choice questions

Solutions start on page 274.

Question 1

Validity is best defined as

A a measurement of how close a result is to the true value.

B the extent to which the experimental procedure tests the actual aim of the experiment.

C the extent to which repeated measurements taken under identical circumstances will yield similar results.

D the extent to which an investigation can be conducted again by the same investigator under the same conditions and generate similar results.

Question 2 ©NESA 2012 SI Q16

A student carried out an investigation to identify the presence of microbes in water from different sources. The student's lab notes are shown.

	Control	*Bottled water*	*Tap water*	*Tank water*
Inoculation of an agar plate with water sample	✗	✓	✓	✓
Incubation at 37°C	✓	✓	✓	✓
Appearance of agar plate				

What can be inferred from these results?

A The inoculation loop was not sterilised properly.

B The water from each of these sources is unsafe to drink.

C These water sources are contaminated with the same microbe.

D The agar plates were contaminated prior to the beginning of the experiment.

Question 3 ©NESA 2019 SI Q8

A group of four students set out to determine the animal species diversity over an area of one hectare in each of five different habitats. Each student graphed their data as shown. Which of the graphs produced is the most suitable to represent animal species diversity in the different habitats?

A

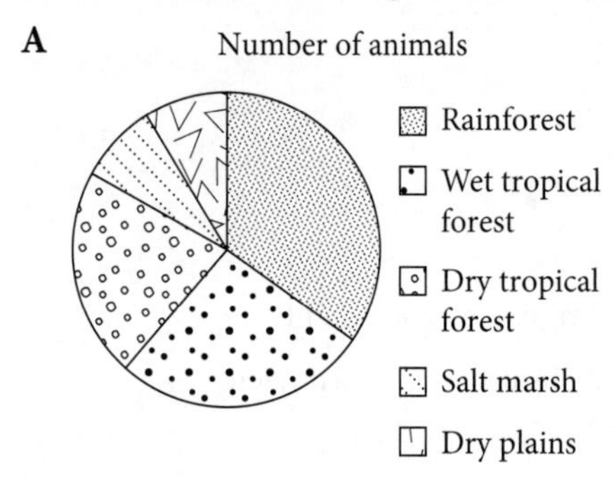

B

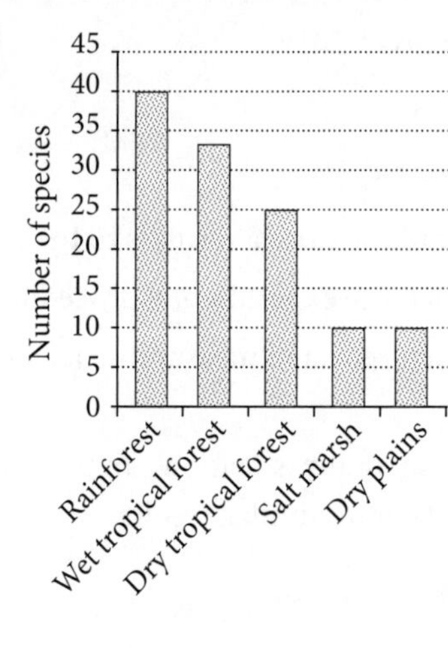

C

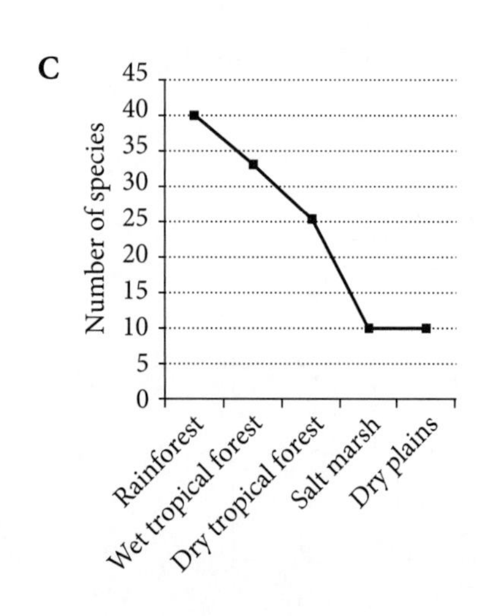

D

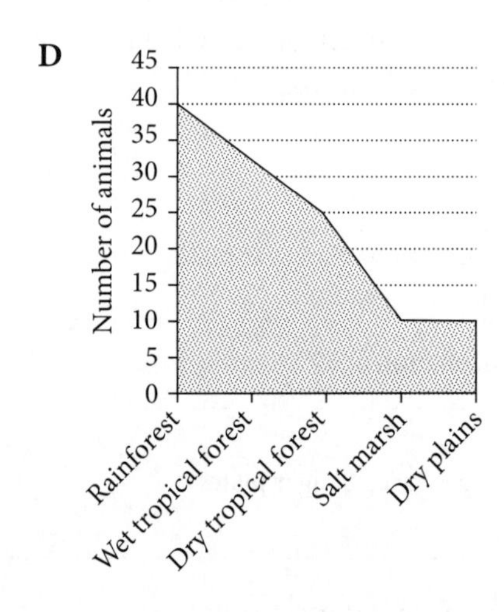

Use the following information to answer Questions 4 and 5.

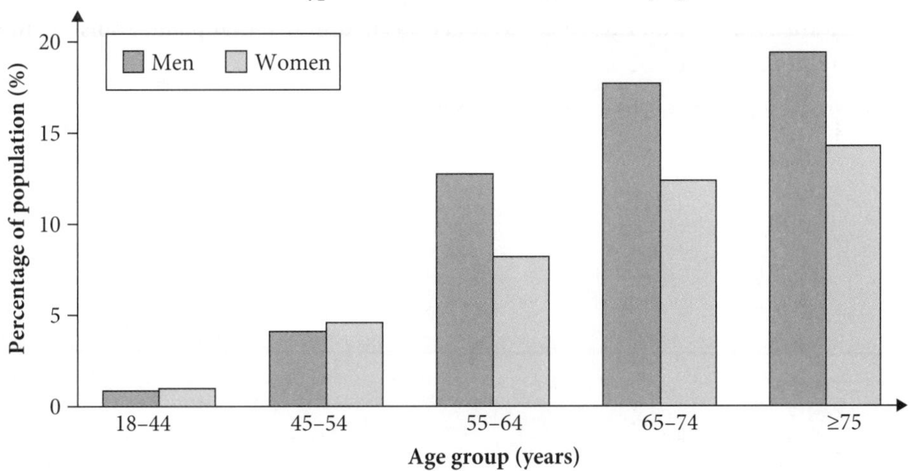

Question 4

The following graph shows a comparison of type 2 diabetes in men and women by age group.

Which of the following correctly identifies the dependent and independent variables?

	Dependent variable	Independent variable
A	age group	percentage of population
B	percentage of population	sex
C	percentage of population	age group
D	percentage of population	age group and sex

Question 5

What trend is shown from the data in the graph?

A In the 45–54 age group, more women than men have type 2 diabetes.

B As age increases, the risk of type 2 diabetes decreases for both men and women.

C As age increases, the risk of type 2 diabetes increases for men and women but to a lesser degree in men.

D As age increases, the risk of type 2 diabetes increases for men and women but to a lesser degree in women.

Question 6 ©NESA 2009 SI Q11 ●●●

Experiments were carried out on plants living in different environments to measure the size of the leaf stomata at different times of the day. Previous investigations had shown that plants transpire more water when the size of the stomata is larger.

Which graph best represents a plant living in a dry environment?

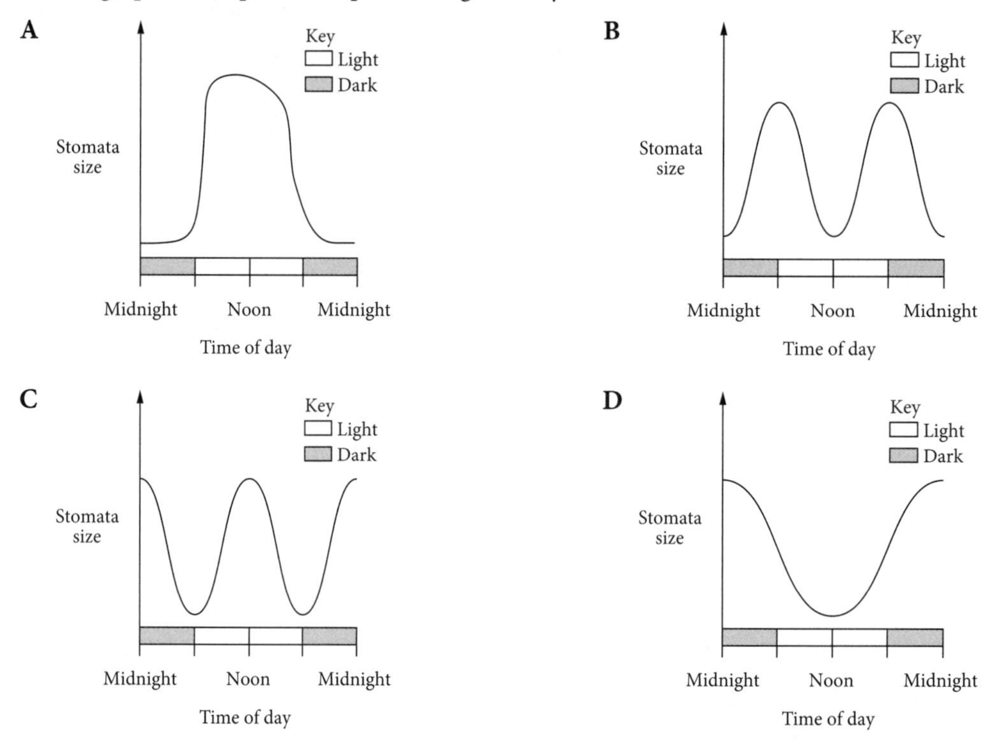

Use the following information to answer Questions 7–9.

Giardia are protozoan parasites that can cause stomach upset and diarrhoea if ingested in large enough quantities. A level of 0–10 cysts L^{-1} in drinking water is safe for consumption.

Chlorine treatment of water at 5 mg L^{-1} is known to kill the parasite.

A scientist wanted to determine whether a half dose of chlorine (2.5 mg L^{-1}) was sufficient to kill the parasite and make the water safe for consumption. Her results are shown in the table below.

	Number of *Giardia* cysts L^{-1} after 48 h		
	Untreated sample	**5 mg L^{-1} chlorine**	**2.5 mg L^{-1} chlorine**
Test 1	116	5	18
Test 2	198	8	9
Test 3	204	0	10
Test 4	3	2	12
Test 5	118	1	21
Average	**159**	**3**	**14**

Question 7 ●●

Test 4 for the untreated water sample had a count of three *Giardia* cysts L^{-1}, yet the scientist determined the overall average to be 159. What is the likely reasoning for this?

A The scientist calculated the average incorrectly.

B The scientist acted unethically and removed the data point.

C The scientist determined that the sample was not prepared correctly, and removed the data point as an outlier.

D The scientist determined that the samples were mixed up and that it actually came from the 5 mg L^{-1} chlorine data set.

Question 8

The scientist wanted to compare the effect of different chlorine levels on *Giardia*. Why did she include the untreated sample?

A The untreated sample is a control.

B The results are more accurate if compared against the untreated sample.

C To ensure that any observed results were due to the different chlorine levels.

D It is always recommended that a third dose be tested to make the results more reliable.

Question 9

What is the most appropriate conclusion for the experiment?

A Water from test 2 for the $2.5\,\text{mg}\,\text{L}^{-1}$ chlorine is safe to drink.

B Water treated with both $2.5\,\text{mg}\,\text{L}^{-1}$ and $5\,\text{mg}\,\text{L}^{-1}$ is safe for consumption.

C Water treated with $5\,\text{mg}\,\text{L}^{-1}$ chlorine has no risk of *Giardia* infection associated with it.

D Treating the water with $2.5\,\text{mg}\,\text{L}^{-1}$ chlorine is not sufficient to make the water safe for consumption.

Question 10 ©NESA 2021 SI Q13

The photographs show an open and a closed stomate on a leaf surface. When open, stomates allow water vapour to pass out of the leaf.

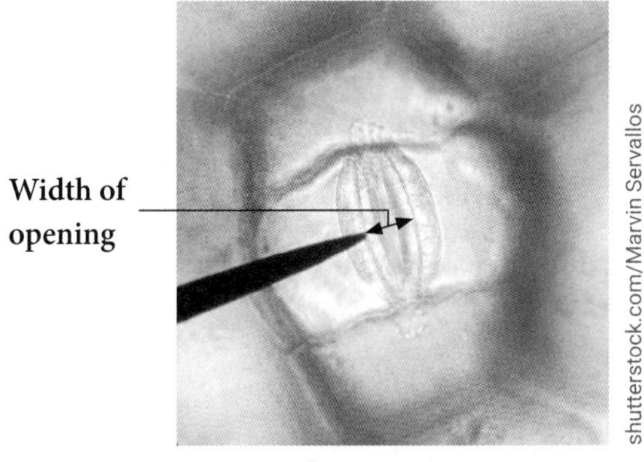

Open stomate

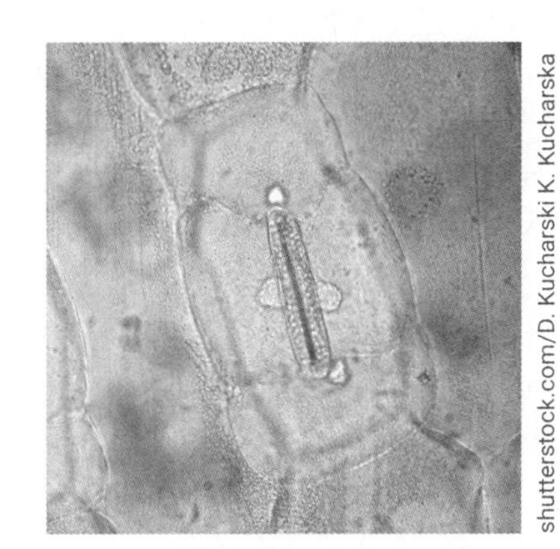

Closed stomate

Regulating stomates is a mechanism by which plants maintain water balance. Which of the following graphs best illustrates this homeostatic mechanism?

A

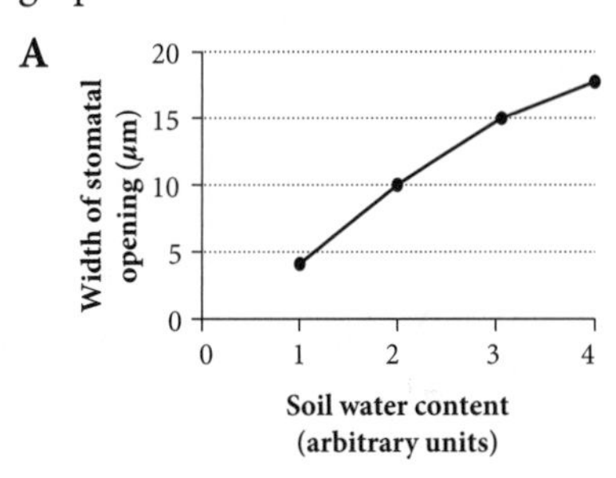

B

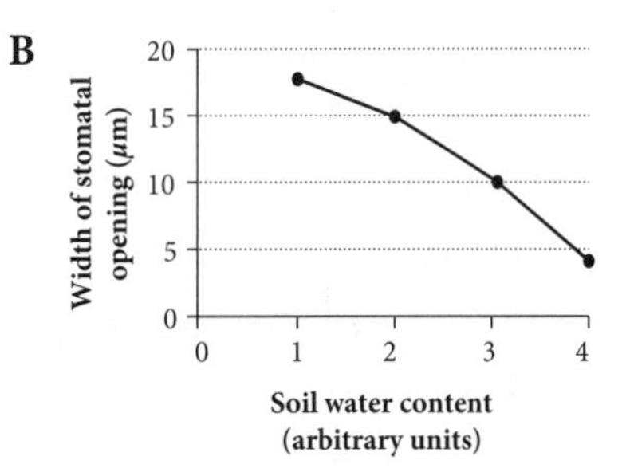

C

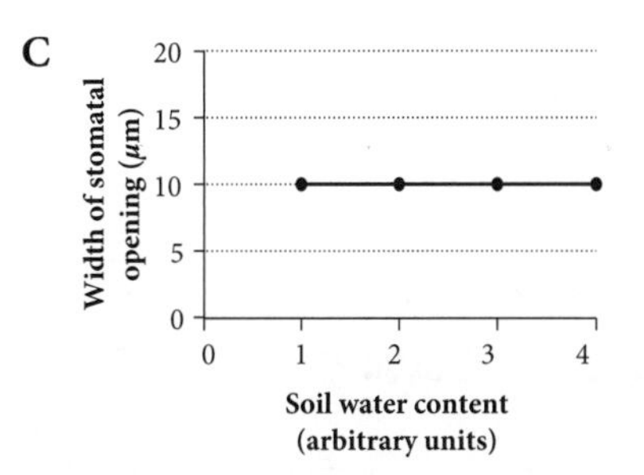

D

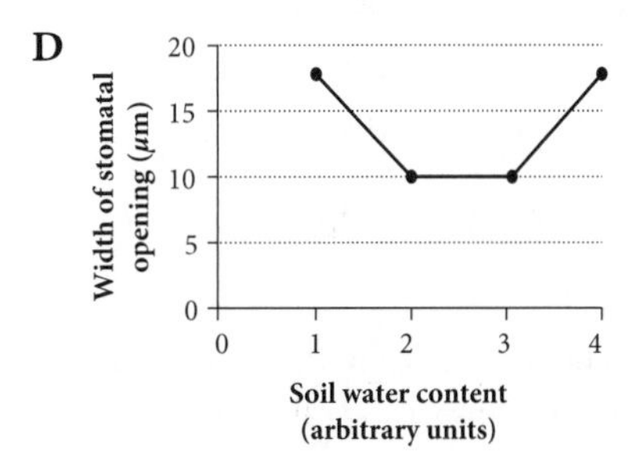

9780170465267

Question 11

Cystic fibrosis is an autosomal recessive disease caused by the F508del mutation in the CFTR gene. The mutation results in a 3 base pairs (bp) deletion and the removal of a single amino acid from the CFTR protein.

The gel electrophoresis at right shows the PCR products of a couple wanting to have a child. What is the likelihood that the child will have cystic fibrosis?

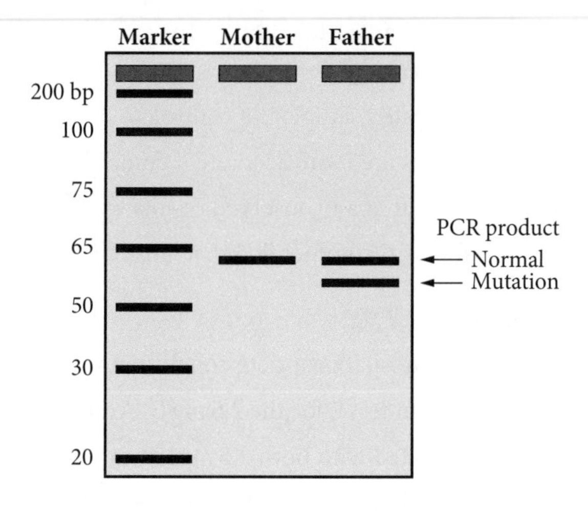

A 0%

B 25%

C 50%

D 0% chance if the child is female and 50% chance if the child is male

Question 12

A student was researching the use of genetically modified organisms. Which of the following would be considered the most valid source of information?

A Are GMOs good or bad for us? https://www.youtube.com/watch?v=6TmcXYp8xu4. 16 961 461 views

B Slide share. Ethical and safety concerns related to GM crops. https://www.slideshare.net/PATELMOHOMMEDFAIZAN/ethical-and-biosafety-issues-related-to-gm-crops. 2312 views

C Stop bashing GMO foods say scientists. *The Sydney Morning Herald* Newspaper. https://www.smh.com.au/technology/stop-bashing-gmo-foods-more-than-100-nobel-laureates-say-20160701-gpw5y6.html

D Fernbach, P.M., Light, N., Scott, S.E. *et al.* (2019). Extreme opponents of genetically modified foods know the least but think they know the most. *Nature Human Behaviour* **3**: 251–256. https://doi.org/10.1038/s41562-018-0520-3. Cited in 172 peer-reviewed articles

Short answer questions

Solutions start on page 276.

Question 13 (10 marks)

A student designed the following procedure to test the effect of three different antibiotics (A, B and C) on the bacterium *Escherichia coli.*

1 Add 25 mL of sterile agar solution to each of five sterile agar plates. Label the plates 1–5.

2 Tape off plate 1 and label 'no exposure'.

3 To plates 2–5 add 5 mL of the *E. coli* culture provided and ensure the culture is evenly spread over the agar plate.

4 Tape off plate 2 and label it 0 μg mL^{-1} antibiotics.

5 To the centre of:

- plate 3 add a 5 μg mL^{-1} antibiotic A disc
- plate 4 add a 10 μg mL^{-1} antibiotic B disc
- plate 5 add a 100 μg mL^{-1} antibiotic C disc.

6 Tape off plates 3–5.

7 Leave plates 1 and 2 at room temperature and place plates 3–5 in an incubator at 30°C. Leave all plates for 48 hours.

8 Use a ruler to measure the zone of inhibition in millimetres from the edge of each disc.

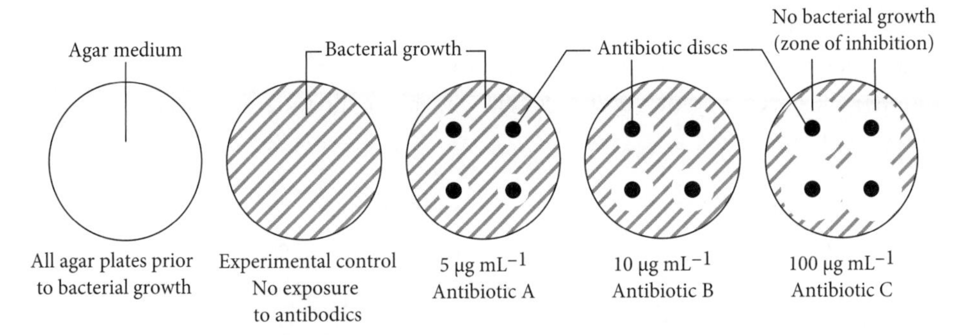

a Identify the independent and dependent variables in this experiment. 2 marks

b Identify a risk and corresponding safety measure associated with this experiment. 2 marks

c Assess the procedure in terms of both reliability and validity. 4 marks

d The student wanted to repeat the experiment to study the effect of three different antiviral drugs on *E. coli*. Explain the likely outcome of the experiment. 2 marks

Question 14 (6 marks) ©NESA 2020 SII Q27

Exposure to arsenic in drinking water has been associated with the onset of many diseases. The World Health Organisation recommends arsenic levels in drinking water should be below 10 $\mu g\,L^{-1}$. An epidemiological study involving 58 406 young adults was conducted over an 11-year period in one country to investigate young-adult mortality due to chronic exposure to arsenic in local drinking water. Each individual's average exposure and cumulative exposure to arsenic over the time of the study were calculated. Age, sex, education and socioeconomic status were taken into account during the analysis of the results. The graphs show survival rates for males and females over the 11-year period associated with different average levels of exposure to arsenic in drinking water.

KEY ---- < 90 $\mu g\,L^{-1}$ .... 90–223 $\mu g\,L^{-1}$ —— > 223 $\mu g\,L^{-1}$

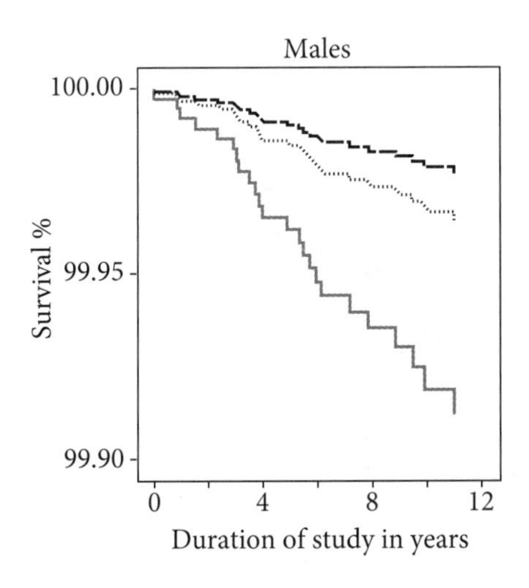

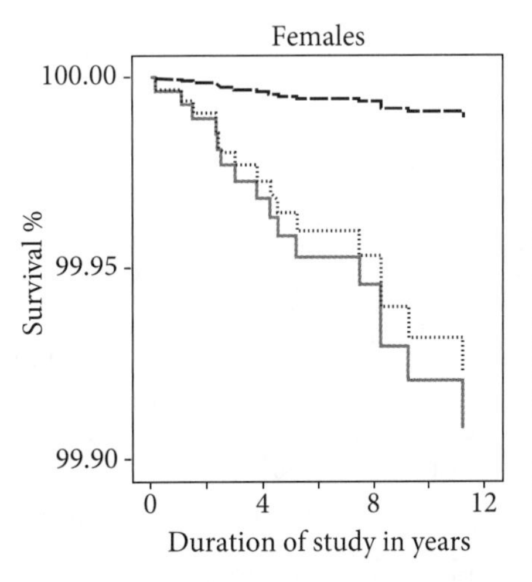

a Identify **two** features of the method used that contributed to the validity of this study. 2 marks

b The hypothesis put forward was that exposure to arsenic in drinking water increases mortality in young adults. Discuss the data presented in the graphs in relation to this hypothesis. 4 marks

Question 15 (14 marks)

A student designed an experiment to examine water loss by transpiration from a plant under four different conditions. They used a potometer to accurately measure the water loss over 30 minutes. The results are shown below.

	Volume of water loss from transpiration (mL)			
Time (min)	**25°C**	**40°C**	**25°C + wind (fan)**	**25°C + high humidity**
0	0	0	0	0
10	0.1	0.25	0.22	0.05
20	0.2	0.45	0.42	0.1
30	0.35	0.7	0.68	0.15

a Identify the independent and dependent variables in this experiment. 2 marks

b Identify two other variables that need to be kept constant. 2 marks

c How could the reliability of the experiment be improved? 2 marks

d Plot the results of the experiment. 4 marks

e Identify a trend shown in the graph. 1 mark

f Plants needs to ensure that their internal water levels remain relatively constant. Draw a negative feedback loop to demonstrate this process, with reference to one of the conditions present in the graph. 3 marks

Question 16 (6 marks)

The following graph shows the changing atmospheric carbon dioxide levels (CO_2) and temperature anomalies over the last 400 000 years. The readings are based on Antarctica ice core drillings and measurements taken at the Mauna Loa Observatory in Hawaii.

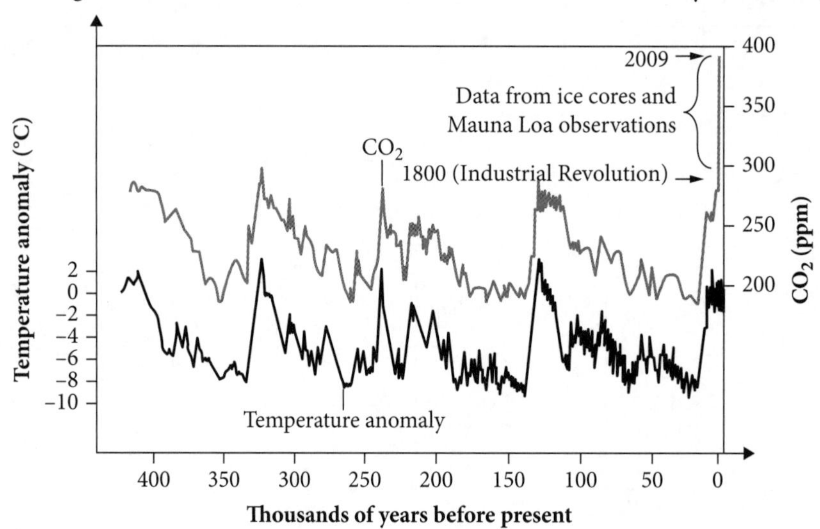

a Identify two trends shown in the graph. 2 marks

b Analyse the data shown in the graph. 4 marks

SOLUTIONS

CHAPTER 1, MODULE 5

Heredity

Multiple-choice solutions

1 A

Budding refers to an outgrowth of a body region that separates from the original organism, resulting in two individuals.

B is incorrect because binary fission is a type of asexual reproduction where the parent cell divides into two approximately equal daughter cells. **C** is incorrect because fragmentation involves a body part detaching and developing into a new organism, and the original organism regenerates the lost body part. **D** is incorrect because external fertilisation relates to the union of a sperm and egg in sexual reproduction.

2 C

The offspring produced are genetically identical and therefore better suited to a stable environment.

A is incorrect because asexual reproduction inhibits variation, making the species more susceptible to environmental changes. **B** is incorrect because there is no allele recombination during asexual reproduction; it only occurs during meiosis in the production of gametes. **D** is incorrect because fertilisation only applies to sexual reproduction.

3 C

For a successful pregnancy, the correct order of events is required, starting with ovulation, then fertilisation, followed by implantation and placental formation.

A is incorrect because implantation can only occur after ovulation and fertilisation. **B** is incorrect because implantation can only occur after ovulation and fertilisation, while placental formation occurs last. **D** is incorrect because fertilisation occurs after ovulation and before implantation.

4 A

The peak of FSH corresponds to the release of the egg, while the peak of progesterone corresponds to the thickening of the endometrium of the uterus.

B is incorrect because both progesterone and oestradiol peak at the same time. **C** is incorrect because both FSH and LH peak at the same time. **D** is incorrect because FSH levels drop as the endometrium thickens.

5 C

A nucleotide consists of one sugar, one phosphate and one nitrogen base. There are six of these nucleotides present.

A is incorrect because there are 2 backbones and 6 nucleotides. **B** is incorrect because there are 3 nucleotide monomers on each backbone. **D** is incorrect because there are a total of 12 molecules (6 sugars and 6 phosphates) in the backbone, which together form part of the 6 nucleotides.

6 B

If 22% of the DNA segment is guanine, it must also be 22% cytosine. As these bases are complementary and form bonds in the DNA molecule, this leaves a total of 56%, which is equally split between adenine and thymine.

A is incorrect because 22% is the amount of each of guanine and cytosine. **C** is incorrect because 44% is the total amount of guanine and cytosine. **D** is incorrect because 78% is the total amount of nitrogen bases other than guanine.

7 B

A distinguishing feature of prokaryotic cells is that they each contain one single circular chromosome, compared to the linear chromosomes of eukaryotic cells.

A is incorrect because the chromosome in prokaryotic cells is circular. **C** is incorrect because only eukaryotic cells have histone proteins. **D** is incorrect because prokaryotic cells have a single circular chromosome in combination with plasmids.

8 B

If the template DNA strand is GAT ATC GAT CTA, the mRNA sequence will be CUA UAG CUA GAU, which will in turn align with the tRNA sequence GAU AUC GAU CUA.

A is incorrect because this is the corresponding mRNA sequence. **C** is incorrect because thymine is not present in tRNA. It is replaced by uracil. **D** is incorrect because this corresponds to the coding DNA strand.

9 C

Using the mRNA table, UAC corresponds to the amino acid tyrosine.

A is incorrect because methionine is produced by AUG. **B** is incorrect because threonine is not coded for by the UAC codon. **D** is incorrect because a 'stop' is not inserted by the UAC codon.

10 A

The second triplet of ACG will have a corresponding mRNA of UGC and code for cysteine. With the guanine replaced by thymine, the DNA triplet becomes ACT, and the mRNA becomes UGA, which will insert a stop.

B is incorrect because threonine requires mRNA of AC(U, C, A, G). **C** is incorrect because tryptophan requires mRNA of UA(U, C). **D** is incorrect because cysteine will only be coded for by an mRNA sequence of UG(U, C).

11 A

For trait A the similarity between siblings is high, no matter what their type, with relatively small difference between groups. Therefore, the family environment has a high impact with little genetic influence. For trait B the identical twins have high similarity, while the other two groups are significantly lower, suggesting a high genetic influence compared to environmental influence. Trait C has few sibling similarities in any of the three groups, suggesting little genetic or environmental influence.

B is incorrect because trait C has low genetic and environmental influence. **C** is incorrect because trait A has low genetic influence and high environmental influence. **D** is incorrect because trait A has low genetic influence and high environmental influence and trait C has low genetic and environmental influence.

12 B

Crossing over does not occur, and therefore the gametes will receive one chromosome from each homologous pair of chromosomes that have replicated to form identical chromatids.

A is incorrect because there are four possible combinations (AB or Ab with cD or Cd), and the gamete must have one chromosome from each pair of the homologous chromosomes. **C** is incorrect because crossing over did not occur and so neither CD nor cd is possible. **D** is incorrect because the gamete will receive only one allele of each gene.

13 C

W is a pair of homologous chromosomes. X represents sister chromatids from one of the homologous pairs of chromosomes, while in Y, crossing over is occurring between non-sister chromatids of the corresponding homologous chromosomes.

A is incorrect because Y represents crossing over and not independent assortment. **B** is incorrect because W and X are labelled the wrong way around. **D** is incorrect because all three options for W, X and Y are incorrect.

14 A

All genetic variation is the result of a mutation at some point in time.

B is incorrect because while fertilisation increases variation within a population of a species, it is not the source of new alleles. **C** is incorrect because while crossing over increases variation within a population, it is not the source of new alleles. **D** is incorrect because polypeptide synthesis is the process by which a gene codes for a specific amino acid sequence.

15 C

The trait being tracked must be recessive, as neither individual 1 nor the partner shows the trait in the phenotype, and they must therefore both pass on a recessive allele to the two children who have it. Therefore 1 must be Tt. If the trait is recessive, individual 2 must therefore be heterozygous and carry the allele for her son to express the trait.

A is incorrect because there is sufficient detail to determine that individual 2 is unaffected and therefore must be heterozygous. **B** is incorrect because if individual 2 was homozygous dominant, then her son would inherit a dominant allele from her and therefore not express the trait. **D** is incorrect because it is not a sex-linked trait as both males and females are shown to be equally affected in the pedigree and male 1 'cannot be unaffected and pass on a recessive allele to his daughter as he only has one copy of the X chromosome', meaning no daughters would be affected if his genotype was X^TY as they would all be homozygous dominant X^TX^T or heterozygous X^TX^t.

16 B

The trait is codominant, and the cross is therefore $I^AI^A \times I^AI^B$. With the study involving 1000 couples, it is expected that the offspring would appear in a 1 : 1 ratio of heterozygous (I^AI^B) to homozygous (I^AI^A) for blood group A.

A is incorrect because the sample size is large, and the best representation is a 1 : 1 ratio of heterozygous (I^AI^B) to homozygous (I^AI^A) for blood group A. **C** is incorrect as the third phenotype produced by I^BI^B is not possible from this cross. **D** is incorrect because a fourth phenotype is not possible from this cross.

17 D

In a population of 200 there are 400 possible alleles. There are 72 TT individuals, 88 Tt individuals and therefore 40 tt individuals. The number of T alleles is 144 + 88 = 232 and the number of t alleles is 80 + 88 = 168. The T% = 232/400 × 100 = 58%. The t% = 168/400 × 100 = 42%.

A is incorrect because 36% is the genotype frequency of TT only. **B** is incorrect because the analysis to obtain this result is incorrect. **C** is incorrect because 80% is the frequency of phenotype expressing the dominant allele.

18 A

The disorder is caused by a recessive allele, which will therefore be represented as a lowercase letter. Of the three (a1, a2 and a3), the a1 allele has the highest frequency. In order to be expressed in the phenotype, an individual must be a1a1.

B is incorrect because the a2 allele has a lower frequency than the a1 allele and would therefore not appear in the phenotypes as often. **C** is incorrect because the dominant A allele would mask the a1 allele. This individual would be a carrier. **D** is incorrect as the 'healthy' A allele is dominant.

19 B

The study group has 73% with the SNP, compared to 27% without it. This indicates that the SNP increases the likelihood of developing the disease.

A is incorrect because 51% of the control group has the SNP. **C** is incorrect because some individuals without the SNP (27%) still develop the disease. **D** is incorrect because there is clearly an increased risk associated with the SNP (73% vs 27%).

20 B

The two bands indicate that both parents are heterozygous, and the cross is therefore Tt × Tt. This results in a 25% chance that one of the offspring will be homozygous recessive.

A is incorrect because the only way the offspring can have zero chance of having the condition is if one parent was homozygous dominant. **C** is incorrect because the only way the offspring can have a 50% chance of having the condition is if one parent had the condition and the other was heterozygous. **D** is incorrect because the only way the offspring can have a 100% chance of having the condition is if both parents were homozygous recessive.

21 A

The suspect must have all bands in common with the evidence sample.

B is incorrect because the suspect does not have all bands in common with the evidence sample. **C** is incorrect because the suspect does not have all bands in common with the evidence sample. **D** is incorrect because the suspect does not have all bands in common with the evidence sample.

Short-answer solutions

22 a Asexual reproduction (1 mark)

b Prepare a suspension of the bacteria *E. coli* K-12.

1 Prepare 12 sterile agar plates with 30 mL agar.

2 Prepare three incubators at temperatures 10°C, 25°C and 40°C.

3 Add 1 mL of the *E. coli* K-12 suspension to nine of the plates.

4 Add three of the plates to each incubator.

5 To the remaining three plates add no *E. coli* K-12 as these are experimental controls. Add one plate to each incubator.

6 Incubate for 48 hours.

7 For each plate, count the number of colonies present.

8 Compare the number of colonies recorded at each temperature.

To receive 4 marks, your procedure must include how the independent variable is changed, how the dependent variable is measured, how all other variables are kept constant/controlled, and how you ensured reliability by repeating the experiment and determining whether the results of each trial were consistent. The method must also follow a logical sequence.

23 ©NESA 2019 ADDITIONAL SAMPLE EXAM MOD 5 Q11 (ADAPTED)

a

Process	Location
Site of egg development	Ovary (1 mark)
Site of fertilisation	Fallopian tube or oviduct (1 mark)
Site of implantation of blastocyst	Uterus (endometrium) (1 mark)

b

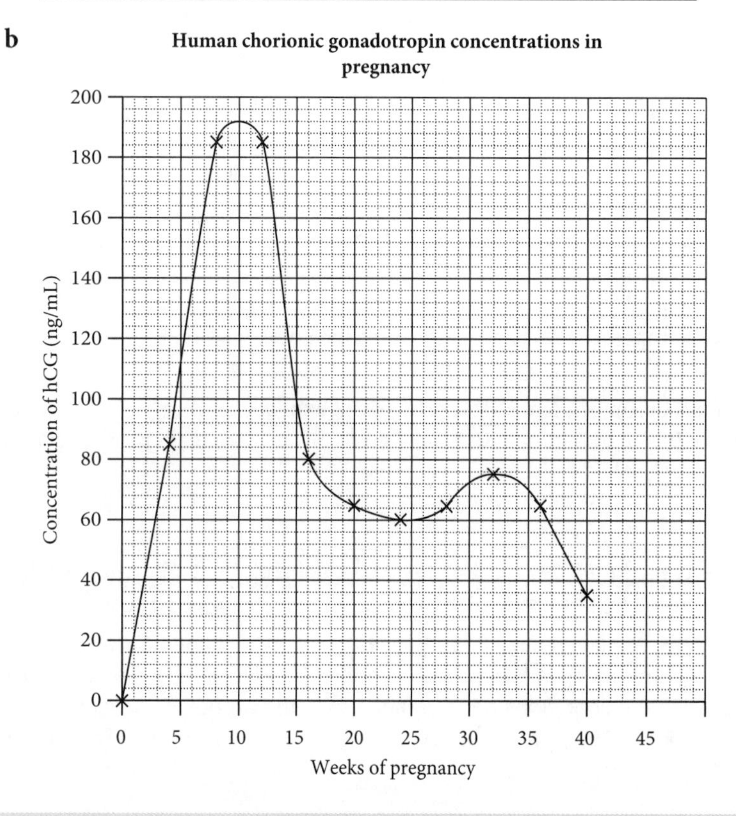

For 3 marks, your graph must include: all axes labelled correctly with correct units, appropriate scale and correctly plotted points. It must be a line graph.

c Progesterone (1 mark) is an important hormone in the control of pregnancy. Progesterone is initially produced by the corpus luteum in the ovary and causes the endometrium to thicken, which helps to support and maintain the pregnancy in the first weeks when the placenta is still developing. The developed placenta then produces progesterone at significantly higher levels to maintain the pregnancy (1 mark). Prior to birth, progesterone levels drop significantly to facilitate labour (1 mark).

Your answer must describe both the role *and* changes in levels of a hormone in pregnancy.

24 Overall, the accumulation of scientific knowledge in the field of genetics has been central to our ability to manipulate both plant and animal reproduction in agriculture. As our understanding of genetics increases, reproductive technologies improve in parallel, allowing for greater manipulation and control over desired traits in agriculture (1 mark).

The knowledge that offspring inherit one gamete from each parent has allowed increased use of artificial pollination and artificial insemination (1 mark). Both techniques increase the likelihood of successful fertilisation; however, neither guarantee that the desired traits will be passed on to the offspring. This is based on our knowledge of chromosomal behaviour during meiosis. A significant benefit of these techniques is that they are relatively cheap, accessible and provide a reliable product source (1 mark).

More recently, gene sequencing technologies (1 mark), combined with the knowledge that DNA is common to all life, has allowed scientists to transfer genes between species (1 mark). Common examples are 'super salmon', which grow larger than wild type fish and therefore produce higher yields more quickly; and a wide range of Bt crops, which have had a gene from the bacteria species *Bacillus thuringiensis* inserted. The gene product is a protein toxin that kills insect pests. This allows for increased crop yields and a reduction in the use of pesticides.

A disadvantage of all of these techniques, and therefore the scientific knowledge that led to them, is that they are typically associated with farming monocultures. There is a significant reduction in biodiversity, which can be problematic for species and ecosystem survival. There are also a range of social and ethical concerns surrounding the use of reproductive technologies, particularly in animal agriculture (1 mark).

25 a For 2 marks. any two of the following:

- Mitosis produces two daughter cells; meiosis produces four daughter cells.
- Mitosis produces genetically identical daughter cells; meiosis produces genetically unique daughter cells.
- Mitosis produces diploid daughter cells; meiosis produces haploid daughter cells.
- Mitosis involves a single cell division; meiosis involves two cell divisions.
- Meiosis includes processes that lead to genetic diversity, e.g. crossing over, random alignment and segregation of homologous chromosomes, independent assortment of alleles; mitosis doesn't.

b The purpose of mitosis is to produce genetically identical daughter cells. They need to be genetically identical because the chromosomes contain the genetic code essential for protein production. If the code is altered, then the amino acid sequence of the protein produced may be changed and its function impaired (1 mark). This can affect the growth of an organism, repair of damaged tissue or impact asexually reproducing organisms. Therefore, mitosis is essential for survival of an individual, which in turn can affect the continuity of species (1 mark).

The purpose of meiosis is to produce genetically unique daughter cells with half the chromosome number of the parent cell. The processes that occur during meiosis, e.g. crossing over and independent assortment, are mechanisms by which variation is maintained within sexually reproducing organisms (1 mark). If conditions change, variation is important to the survival of species that reproduce sexually. Therefore, meiosis is essential to the continuity of species that reproduce sexually (1 mark).

c ©NESA 2012 EXAM SAMPLE ANSWERS Q29a

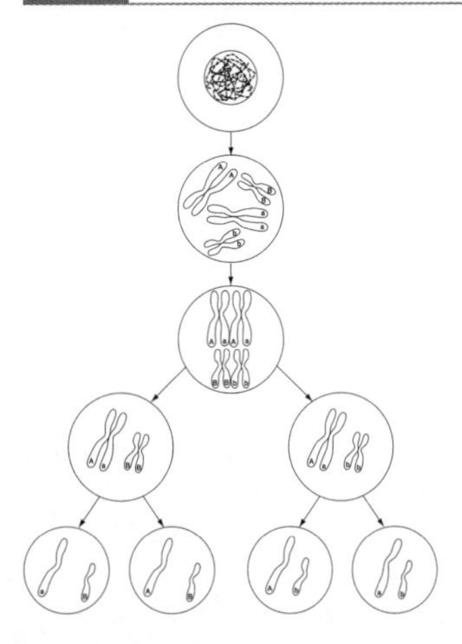

For 3 marks, you must include: correct allele arrangement in all cells, sister chromatids joined at the centromere, alleles in correct location on chromosome.

d During meiosis, chromosome pairs line up independently of each other. The orientation of each tetrad (each homologue has two sister chromatids) is independent of the orientation of the other tetrads (1 mark). Any maternally inherited chromosome may face either pole and any paternally inherited chromosome may face either pole. The random alignment of tetrads on the metaphase plate produces a unique combination of maternal and paternal chromosomes (1 mark).

26 ©NESA 2005 EXAM SI Q10 (diagram only)

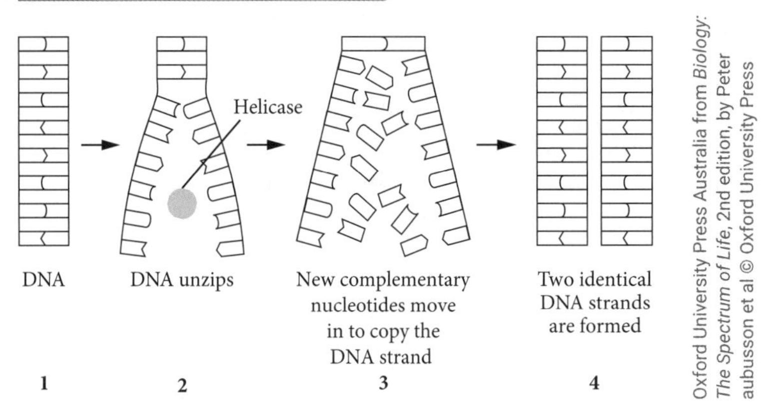

For 6 marks, you must include the following:

- a flow chart with arrows showing direction
- initial DNA helix
- DNA helix unwinding
- the hydrogen bonds linking the two strands of the molecule unzipping under control of DNA helicase
- DNA nucleotides lining up on complementary bases on the exposed strands (A-T and C-G)
- two new identical double-stranded molecules, each forming a double helix.

27

	Prokaryotic cell	Eukaryotic cell
Building blocks of DNA	Nucleotides with one of four nitrogen bases (A, T, G, C)	Nucleotides with one of four nitrogen bases (A, T, G, C)
Chromosome structure	Single circular	Multiple linear chromosomes
DNA packaging	No histone proteins	Tightly wound around histone proteins
Plasmids	Present	Not present

- Table must be correctly set up to compare prokaryotic and eukaryotic cells (1 mark)
- Must include at least three distinct points (1 mark)
- Must include one similarity (1 mark)

28 a The sequence of bases on the template strand of the section of DNA forming the INS gene determines the exact sequence of amino acids in the polypeptide molecule that makes insulin (1 mark).

Transcription: Under the control of DNA helicase, the relevant DNA sequence unzips to allow free mRNA nucleotides to line up on complementary bases of the template strand, and then RNA polymerase joins the forming nucleotide chain together into mRNA (1 mark). The mRNA exits the nucleus and moves to a ribosome, where translation occurs (1 mark).

Translation: The mRNA moves through the ribosome and in sets of three nucleotides called codons, the mRNA attracts a complementary tRNA molecule that binds to the codon via the anticodon sequence. At the other end of the tRNA molecule is an amino acid coded for by the codon (1 mark).

At the ribosome each triplet of bases, or codon, attracts a tRNA molecule carrying the complementary anticodon. As successive tRNA molecules are attracted to the ribosome they release their amino acids, which join to form a polypeptide whose amino acid sequence is joined by peptide bonds (1 mark). Once complete, the polypeptide folds to form the final protein structure. Therefore, the specific sequence of nucleotides on the INS determines the exact sequence of amino acids in insulin (1 mark).

b Gene expression refers to the product produced from a specific genetic code (1 mark). For example, there is a specific sequence of nucleotides on the *INS* gene that will determine the specific sequence of amino acids in the final product, in this case insulin (1 mark). The correct sequence of amino acids is essential for forming the correct secondary and tertiary structures necessary for the protein to perform its role (1 mark). If the structure is changed, the function can be impaired and the insulin would be unable to bind to receptors on muscle and fat cells, decreasing their ability to absorb glucose from the blood. Correct gene expression is therefore essential to a functioning organism (1 mark).

29 While genes contain the genetic code responsible for producing the phenotype, the environment plays an important role in controlling gene expression and determining the phenotype (1 mark). For example, in several species of reptiles, e.g. crocodiles and some turtle species (1 mark), the sex of the offspring is determined by the temperature of the eggs. During embryonic development, the thermosensitive period (TSP) affects the expression of the *Sox9* gene, which is involved in sex determination. While both sexes possess the *Sox9* gene, its expression in determined by temperature (1 mark). For example, saltwater crocodiles have an egg incubation period of 80–90 days. The TSP is 25–50 days, during which eggs kept at 32°C or above will hatch as male offspring while those kept at 31°C or below will hatch as females (1 mark).

30 a Three

b AaBbCcDDEe

c AbCDE, AbcDe, aBCDE, aBcDe

31 ©NESA 2018 MARKING GUIDELINES SII Q29 (ADAPTED)

a DNA replication involves separating DNA strands and adding complementary nucleotides to each strand until two identical sequences are produced (1 mark). They remain joined at a centromere and each is referred to as a sister chromatid (1 mark). Note: Crossing over has not yet occurred.

b The model represents homologous pairs of chromosomes as being the same size and identifies the paternal or maternal origin with different shading (1 mark).

In the first division of meiosis, individual chromosomes form homologous pairs. These carry the same genes but the alleles may not be identical (1 mark). When paired, crossing over occurs as genetic material is exchanged between non-sister chromatids, resulting in new combinations of genetic material, as shown by the shading (1 mark).

The two daughter cells formed after this division have half the number of chromosomes (haploid cells) (1 mark), one from each homologous pair assigned at random. This increases variation in the daughter cells, as shown after the first division (1 mark).

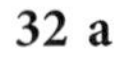

32 a

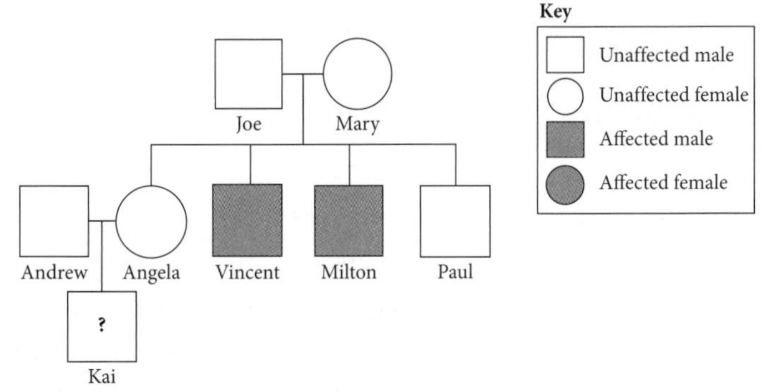

b It cannot be determined whether Angela is X^TX^t or X^TX^T (1 mark). If Angela is a carrier, then there is a 50% chance that Kai will be colour blind. If Angela is not a carrier, Kai cannot be affected. This is because Kai is XY, and his Y chromosome comes from his father, Andrew. He only needs one copy of the X^t to be colour blind (1 mark).

To receive full marks:

- Correct key 1 mark
- Pedigree correctly drawn 1 mark
- Names correctly indicated 1 mark
- Individuals with trait correctly shaded 1 mark

33 ©NESA 2019 MARKING GUIDELINES SII Q30

Graph *A* shows a ratio of 3 : 1 for the two phenotypes. This is typical of dominant/recessive alleles (1 mark).

The pure-breeding parents must have been homozygous for seed shape, e.g. RR for round and rr for wrinkled. The first generation would all have been round heterozygous Rr. When this generation was self-pollinated to provide the second generation, the following Punnet square can be produced (1 mark).

	R	r
R	RR	Rr
r	Rr	rr

(1 mark)

KEY:
R = round seed
r = wrinkled seed

Graph *B* shows a ratio of 1 : 2 : 1 for the three phenotypes. This is typical of either codominant or incomplete dominant alleles (1 mark).

Neither allele is dominant and both are expressed. In the F_1 generation, the chickens express both alleles and are heterozygous. When they breed together, the following Punnett square is produced. For example, if codominant:

	F^B	F^W
F^B	F^BF^B	F^BF^W
F^W	F^BF^W	F^WF^W

(1 mark)

KEY:
F^B = black feathers
F^W = white feathers

This results in 50% with black and white feathers, 25% black feathers and 25% white feathers.

Remember to provide a suitable key for your Punnett square. Answers could also include: Incomplete dominance for graph B where both alleles are blended.

34 A SNP is a point mutation (1 mark) that is present in more than 1% of the population (1 mark).

35 a DNA sequencing is a technique used to determine the exact nucleotide sequence of a segment of DNA (1 mark), whereas DNA profiling is a technique by which a specific DNA pattern (profile) of an individual is generated (1 mark).

b Short tandem repeats or STR (1 mark).

c PCR and gel electrophoresis are two technologies used in inheritance studies (1 mark).

PCR is a technique used to amplify specific segments of DNA so that a large enough quantity is present to be run in gel electrophoresis. There are three key steps. In denaturation, the sample is heated to 95°C to separate the DNA into two single complementary strands. Annealing involves cooling the sample to 55°C, where DNA primers attach to the target sequences, promoting replication from the point of attachment. Finally, elongation involves heating the sample to 72°C, where Taq polymerase binds to the primers and copies the target strand of DNA by adding free nucleotides. After each cycle, the quantity of DNA is doubled (2 marks).

Gel electrophoresis is a technique used to separate a mixture of DNA fragments based on their molecular size after they have been amplified via PCR. Gel electrophoresis can be used to generate a DNA profile of parents who may be both carrying a recessive allele for a genetic disease, e.g. cystic fibrosis. The gene sequence must be known so that DNA primers can be made specific to the target gene. In the case of cystic fibrosis, the main causative mutation results in the deletion of three nucleotides, resulting in shorter/lighter DNA fragments that travel further through the gel (2 marks).

In conclusion, if both parents are heterozygous (carriers), then PCR followed by gel electrophoresis will show two bands due to the difference in molecular weight of each allele (1 mark).

36 The Human Genome Project (HGP) was an international collaboration involving many laboratories around the world. The project, completed in 2003, decoded the entire sequence of the human genome. It was found that the human genome consists of 3 billion DNA base pairs and an estimated 20 000 to 252 000 genes. The Sanger method of DNA sequencing was used. This involves denaturing the DNA and using chain-terminating nucleotides (ddATP, ddTTP, ddCTP or ddGTP). The resulting DNA fragments are denatured into single-stranded DNA, which can then be separated by electrophoresis to determine the sequence (2 marks).

The HGP was next used to develop the HapMap of the human genome. This is a computer database that can be used to look for specific trends, patterns and relationships within the DNA sequences linked to specific diseases or disorders. It can also be used to predict the response of different gene combinations to medications and environmental factors. To use the data, scientists conduct genome-wide association studies (GWAS), where the HapMap is scanned to identify genetic variations between a group of individuals with a particular disease or disorder and a control group of individuals without the condition (2 marks).

Data trends involve a general direction in the data set. For example, several different GWAS have used the data obtained from the HGP and HapMap to identify an increased likelihood of developing certain cancers, such as prostate cancer and breast cancer, if an individual has certain gene variants (1 mark).

Data patterns are slightly different in that they do not necessarily show direction, but rather repeating observation. For example, GWAS studies have found that certain combinations of gene variants are more common in individuals who have certain disorders or diseases. For example, prostate cancer has been linked to 170 common gene variants (1 mark).

Finally, relationships tend to be more numerical. For example, a GWAS study found that individuals with a particular gene variant of the *BRCA1* gene have a 60% chance of developing breast cancer (1 mark).

CHAPTER 2, MODULE 6

Genetic change

Multiple-choice solutions

1 A

X-rays have a short wavelength and high energy, which can displace electrons from atoms (ionise). This can break hydrogen bonds between complementary bases and covalent bonds in the sugar–phosphate backbone.

B, **C** and **D** are incorrect because they are all forms of electromagnetic radiation that have a long wavelength and low energy. They cannot enter cells and interact with DNA in the nucleus.

2 D

Gamma rays, X-rays and particle radiation are forms of ionising radiation. They remove electrons from atoms (ionise) in DNA, which can break hydrogen bonds between complementary bases, or covalent bonds between adjacent nucleotides on the same strand of DNA.

A is incorrect because ionising radiation does not cause the addition of atoms to DNA. **B** is incorrect because ionising radiation can break hydrogen bonds, not add hydrogen bonds, between complementary bases. **C** is incorrect because ionising radiation does not add chemical groups to DNA bases.

3 C

Base analogues are also known as base 'mimics' because their chemical structure is very similar to naturally occurring bases. DNA polymerase is unable to distinguish between a base analogue and a real base and incorporates it into the DNA strand during DNA replication.

A is incorrect because base analogues are chemical mutagens and are not a form of high-energy ionising radiation. **B** is incorrect because this describes a type of mutation caused by the addition of chemical groups to bases. **D** is incorrect because this is describing the action of another chemical mutagen, called an intercalating agent.

4 C

A germ-line mutation must occur in a gamete, and a mutation in a non-coding region will not affect a protein-coding gene.

A is incorrect because mutations originate in DNA, not RNA. **B** is incorrect because mutations originate in DNA, not polypeptides. **D** is incorrect because the question states that it is a germ-line mutation, which means it must be in the DNA sequence of the gametes.

5 D

A base substitution (point mutation) that causes an amino acid to change to a different amino acid is called a missense mutation. In this instance the alanine (Ala) changes to glycine (Gly).

A is incorrect because a nonsense mutation is when a point mutation creates a stop codon. **B** is incorrect because a frameshift mutation is when the insertion or deletion of a nucleotide alters the open reading frame of the DNA sequence, and the polypeptide sequence from that point on is altered. **C** is incorrect because there is no evidence that a new nucleotide has been incorporated into the DNA strand.

6 B

Aneuploidy is the presence of one or more additional chromosomes, or the absence of one or more chromosomes, in a cell. The karyotype is from a person with Down syndrome (trisomy 21). The person is male (XY) and is diploid for all chromosomes except chromosome 21.

A is incorrect because polyploidy is the presence of two or more whole sets of chromosomes in a cell. **C** is incorrect because duplication refers to an additional copy of a gene or chromosomal region. **D** is incorrect because a translocation involves the movement of a large chromosomal region to a new region of the same chromosome or to a non-homologous chromosome.

7 C

Somatic mutations are only present in somatic cells of the body. They are not present in the gametes (egg and sperm) and cannot be inherited by offspring.

A is incorrect, because they do occur in somatic cells, but they are not present in gametes and therefore cannot result in mosaicism in eggs or sperm. **B** is incorrect because somatic mutations only affect an individual, but they cannot be passed onto offspring. **D** is incorrect because somatic mutations are not present in gametes and they cannot be inherited.

8 D

Genetic recombination occurs during meiosis I, after DNA replication. Due to DNA replication, each homologous chromosome (homologue) consists of two sister chromatids. Genetic material is exchanged between the chromatids of the different homologues, i.e. the non-sister chromatids.

A is incorrect because genetic material is exchanged between homologous chromosomes. **B** is incorrect because even though genetic material is exchanged between homologous chromosomes, this does not occur between the sister chromatids. **C** is incorrect because genetic material is not exchanged between sister chromatids of non-homologous chromosomes.

9 C

There are 10 individuals in the final population, which equates to 20 alleles. There are 16 P alleles (2 in the Pp heterozygotes + 14 in the PP homozygotes) and 4 p alleles (2 in the Pp heterozygotes + 2 in the pp homozygotes). Frequency of P allele = 16/20 = 0.8, frequency of p allele = 4/20 = 0.2.

A, B and **D** are incorrect because these calculations all give the wrong allele frequencies.

10 C

Fermentation was used by ancient civilisations thousands of years ago to produce food products.

A is incorrect because fermentation is a biochemical reaction that occurs in microorganisms. It is not a method of reproduction or a cloning technique. **B** is incorrect because CRISPR-Cas9 was only discovered in 2012. **D** is incorrect because recombinant DNA technology was first used in the 1970s.

11 C

Some people argue that humans are 'crossing the line' or 'playing God' by altering the genome of other species and interfering in the process of evolution. Other people believe it should be done so that all humans can access sufficient food for survival.

A, B and **D** are all incorrect because these are the outcomes of genetically modifying plants for agriculture and how they can benefit farmers.

12 A

This is an ethical issue because it is a debate about what two people feel is right or wrong based on their moral principles.

B is incorrect because political topics such as government policy or legislation are not mentioned in the debate. **C** is incorrect because scientific methods involved in ES cell research are not a topic in the debate. **D** is incorrect because money, or the consumption of goods and services, is not a topic in the debate.

13 D

Both processes involve the selective transfer of male gametes to female gametes without the alteration of DNA.

A is incorrect because this statement is true. Not all individuals have the same opportunity to reproduce. The genetic material from one individual with a desirable trait is inherited by many offspring. **B** is incorrect because this is a true statement. Scientists and farmers use male gametes from individuals with traits they would like to pass onto many offspring. **C** is incorrect because this statement is true. Humans manually transfer gametes to assist reproduction.

14 D

Restriction endonucleases, also known as restriction enzymes, recognise a specific target sequence in a plasmid and digest (cut) the DNA. This allows for the insertion of a gene or piece of DNA into the plasmid.

A is incorrect because DNA ligase is an enzyme that fuses the insert and plasmid together after the restriction endonuclease digestion. **B** is incorrect because DNA polymerase is the enzyme that replicates DNA. **C** is incorrect because reverse transcriptase is an enzyme that converts RNA into cDNA.

15 A

Recombinant DNA is a molecule that consists of two pieces of DNA from different species that have been joined together. In this case a gene from one species has been combined with a plasmid from a bacterium.

B, C and **D** are incorrect because scientists cannot clone mRNA, protein or polypeptide in a plasmid.

16 A

The first step is to digest a plasmid with a restriction endonuclease (digestion). The restriction endonuclease sites on the ends of the gene match those at the plasmid cut site. A gene is inserted into the linearised plasmid with DNA ligase, which fuses the end of the genes with the ends of the linearised plasmid (ligation). The recombinant plasmid is then introduced into bacteria using heat shock or an electrical pulse (transformation). Finally, bacteria are grown on agar containing an antibiotic such as ampicillin. Only bacteria containing the plasmid (which contains the cloned gene and an antibiotic resistance gene) will grow (selection).

17 D

Vitamin A deficiency is a common problem in low-income countries because of limited access to food containing essential nutrients. It results in blindness and can make people more vulnerable to infectious diseases. The addition of the beta-carotene pathway – a precursor of vitamin A – to rice is a way to prevent malnutrition in regions where rice is a staple food.

A is incorrect because patents are an implication of making transgenic plants. Patents financially benefit the scientists or companies that invented the transgenic plant. **B** is incorrect because the rice plants have not been modified to be more resistant to disease. **C** is incorrect because the rice plants have not been modified to have a faster growth rate.

18 D

Transgenic organisms may be able to reproduce with closely related species, which may affect genetic variation in wild populations.

A is incorrect because this is a benefit of making transgenic crops that are better adapted to biotic and abiotic conditions. **B** is incorrect because this is a benefit of making a transgenic animal with a gene that regulates growth. **C** is incorrect because this is a benefit of using transgenic crops that are resistant to insect pests.

19 A

The farmer is introducing a new allele into the population of cows.

B is incorrect because this is not a chance event. **C** is incorrect because natural selection is a naturally occurring process in which individuals within a population survive to reproduce if they have the adaptations to do so. **D** is incorrect because selective breeding involves natural mating of individuals based on their physical traits.

20 B

A guide RNA directs the CRISPR-Cas9 complex to a specific target site in the DNA strand and the active site in Cas9 cuts it.

A is incorrect because Cas9 forms a complex with a guide RNA (gRNA), not DNA. **C** is incorrect because even though the gRNA directs the CRISPR-Cas9 complex to the target site on the DNA, the gRNA may have uracil in its sequence. **D** is incorrect because RNA is unable to cut DNA.

Short-answer solutions

21 a A point mutation is a change in the nucleotide at a specific position in the DNA sequence. This results in the substitution, insertion or deletion of a base (1 mark)

b A frameshift mutation (1 mark) is the insertion or deletion of a single nucleotide, which changes the polypeptide sequence from the mutation site onwards. They often result in stop codons within the coding sequence (1 mark). This will produce a non-functional polypeptide that is degraded by the cell (1 mark).

Other responses include:

- silent mutation – doesn't alter the amino acid encoded by the codon; no effect on the polypeptide sequence
- missense mutation – changes the amino acid encoded by the codon; can alter the function of the protein.

22 a The normal mRNA sequence is AUU GCA UGG UUU GCU AUC UGU GAU UUA AUC. The normal polypeptide sequence is Ile-Ala-Trp-Phe-Ala-Ile-Cys-Asp-Leu-Ile. The mutant mRNA sequence is AUU GCA UGG UUU GCU AUC UGA GAU UUA AUC. The mutant polypeptide sequence is Ile-Ala-Trp-Phe-Ala-Ile-Stop. (2 marks)

b A nonsense mutation. The codon changes from TGT (Cys) to TGA (stop). (1 mark)

c The polypeptide sequence after the stop codon will not be translated. The polypeptide will not be able to function because it may not fold properly, it may not localise to the correct position in the cell, it may not have enzymatic activity, it may not bind with other proteins to form functional protein complexes, or the cell may recognise that it has an abnormal polypeptide and degrade and recycle the amino acids. (1 mark)

23 a

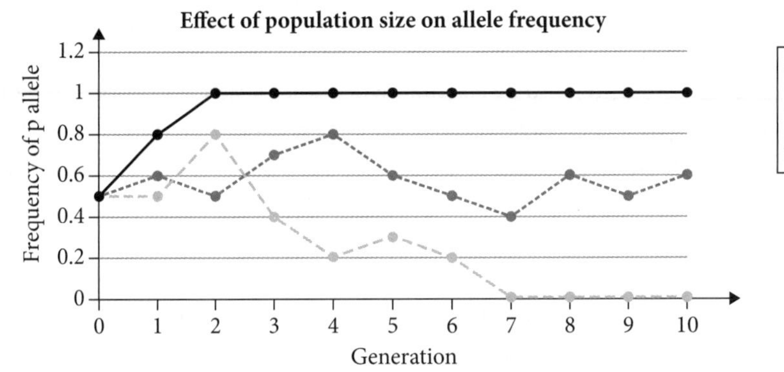

(2 marks: ½ mark for labelled axes, ½ mark for legend, 1 mark for plots)

b Generation 2 (1 mark)

c Genetic drift is a process that causes a change in the frequency of alleles in a population. It is caused by the random inheritance of alleles during each generation because of a chance event. A small population of organisms is more likely to have a smaller number of genotypes and alleles than a larger population. In both small and large populations, some individuals may not mate, which means their alleles are not passed on to the next generation. These non-mating individuals make up a higher percentage of a smaller population than a larger population, so the chance that alleles are lost or become fixed (100% of alleles at a locus in a population) is higher over time for a smaller population. (3 marks)

24 a Gamma rays and particle radiation. These two types of ionising radiation are emitted during the decay of radioactive elements. (1 mark)

b The thyroid gland naturally absorbs iodine, so exposure to radioactive iodine-131 resulted in an increased frequency of thyroid cancer. The high energy in the gamma rays and particle radiation would have been absorbed by the atoms in DNA, which would have displaced electrons in a process called ionisation. This would have caused changes in the chemical structure of bases, broken hydrogen bonds between bases on complementary DNA strands and broken the chemical bonds in the sugar–phosphate backbone. Some of these changes would have affected genes that control cell division, resulting in cancer. (2 marks)

c The incidence of thyroid cancer in children aged 0–14 years sharply increased approximately four years after the Chernobyl accident. The incidence of cancer was close to zero until 1989 and increased to the highest incidence of 4/100 000 in 1996 (ten years after the disaster). The incidence of cancer in this age group then declined for the twenty years after the accident. In 2006 the incidence of cancer was at a similar level to that before the accident. (2 marks)

The initial increase in cancer was caused by the accumulation of mutations in people exposed to high levels of radiation. New babies would have been born in the two decades after the accident at a time when radiation levels would have declined, and these levels may not have been high enough to cause cancer. This would account for the overall decline in the incidence of cancer in this age group. (2 marks: ½ mark for labelled axes, ½ mark for legend, 1 mark for plots)

- 2 marks are allocated for describing the trend and citing specific data in the graph.
- 2 marks are allocated for explaining the trend based on your knowledge of how mutagens operate.

25 a Exons are the only coding regions of the genome (½ mark). Examples of non-coding regions will vary (½ mark). Non-coding regions include introns, intergenic regions, centromeres and telomeres.

b An intronic mutation will only affect a gene if it occurs at a splice site or within a regulatory element such as an enhancer. Splicing mutations result in the loss of exons during the splicing process, while enhancer mutations may reduce the amount of transcription from a gene. (1 mark)

- There are several examples of non-coding regions but only one is required.

26 ©NESA 2019 MARKING GUIDELINES SII Q26 The map shows that ability to digest lactose varies in adult populations around the world. For example, it is much lower in Australia than Northern Europe (1 mark). This variation is likely to be due to natural selection, where the presence of milk in the diet is the selective pressure. A mutation to a gene is likely to cause the continued production of lactase past the age of five years (2 marks). Adults who possess the mutation are likely to have an increased chance of survival as they have increased nutrition in their diets. When they reproduce, they pass the mutation on to their offspring, making it more common in the population. Populations that have remained largely lactose intolerant are less likely to have had milk available, and therefore a mutation of the lactase gene would not offer any advantage and would not become more common in the population. (2 marks)

27 A mutation in a germ cell can be passed on to future generations. It will be present in every cell of an offspring if it is inherited. A mutation in a somatic cell will only affect the person in which it occurred and cannot be inherited by offspring (1 mark). Mutation *B* will be present in every cell of twin 1's body because it has occurred in the embryo before differentiation into germ cells and somatic cells. The twin will display somatic mosaicism because the mutation occurred after twin formation when there were more than two cells in the embryo. The mutation has a 50% chance of being inherited by offspring. If it is inherited, it will be present in every cell of their body, and they may develop a disease (2 marks). Mutation *C* will only be present in the daughter cells of the somatic cell of twin 2. Twin 2 will also display somatic mosaicism, but the mutation will not be passed on to the next generation. Their offspring will be unaffected. (2 marks)

28 If 16% of the population are homozygous recessive, this means that $q^2 = 16\% = 0.16$. Therefore, $q = 0.4$.

If $P + q = 1$

then

$P + 0.4 = 1$

$P = 1 - 0.4$

$P = 0.6$

Genotype	Genotype frequency	Number of individuals
Homozygous dominant (TT)	$= 0.6 \times 0.6$ $= 0.36$ (½ mark)	$= 0.36 \times 10\,000$ $= 3600$ (½ mark)
Heterozygous (Tt)	$= 2 \times p \times q$ $= 2 \times 0.6 \times 0.4$ $= 0.48$ (½ mark)	$= 0.48 \times 1000$ $= 4800$ (½ mark)
Homozygous recessive (tt)	$= 0.4 \times 0.4$ $= 0.16$ (½ mark)	$= 0.16 \times 10\,000$ $= 1600$ (½ mark)

29 ©NESA 2020 MARKING GUIDELINES SII Q29 A gene pool is the total genetic diversity of a population – it results in variation of phenotypes and provides the basis for natural selection (1 mark). When the gene pool of a population changes, evolution has occurred. Gene pools may change as a result of mutation, gene flow and genetic drift.

Gene flow is the movement of alleles into or out of a population (2 marks). For example, a migrant animal may add new alleles when it reproduces with individuals in the population. Genetic drift is a change in allele frequency because of random selection of alleles. This is especially marked in a small, remnant population. The few remaining individuals that survive carry a small sample of the alleles in the original population (2 marks).

30 a Xenotransplants would be beneficial when there are shortages of human organs for transplantation. This would potentially save the lives of people who need organ transplants, and therefore the people can continue to contribute to society. (1 mark)

b There is a possibility that pig retroviruses could infect humans. If a virus jumped between species, it may not only affect the health of the person with the transplanted organ, but it may spread to other people and result in a disease outbreak that could be difficult to prevent or treat. (1 mark)

c Ethical issue in support of transplantation: (1 mark)

- The use of pigs for xenotransplantation has the potential to save people's lives. The human lives saved by this technology outweigh the deaths of the animals that provide the organs.

Ethical issue against transplantation: (1 mark)

- The value of every life is equal. It is therefore unethical to kill pigs to save human lives.

Other responses include:

Ethical issues against transplantation: (1 mark)

- The creation of a transgenic animal is done without the animal's consent.
- The creation of transgenic organisms interferes in the process of evolution.

31 a mRNA vaccines (1 mark)

b mRNA vaccines are a novel type of prevention for diseases caused by viral infections. This technique takes advantage of natural processes (transcription, translation and the immune response) that take place in eukaryotic cells (1 mark). It involves packaging a piece of mRNA that encodes a protein on the surface of a virus (e. g. the spike protein on the surface of COVID-19) into lipid vesicles. These vesicles prevent the breakdown of the mRNA before it is taken up by cells in our body. The mRNA is injected into a person, the mRNA is absorbed by cells that translate it into a polypeptide, and the protein is presented on the surface of the cell. The immune system recognises the foreign protein and creates antibodies against it. If the person is infected by the virus in the future, the adaptive immune system will be able to recognise the virus and eradicate it before it causes serious illness (1 mark).

c mRNA vaccines are relatively quick to design and mass-produce and can be adapted for use with other viruses. Vaccinations can be provided to eligible people in the community, which prevents the spread of infectious diseases caused by viruses. mRNA vaccines prevent serious illness and death as demonstrated with the Pfizer and Moderna mRNA vaccines and COVID-19 (1 mark). This also means the healthcare system is not overwhelmed by patients with viral infections and can continue to treat people with other health conditions (1 mark).

mRNA vaccines are provided as an example answer for this question. Other applications from section 2.2.1 in Module 6 could also be used.

32 a Step 1: The isolation of the desired genes from the two species of fish. The growth hormone gene from the Chinook salmon has been combined with the promoter of the Ocean pout's antifreeze protein gene. (1 mark) Step 2: The gene construct has been cloned into a plasmid. The plasmid was digested with a restriction enzyme and the ends of the linearised plasmid DNA would have been compatible with the ends of the gene insert. The plasmid and gene were joined with the enzyme DNA ligase. (1 mark) Step 3: Bacteria were transformed with the plasmid containing the transgene. The bacteria were grown on agar plates, and then a positive colony was transferred to culture broth. The bacteria replicated by binary fission to produce enough of the plasmid to be used in experiments to make the transgenic fish. Step 4: The transgene was isolated from the plasmid DNA using restriction enzymes. Step 5: The transgene was injected into a salmon egg where it may randomly incorporate into the fish's genome to make a transgenic fish. This will be performed on many eggs to produce some offspring with the desired trait. (6 marks)

b The three levels of biodiversity are genetic diversity, species diversity and ecosystem diversity. Genetic diversity is maintained by breeding the transgenic male with the wild-type, non-transgenic females (3 marks). The transgenic salmon could outcompete wild-type salmon and other species if they escaped into the wild, which could affect species diversity. Species diversity is protected by pressure shocking the eggs to produce sterile adults and by growing the fish in inland tanks. This reduces the chance of fish escaping into rivers, and even if the eggs escaped from the tanks the mature fish couldn't breed in the wild (3 marks). By protecting and preserving genetic and species diversity, these techniques also minimise ecological damage. This therefore protects ecosystem diversity as part of wider management of the habitat with other organisations (3 marks).

33 ©NESA 2020 MARKING GUIDELINES SII Q28

a The paired homologous chromosomes are incorrectly drawn. In a pair of chromosomes, one is paternal and the other is maternal. Prior to crossing over, each chromosome duplicates itself forming two chromatids and they should be identical, that is both chromatids should be either maternal or paternal and not different as shown in the model. (3 marks *for accurately explaining the misunderstanding of meiosis shown in the model*)

b In meiosis, homologous chromosomes are lined up in Metaphase I in random order and orientation (independently assorted). They separate in Meiosis I, resulting in different combinations of parental chromosomes in the gametes. Crossing over is the exchange of genetic material between the chromatids of homologous chromosomes during Meiosis I. This leads to a new combination of alleles on each chromatid. (3 marks *for explaining the processes in meiosis that lead to genetic variation*)

34

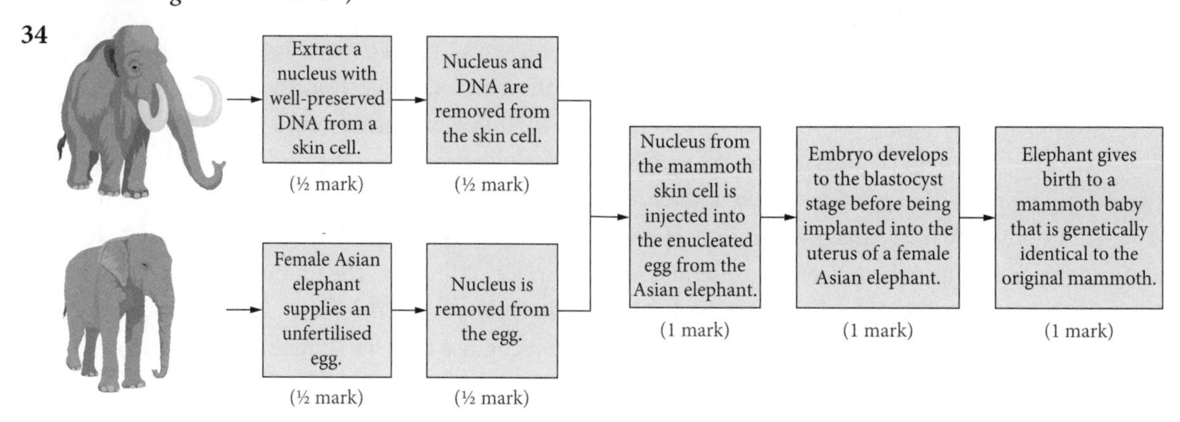

35 a Cattle (1 mark)

b A fertilised egg develops into a morula/blastocyst before being implanted into a recipient female. The embryo develops into a foetus and continues to grow and develop until the mother gives birth. There is a general decline in the success rate at each stage of development. The most significant decline in the success rate is the number of embryos that develop into blastocysts. There continues to be a decline in the number of foetuses that develop from the blastocysts and again in the number of live births. (2 marks)

CHAPTER 3, MODULE 7

Infectious disease

Multiple-choice solutions

1 C

Bacteria are prokaryotic. They contain no membrane-bound organelles and have a cell wall.

A is incorrect because prions are non-cellular pathogens. **B** is incorrect because fungi are eukaryotic and therefore have membrane-bound organelles. **D** is incorrect because protozoa are eukaryotic and therefore have membrane-bound organelles.

2 C

The female *Aedes aegypti* mosquito is the vector that transmits the pathogenic virus when it bites a human.

A, B and **D** are incorrect because the virus is a pathogen, and the mosquito is a vector.

3 B

Quantitative data is numerical. Qualitative data is descriptive. The data being collected is the appearance of the broth – whether it is clear or cloudy. No quantitative data is being collected in the experiment. Option **B** is therefore correct.

A is incorrect because the data is not quantitative. **C** is incorrect, as modelling a past experiment does not mean it is qualitative. **D** is incorrect because the data is not quantitative.

4 C

The experiment was not valid. The presence of microbes in the broth indicates that it was not sterilised before the experiment.

A is incorrect because the nature of the broth is not relevant to the experiment and insufficient information is provided to make this statement. **B** is incorrect because the nutrient broth will only go cloudy in response to microbial growth. **D** is incorrect because exposure to oxygen will not cause the nutrient broth to go cloudy. It requires microbes in the air to access the broth.

5 C

The organism shown has membrane-bound organelles, which therefore indicates that it is a eukaryotic organism. Protozoans are the only viable option.

A is incorrect because a bacterium is a prokaryote and therefore does not have membrane-bound organelles. **B** and **D** are incorrect because both prions and viruses are non-living and therefore do not have organelles.

6 C

Mast cells release histamines to increase the permeability of blood vessels, which in turn provides easier access to the infection site for phagocytes.

A is incorrect because perforin released by killer T cells or natural killer cells lyses cells. **B** is incorrect because other chemoattractants such as IFN-γ and cytokines are involved in attracting other cell types. **D** is incorrect because memory B cells are produced as part of the adaptive immune response.

7 A

Skin is a physical barrier and stomach acid is a chemical barrier.

B is incorrect because peristalsis is not a chemical barrier. **C** is incorrect because cilia are not a chemical barrier. **D** is incorrect because stomach acid is not a physical barrier.

8 B

The image shows phagocytosis, with a macrophage engulfing and destroying a bacterium.

A is incorrect because apoptosis is programmed cell death. **C** is incorrect because inflammation is the result of histamine release, which causes an increase the permeability of blood vessels in the area of infection. **D** is incorrect because chemoattraction is the release of chemicals to attract phagocytes and T lymphocytes.

9 D

Killer T lymphocytes release perforin to lyse infected 'self' cells.

A is incorrect because phagocytes are part of the non-specific innate immune response. **B** is incorrect because B lymphocytes are responsible for the antibody-mediated response. **C** is incorrect because natural killer cells are part of the innate immune response.

10 C

A and **B** are incorrect because a B cell is not activated by a helper T cell until the helper T cell itself is activated. **D** is incorrect because the production of memory B cells only occurs after exposure to the pathogen.

11 D

The memory B cells that remain circulating after the primary exposure are activated when the rubella virus is detected, leading to a rapid production of high levels of antibodies.

A is incorrect because the axes are back to front and the first exposure will not produce more antibodies than the second exposure. **B** is incorrect because the second exposure should produce higher levels of antibodies and more rapidly than the first exposure. **C** is incorrect because the first exposure will not produce more antibodies than the second exposure.

12 A

An antigen is any molecule that is recognised by the immune system as being foreign and causes an immune response in the body.

B is incorrect because an antibody, produced by B (plasma) cells, binds to the antigen to neutralise it. **C** is incorrect because an antibiotic is a drug used to either kill or inhibit bacteria. **D** is incorrect because an anticodon is the component of a tRNA molecule that binds to the mRNA codon during polypeptide synthesis.

13 B

When exposed to the same pathogen, the immune response in person 1 is more rapid and produces more antibodies than in person 2.

A is incorrect because person 1 has been previously exposed to pathogen P as indicated by antibody levels above zero at the time of the exposure. **C** is incorrect because person 2 has not been previously vaccinated against pathogen P as there are no circulating antibodies at the time of the exposure and the response is slower and less dramatic. **D** is incorrect because exposure to a pathogen is specific. A similar pathogen would not induce a faster response.

14 D

Dendritic cells are antigen-presenting cells that present the antigen to helper T cells via the MHC class II molecules. This occurs in the lymphatic system.

A is incorrect because phagocytes do not present antigens to B cells. **B** is incorrect because dendritic cells present antigens to helper T cells via the MHC class II molecules and not MHC class I. **C** is incorrect because antigen-presenting cells do so via the MHC class II molecules.

15 A

Snow observed and recorded the number of cholera cases associated with the water pumps.

B and **C** are incorrect because Snow had no knowledge of *Vibrio cholerae*. It had not yet been discovered. **D** is incorrect because vaccinations had not been developed for cholera.

16 D

Passive vaccinations involve the introduction of antibodies rather than an attenuated pathogen. They therefore do not induce a memory of the pathogen.

All other options are incorrect because passive vaccinations do not produce memory cells.

17 B

The quarantining of animals or animal-based products until the contagious period has passed means that non-infected individuals will not come in contact with the pathogen.

A is incorrect because monitoring locally farmed animals does not prevent entry of the pathogen. If detected at this point, it has already entered the country. **C** is incorrect because inspections will only detect symptomatic animals. It is possible that they have only recently contracted the disease and are not yet showing symptoms. **D** is incorrect because border checks may prevent some spread between states. If detected, it has already entered the country and some animals may not yet show symptoms.

18 D

Incidence is the number of new cases of a disease measured over a specific time. Prevalence is the total number of cases in a population at a given time.

A and **B** are incorrect because the number of deaths caused by a disease is mortality. **C** is incorrect because the definitions are back to front.

19 A

Overall, antibiotic W has a significant effect on both bacterial species.

B is incorrect because antibiotic X has the least impact on bacteria B. **C** is incorrect because antibiotic Y has less impact than antibiotic W on both species of bacteria. **D** is incorrect because antibiotic Z has the least impact on bacteria A.

20 B

$$\frac{150}{7500} \times 1000 = 20$$

A, **C** and **D** are incorrect because the answer has been derived incorrectly.

9780170465267

Short-answer solutions

21

Pathogen	Distinguishing characteristics of the pathogen	Disease caused by the pathogen
Virus	Non-cellular; contain nucleic acids	Chickenpox (1 mark)
Bacteria	Prokaryotic	Cholera (1 mark)
Fungi	Eukaryotic; cell wall	Tinea (1 mark)
Protozoan	Eukaryotic; no cell wall	Malaria (1 mark)

22 Part a:

1 Prepare four identical servings of a chosen hot meal.

2 Prepare four incubators at temperatures 30°C, 40°C, 50°C and 60°C. (1 mark)

3 Bring each hot meal to the allocated temperature and leave for 4 hours.

Part b:

4 Prepare 12 sterile agar plates with 30 mL agar. (1 mark)

5 Use a sterile inoculation loop to add a sample of the 30°C food to three of the plates. (1 mark)

6 Repeat step 5 for each of the other temperatures.

7 Leave three plates unexposed to any food (experimental control).

8 Add all plates to an incubator kept at 40°C.

9 Incubate for 48 hours.

10 For each plate, count the number of colonies present.

11 Compare the number of colonies recorded at each temperature. (1 mark)

23 Like all viruses, HIV replicates within a host cell. It infects T lymphocytes, which are involved in the adaptive immune response. For example, T helper cells bridge the communication between the innate and adaptive immune systems by receiving an antigen from an APC (1 mark). The T helper cells activate killer T cells, which destroy 'self' cells infected with pathogens (1 mark). As the HIV virus replicates, it destroys increasing numbers of T lymphocytes (1 mark). This places an infected person at risk of diseases from other pathogens as the adaptive immune system is unable to respond to an infection (1 mark).

24 ©NESA 2016 MARKING GUIDELINES SII Q25

a Light microscopes were not powerful enough (1 mark) to see the viruses, which are microscopic (1 mark).

b Any number of microorganisms in the dog saliva may be causing rabies. Pasteur needed to identify the particular microorganism he suspected caused the disease and culture it (1 mark). The microorganism had to be introduced into a healthy animal (1 mark) – if the animal developed rabies and the same microorganism was identified in the saliva (1 mark) then Pasteur could conclude that the particular microorganism caused rabies (1 mark).

25 Because the bananas are seedless, they cannot reproduce sexually, which means no new genetic variation can be introduced into the population, unless by mutation (1 mark). 'Suckers' can be split from parent plants in order to produce a new fruit-bearing tree. This is the main asexual reproductive technique used in bananas. Because of this, all plants are genetically identical (1 mark).

Like many crops, the Cavendish banana is grown mainly in monoculture in order to produce large yields and meet market demand (1 mark). The combination of monoculture and genetically identical plants means that a change in environmental conditions, including a disease outbreak, puts the entire world crop at risk (1 mark). For example, if a previously unencountered fungal disease outbreak occurred in one part of the world it would likely spread quickly around the world, due to large movement of product involved in the export market (1 mark). Because there is a lack of genetic diversity within the population, the disease would have a significant impact on the species (1 mark).

For 6 marks, you must demonstrate knowledge of asexual reproduction and link to offspring being genetically identical (2 marks) *and* monoculture practices used in agricultural production places genetically identical crops at greater risk of disease (2 marks). You must also provide an assessment of the likely cause and effect of disease outbreak (2 marks).

26

Pathogen	Adaptation for entry	Adaptation for transmission
HIV	HIV uses the host's innate immune system's antigen-presenting macrophages, situated in the mucous membrane of the reproductive tract and rectal passage. When detected, HIV is presented to T lymphocytes, the virus's primary target.	Infected individuals do not immediately show symptoms. They therefore can unknowingly transmit the virus, which has been replicating in T cells and is passed on via body fluids.
Plasmodium falciparum	Has evolved to exit the *Anopheles* mosquito only once the mosquito has pierced the skin of the human host and injected an anticoagulant protein.	Avoids detection by the immune system by alternating the expression of different protein antigens on the surface of the red blood cells. Ensures survival and uptake during mosquito blood meal.

For 5 marks, you must include a table (1 mark) containing different adaptations for entry (1 mark) and transmission (1 mark) for two pathogens (2 marks).

27 a All eucalyptus trees secrete an oil, which they store in subdermal secretory glands (1 mark). The oils have antimicrobial properties (1 mark).

b Eucalyptus rust is a fungal disease caused by *Puccinia psidii*. The disease is common in many Australian plant genera including *Eucalyptus*; for example, *Corymbia citriodora* (spotted gum) (1 mark). In significant infections the disease leads to deformity of the leaves and flowers, heavy defoliation, stunted growth and often death.

The tree will activate the localised hypersensitive response (HR) in an attempt to prevent further spread of the pathogen to other parts of the plant. The HR involves intentional plant cell suicide at the site of infection (1 mark). This causes the cells in the region to produce specific chemicals that alter the cell wall structure (1 mark). The pathogens are trapped inside the host cell, where they are killed upon apoptosis (1 mark).

28 An antigen-presenting cell, such as a dendritic cell, phagocytoses a pathogen and then expresses an antigen on its surface via an MHC II marker (1 mark). The dendritic cell migrates to the lymphatic system, where it presents the antigen to a specific T helper cell (1 mark). The T helper cell then communicates with B cells, which begin antibody production specific to the pathogen (1 mark) and T killer cells, which seek and kill 'self' cells infected with the pathogen (1 mark).

29 Mast cells detect the presence of the pathogens and release histamines (1 mark). The histamines change the chemical environment of the surrounding tissue and cause increased permeability of the surrounding blood vessels. As a result, this leads to inflammation and an increased presence of phagocytes in the infected tissue (1 mark). The phagocytes engulf and destroy the pathogens and release cytokines (1 mark). These chemicals attract other phagocytes and natural killer cells to the infected area to assist in fighting the pathogen (1 mark).

30 a

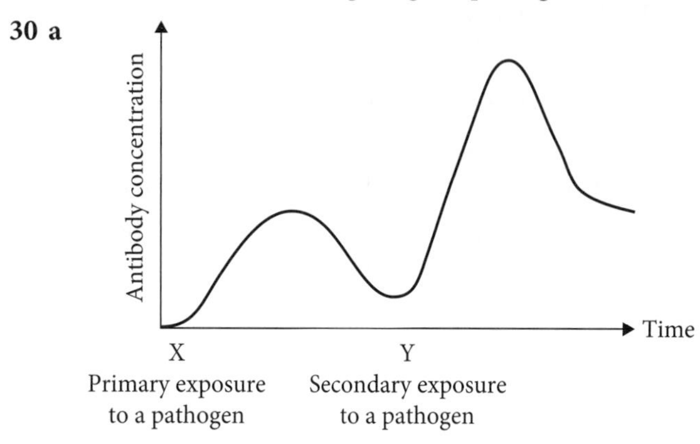

For 2 marks, your graph should correctly show distinct peaks for primary and secondary exposure (1 mark) with secondary exposure peak and remaining antibodies higher than after primary exposure (1 mark). The slope of the second peak is steeper, because production of antibodies is faster after the second exposure.

b

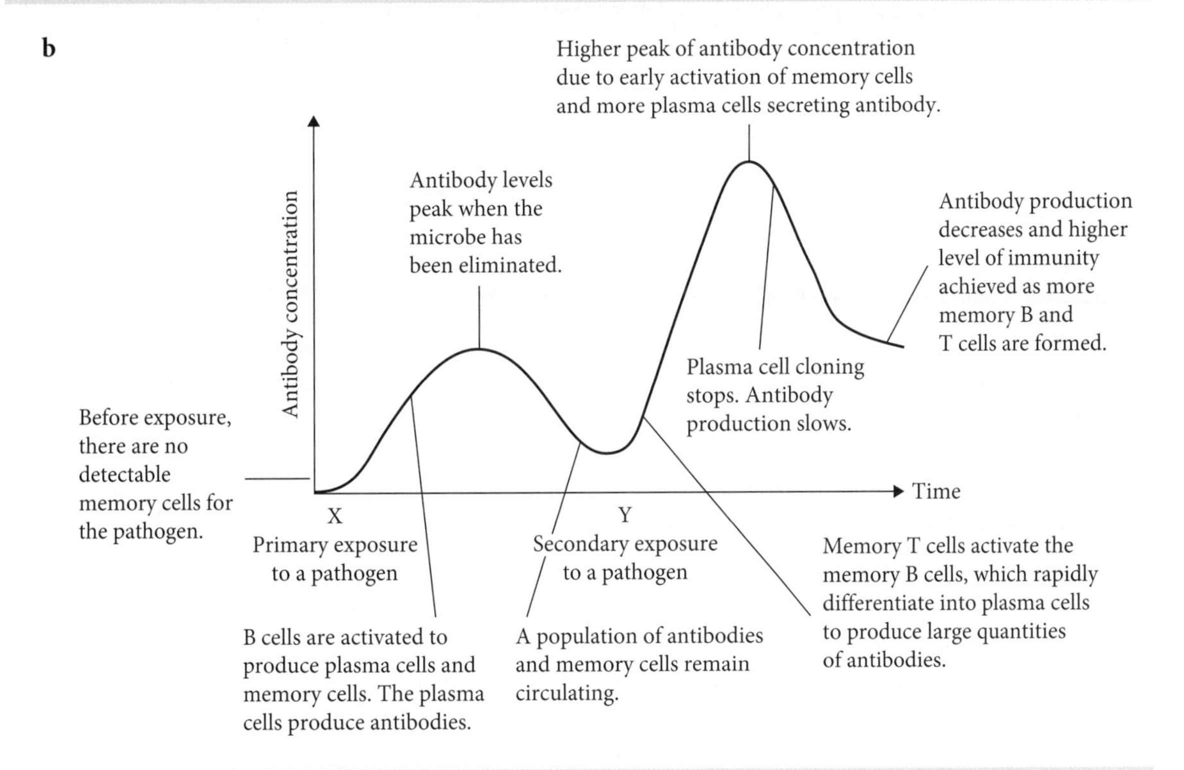

For 4 marks, your graph must show a minimum of six correctly annotated points relating to the shape of the graph.

31 The immune system is able to recognise its own cells due to the presence of self-antigens. Any cells that do not possess these self-antigens are recognised as foreign and an immune response will be initiated against the antigen (1 mark). Blood group O individuals do not possess A or B antigen molecules (1 mark). Therefore, if they receive blood from an A, B or AB individual, these antigens will be recognised as foreign, and an immune response will begin, which will include the production of antibodies and agglutination of the 'foreign' blood cells. The volume of blood received will determine the impact that the immune response has on the individual (1 mark).

32

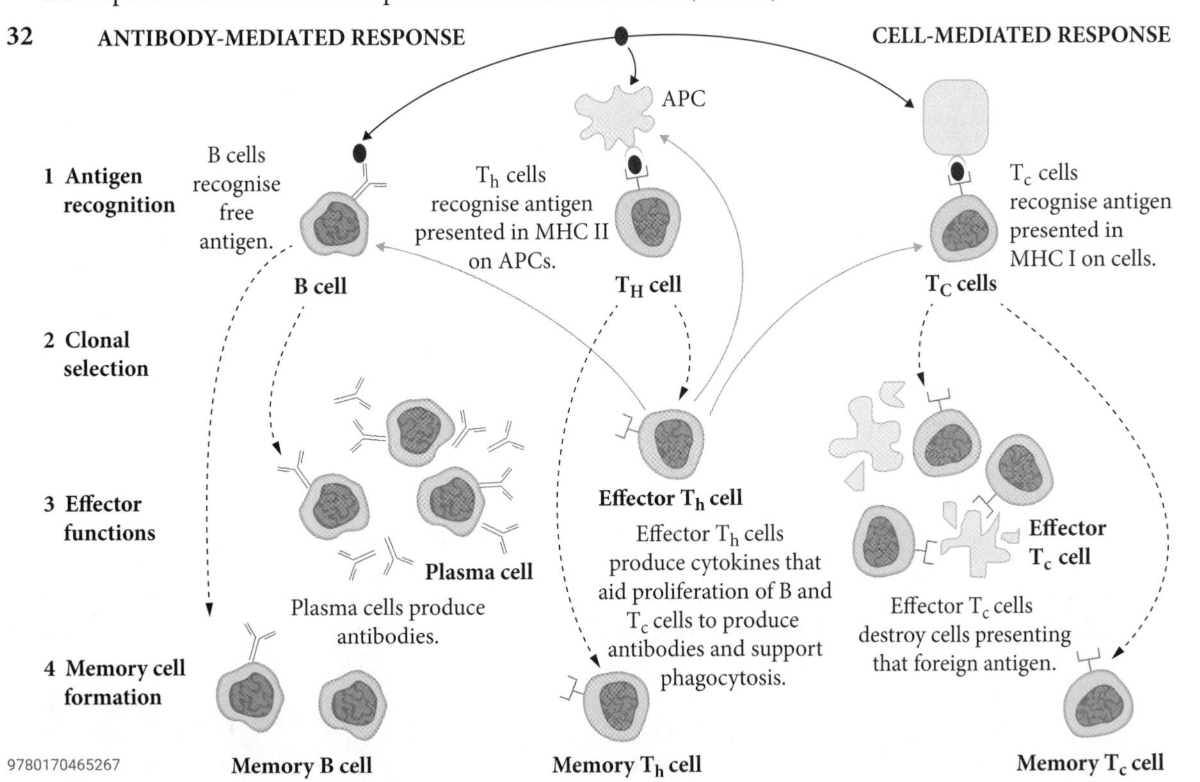

For 5 marks, your diagram must include antigen-presenting cell from innate immune system (1 mark) presenting antigen to helper T cell (1 mark), helper T cell activating both antibody-mediated response (B cell production) and cell-mediated response (T cell production) (1 mark), B cells differentiated to plasma cells and producing specific antibodies, killer T cells destroying infected self-cells (1 mark), and memory B and T cells produced (1 mark).

33 Passive vaccines contain antibodies that are specific to a pathogen; for example, tetanus vaccines (1 mark). The antibodies act directly on the pathogen if encountered, but because there is no activation of the adaptive immune system memory cells are not produced and therefore there is no long-term immunity (1 mark). In contrast, active immunisation involved the introduction of a weakened form of a pathogen (1 mark). This acts as a primary exposure and leads to the production of antibodies and memory B and T cells. If the pathogen is encountered, an immune response is initiated more rapidly (1 mark).

34 Antibiotics are effective against bacteria only. They are relatively cheaper and broader spectrum than antivirals due to the mechanism by which they target the cell wall of bacteria (1 mark). They therefore have far fewer side effects, as human cells do not contain a cell wall. However, overuse and misuse of antibiotics has led to resistant strains of bacteria, and this is becoming increasingly problematic due to the rapid turnover of generations (1 mark).

Because viruses replicate inside host cells, antiviral drugs cannot eliminate a viral infection. Any drugs that interfere with the virus will also harm the host organism's cells (1 mark). They only slow the replication rate of the virus. They also need to be very specific to each virus, so they are relatively expensive to produce and run the risk of becoming obsolete if the virus mutates (1 mark). Effective antivirals, however, can allow infected individuals to lead relatively normal lives as the viral replication rate is slowed significantly. Relative to antivirals, antibiotics have had much greater success (1 mark).

35 Washing hands after going to the toilet and before and after cooking food kills any possible pathogens (1 mark) and therefore reduces the likelihood of transmitting pathogens to non-infected individuals (1 mark).

Correct water treatment (e.g. chlorination) kills any water-borne pathogens (1 mark). It is also essential to keep drinking water and sewerage infrastructure separate to avoid contamination. These practices minimise the risk of transmitting disease (1 mark).

An example of genetic engineering to prevent disease spread is Bt cotton. The GMO crops have a gene inserted that produces a toxin that kills insect pests, which often act as disease vectors (1 mark).

Note that the question requires hygiene practices (plural). You must therefore provide at least two.

36 Prior to the introduction of a COVID-19 vaccine a combination of both environmental management strategies and quarantine were essential in controlling the spread of the disease (1 mark). For example, lockdowns and mandatory mask wearing in public reduced the opportunity for infected individuals to pass on the pathogen to non-infected individuals (1 mark). However, given that essential travel and shopping were still permitted, these environmental management strategies did not stop the spread of COVID-19 entirely (1 mark). Instead, mass public testing identified infected individuals who were then required to quarantine for 14 days. This quarantine period was longer than the contagious period of the disease. This again minimised the risk of infected individuals coming in contact with non-infected individuals (1 mark).

While environmental management strategies and quarantine had a major impact on reducing the spread of COVID-19, they were not 100% effective (1 mark) as there was still movement within a population occurring, and quarantining required that individuals be tested in the first place and that they follow the health orders once identified. In some cases, this did not occur (1 mark).

For longer-response questions, a mark will often be awarded for concise writing. Also note that any disease could have been chosen. COVID-19 is an obvious choice.

37 a Bush medicine is the combination of practices and beliefs in which plants or other materials are used in the maintenance of good health (1 mark).

b In order to make a commercially viable medicine, pharmaceutical companies need to ensure their product is reliable, with a consistent amount of active ingredient (1 mark). They perform the research and develop products with a view to making money (1 mark).

For 2 marks, you must outline more than one reason.

c Smoke bush, which is endemic to Western Australia, has been used for thousands of years by local Indigenous Australian peoples for its natural healing properties (1 mark).

In the 1960s various governments and pharmaceutical companies negotiated with the Western Australian government for exclusive research rights to smoke bush. In the 1980s it was found to have the active ingredient conocurovone, which is known to destroy HIV in low concentrations.

At the time of issuing research licences, the government chose not to acknowledge or include local Indigenous Australians in any negotiations, royalties or compensation agreements. A pharmaceutical company paid over $1.5 million for the exclusive research rights to smoke bush. If successful in developing an anti-HIV drug they would potentially earn over $100 million per annum from their investment (1 mark).

While it is important that pharmaceutical companies obtain a return on the research investment (1 mark), it is essential that Indigenous cultural and intellectual property (ICIP) laws are in place to protect the exploitation of traditional arts and culture of Indigenous Australian people and ensure the appropriate balance is met between the competing interests (1 mark).

38 ©NESA 2019 MARKING GUIDELINES SII Q32 (ADAPTED)

a The distribution of dengue fever appears to have increased markedly since 1950. Many more parts of the world, such as South America and Africa, are now affected (1 mark). The number of countries with reported cases of malaria decreased significantly between 1900 and 2010, from 140 to 88 (1 mark). However, a growing number of people are at risk, though they represent a smaller percentage of the global population. The actual numbers of at-risk people has increased from 0.9×10^9 to 3.4×10^9, which equates to an actual drop from 75% to 50% (1 mark).

It is essential that you respond with the necessary depth associated with the verb in the question stem. In this case 'analyse' requires you to identify different components and relationships shown in *both* the table *and* the image and discuss the implications.

b The global distribution of both dengue fever and malaria is likely to be associated with mosquito vectors (1 mark). Airline travel has increased markedly over the last century, offering opportunities for both infected people and the mosquito vectors to be transported around the world. In addition, the world population has increased as shown in the table, from 1.2 to 6.8 billion, which will increase the density of potential hosts of both diseases (1 mark). With increasing urbanisation of a larger population, new urban habitats for the mosquitoes could have emerged. It could be argued that these factors have led to the increased distribution of dengue fever, shown with the greater shading in the world map in 2010, and the increased population at risk of malaria (0.9×10^9 to 3.4×10^9), but the number of countries (i.e. the distribution of malaria) has reduced (1 mark). This suggests that the mosquito vector for malaria has been contained within known areas. This could have been achieved by spraying pesticides on water bodies to kill the mosquitoes, or by strict quarantine regulations whereby mosquitoes are eliminated before they can establish populations in new areas. To prevent spread of these diseases, both the vector and the hosts need to be contained if possible (1 mark).

Medical advances have the capacity to prevent or control these diseases (e.g. development of vaccines), especially for a viral disease such as dengue fever. From the data it would seem that medicines have had a greater effect in containing malaria than dengue fever (1 mark). It may be that application of a vaccine in remote areas can limit the spread, given that the higher the proportion of the population vaccinated, the more effective the control of the disease. Also, antiviral drugs may assist in treatment of the disease. However, the evidence suggests that there are no suitable vaccines or drugs for dengue fever or that these have not been widely applied. It is possible too that the virus evolves quickly, making vaccines ineffective after a short period of time (1 mark). Vaccines and pharmaceuticals, such as antibiotics and antimalarial drugs, appear to have been more effective for malaria than for dengue fever (1 mark).

With longer-response questions it is highly recommended that you spend time breaking down the question to determine where the marks will be allocated and how best to structure your response. With this question, remember that the syllabus *does not* require a specific knowledge of malaria or dengue fever. The focus is instead on your interpretation and analysis of the data relating to the incidence and prevalence of infectious disease in populations. To ensure you address all the necessary details, you could consider constructing a table and addressing each disease separately.

CHAPTER 4, MODULE 8

Non-infectious disease and disorders

Multiple-choice solutions

1 C

Endotherms regulate their internal body temperature through a negative feedback pathway. They respond to external temperature stimuli and have many effector responses that maintain a stable body temperature.

A and **B** are incorrect because the word 'endotherm' refers to inside (endo) and temperature (therm). **D** is incorrect because the external temperature refers to the surrounding environment such as the atmosphere or water.

2 A

A negative feedback loop responds to stimuli to maintain the body in homeostasis. It stabilises rather than destabilises the body.

B is incorrect because a negative feedback loop responds to stimuli. **C** is incorrect because a negative feedback loop decreases a function to return the body back to homeostasis. **D** is incorrect because a negative feedback loop reacts to a deviation from a set point such as a change in temperature or glucose levels.

3 B

A behavioural adaptation is a physical action that an organism consciously performs to help it survive.

A, **C** and **D** are incorrect because they are physiological responses.

4 B

When water is scarce, the stomata in leaves close to reduce water loss. This in turn reduces the amount of CO_2 taken in through stomata for photosynthesis.

A is incorrect because plants take up CO_2 through their stomata, which will be closed when water levels are low in the plant. **C** and **D** are incorrect because plants close their stomata when water levels are low, to prevent further water loss by evaporation.

5 D

Your interpretation of the graph should be based on comparing blood glucose levels at each time point relative to time = 0. After consuming the glucose, cells of the body in a person with type 2 diabetes do not respond adequately to insulin. This means blood glucose levels remain elevated compared to a person without diabetes.

A is incorrect because the blood glucose level in the person with type 2 diabetes increases and declines rapidly. This would not occur because their cells do not efficiently absorb glucose to remove it from the bloodstream. **B** is incorrect because glucose levels will increase in any person after they consume glucose. **C** is incorrect because there is no change in the blood glucose level in the person with type 2 diabetes. Blood glucose will always increase immediately after glucose is consumed.

6 D

Diseases such as cystic fibrosis and haemophilia A are caused by germ-line mutations and most cancers are caused by mutations in somatic cells. A genetic disease is caused by mutations in DNA. A mutation can occur in the germ line and be inherited by offspring. If this occurs, the mutation will be present in every cell of the body. Cancer is the main disease caused by somatic mutations. A somatic mutation occurs in a somatic cell of the body and only affects the individual with the mutation.

A is incorrect because there are many types of disease caused by germ-line mutations. **B** and **C** are incorrect because an inherited mutation occurs in the germ line but diseases such as cancer are caused by somatic mutations.

7 D

Metastasis means cancer cells move from the site of the primary tumour to other sites in the body via the circulatory and lymphatic system.

A is incorrect because benign tumours remain localised. **B** is incorrect because metastatic means cancer cells move to other sites in the body. **C** is incorrect because cancer cells evolve mechanisms to evade the immune system.

8 A

Nutritional diseases are caused by deficiencies or excesses in the diet.

B is incorrect because type 1 diabetes is an autoimmune disease. **C** is incorrect because mesothelioma is a disease caused by an environmental exposure (asbestos). **D** is incorrect because skin cancer can be classified as a genetic disease and a disease caused by an environmental exposure (UV light).

9 D

Non-infectious diseases make up just over 62% of the total disease burden in 2017, compared to 28% for infectious/nutritional diseases and 10% for injuries.

A is incorrect because injuries have remained stable at about 10% in 1990–2017. **B** is incorrect because the proportion of infectious/nutritional diseases has declined from 47% in 1990 to 28% in 2017. **C** is incorrect because non-infectious diseases are not caused by pathogens. Also, the graph provides no information about pathogens, so it is not possible to conclude from this data that pathogens are the cause of the diseases.

10 C

Some non-infectious diseases are caused by germ-line mutations that are inherited. An inherited mutation exists in every cell of a person's body and will affect them for all their life.

A is incorrect because transmission requires the disease to be caused by a pathogen. **B** is incorrect because non-infectious diseases are not caused by pathogens. **D** is incorrect because non-infectious diseases can be treated using a variety of methods such as lifestyle changes and medication.

11 A

Injuries account for the lowest proportion of total burden (8.4%) and most injuries are fatal.

B is incorrect because almost all cases of mental and substance use disorders are non-fatal. **C** is incorrect because most cases of cancer are fatal. **D** is incorrect because cardiovascular disease is the second-highest contributor to total burden (13%) along with musculoskeletal conditions and mental and substance use disorders.

12 D

The study found a statistically significant association between obesity and cardiovascular disease.

A is incorrect because the study only showed an association between obesity and cardiovascular disease. **B** is incorrect because the study did not investigate the physiology of the disease. **C** is incorrect because there is no mention of physical activity being investigated in this study.

13 D

More women than men have dementia in every age group (except 30–59) and the number of women with dementia increases in each age group.

A is incorrect because there are men with dementia in each age group. **B** is incorrect because more women than men have dementia in every age group (except 30–59). **C** is incorrect because the graph only shows the prevalence of dementia in different age groups of a population of humans; it does not show the cause of the disease.

14 B

Prospective studies follow people through time after the start of the study.

A is incorrect because a retrospective study identifies risk factors that occurred before the start of a study. **C** is incorrect because prospective means to follow participants through time. **D** is incorrect because a retrospective study only identifies risk factors that occurred before the start of a study.

15 B

Plain packaging of cigarettes includes graphic images of diseases caused by smoking and no brand labels. This is aimed at discouraging smoking and is therefore a preventative approach.

A is incorrect because diagnosis identifies a type of disease a person has and there is no way the packaging can identify which people smoke. **C** is incorrect because treatment refers to medication to alleviate the symptoms of a disease. **D** is incorrect because screening aims to identify the early stages of a disease in a population and there is no way the government can identify a person based on a packet of cigarettes they bought.

16 C

Conductive hearing loss occurs when there is a problem with the outer or middle ear (usually a temporary blockage) that prevents sound waves reaching the inner ear.

A is incorrect because it is not a formal name for a type of hearing loss. Also, the cochlea is not affected in the situation described. **B** is incorrect because it is not a formal name for a type of hearing loss. **D** is incorrect because sensorineural hearing loss occurs when there are problems with the inner ear.

17 D

Cochlear implants bypass the outer and middle ear and directly stimulate the auditory nerve.

A is incorrect because cochlear implants do not amplify sound like a hearing aid. **B** is incorrect because the pinna is the fleshy part of the outer ear that captures sound waves and directs them into the ear canal. **C** is incorrect because cochlear implants are used by people who have inner ear damage. Stimulating the eardrum would not be beneficial.

18 A

Light is first refracted by the cornea when it enters the eye.

B is incorrect because the retina is located at the back of the eye. **C** is incorrect because the retina does not focus light. It receives light and converts it into neural signals that are transmitted to the brain via the optic nerve. **D** is incorrect because the cornea is located at the front of the eye, and light therefore needs to pass through it to other structures in the eye.

19 B

This refractive condition occurs when the focal point is behind the retina.

A is incorrect because hyperopia is a refractive eye condition. **C** is incorrect because this describes myopia (short-sightedness). **D** is incorrect because this describes a problem with the retina, but hyperopia is a refractive eye condition.

20 D

Nephrons are the functional unit of the kidney; they filter blood, remove waste and excess liquid, and produce urine, which is released into the ureter.

A is incorrect because the nephron releases urine directly into the ureter, not the urethra. **B** is incorrect because nephrons filter blood, not urine. **C** is incorrect because nutrients are absorbed by the digestive system and transported to cells by blood.

Short-answer solutions

21 a Osmoregulation (1 mark)

b Osmoregulation maintains the balance of solutes such as salts and sugars in the body (1 mark). This is an example of homeostasis because the balance is maintained despite external factors such as temperature, diet and the weather. Osmoregulation is important because it keeps solutes at the ideal concentration for optimal cell, tissue and organ function (1 mark). In the absence of osmoregulation (e.g. caused by kidney failure), water and waste would accumulate in the body. This results in a range of symptoms such as nausea, vomiting, high blood pressure and shortness of breath (1 mark).

22 ©NESA 2019 MARKING GUIDELINES SII Q21

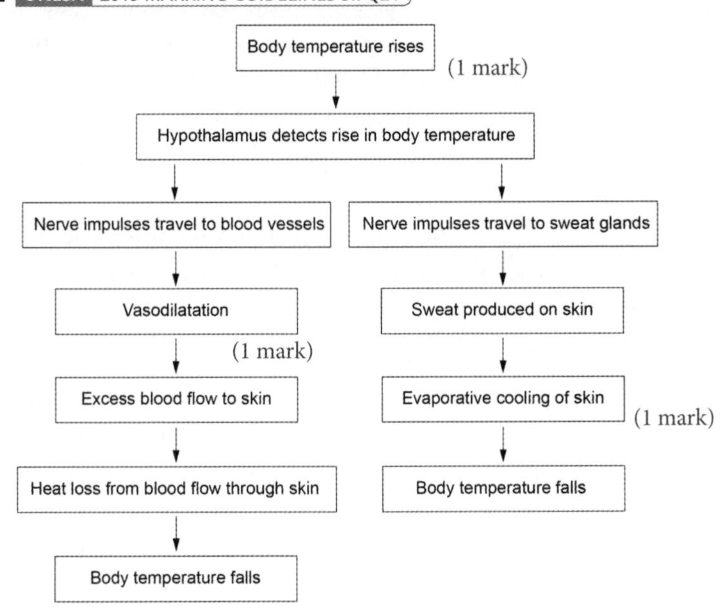

23 ©NESA 2019 MARKING GUIDELINES SII Q29 (ADAPTED) Plants regulate the loss of water by transpiration through stomata on the underside of their leaves (1 mark). Plants can control movement of water out of the leaf to keep a stable internal environment with respect to water, by altering the size of the opening of the stomates (1mark). The opening and closing is under hormonal control; for example, abscisic acid is a stress hormone produced by plants when internal water is low, causing stomata to close, thus reducing water loss (1 mark).

24

Non-infectious diseases			
Type	**Example**	**Cause**	**Effect**
Genetic	Cystic fibrosis (½ mark)	Germ-line mutations in the *CFTR* gene (½ mark)	The accumulation of thick mucus in the lungs which results in persistent coughing and makes it difficult to breath (1 mark)
Nutritional	Type 2 diabetes (½ mark)	Obesity caused by excessive consumption of processed food and a sedentary lifestyle (½ mark)	Insulin is not used efficiently in the body, resulting in high blood glucose levels (1 mark)
Cancer	Breast cancer (½ mark)	Genetic mutations that inactivate tumour suppressor genes and activate proto-oncogenes in somatic cells in the breast (½ mark)	Primary tumour begins in the breast. Cancer cells spread to other regions of the body where they form secondary tumours, causing organ failure and death (1 mark)

25 a Mesothelioma (1 mark)

Mesothelioma is provided as an example answer. Tobacco smoking, which causes lung cancer, is another possible answer to this question.

b Mesothelioma is caused by exposure to asbestos fibres. In the past this occurred in people who mined asbestos or used it in the construction industry. Microscopic asbestos fibres are inhaled and travel into the respiratory system (1 mark). The fibres settle in the pleural lining of the lungs and chest wall. Macrophages attempt to engulf the fibres, which results in local inflammation and the release of reactive oxygen species (a type of naturally occurring mutagen). This causes genetic mutations and lung cancer decades after the initial exposure (1 mark).

26 a The A1C test shows an average reading of 7% (which is above 6.5%) over the time period (horizontal line). This person would be diagnosed with diabetes based on this test. (1 mark)

b Blood glucose levels increase after meals. For example, on Sunday the fasting blood glucose level was 145 mg dL^{-1}, which increased to a maximum of 170 mg dL^{-1}) (1 mark). In a person without diabetes, this stimulates the pancreas to release insulin, which results in the absorption of glucose by cells and a decline in the amount of glucose in the blood. The A1C test diagnosed this person with diabetes. The blood glucose levels increased after each meal, but a control person is needed to determine whether they remain elevated for longer than a person without diabetes (1 mark).

27 Incidence is the number of new cases of a disease in a population in a specified time period. It is presented as a rate to allow comparison of data sets between populations, time periods or locations (1 mark). It is different from prevalence, which is the proportion of a population who have a specific characteristic at a given time period, regardless of when they first developed the characteristic. It is usually presented as a proportion of the population (1 mark).

28 a Type 2 diabetes is an example of a non-infectious disease caused by the body's inability to efficiently use insulin to absorb glucose (1 mark).

The disease can be managed through lifestyle changes such as better nutrition and physical activity. Eating habits can include a high-fibre diet that is low in high-calorie, processed foods, and regular meals with a smaller portion size. Physical activity can include aerobic and strength exercise, and reducing longer periods of sedentary activity. This reduces body weight and results in a natural regulation of blood glucose levels. If these prevention strategies fail, then the disease can be treated with drugs such as metformin, which reduces glucose production in the liver and makes the body more sensitive to insulin. Insulin therapy may also be used, and weight loss surgery can be performed in extreme cases (2 marks).

b A future direction of research could be the development of an artificial pancreas (1 mark), which is an electronic device that automatically senses glucose levels in the blood and releases insulin into the body (1 mark).

Another possible example is beta-cell transplants (1 mark). People with type 2 diabetes produce less insulin, so the introduction of these cells back into the body may be a way to increase insulin production and eliminate the need for ongoing insulin therapy (1 mark).

29

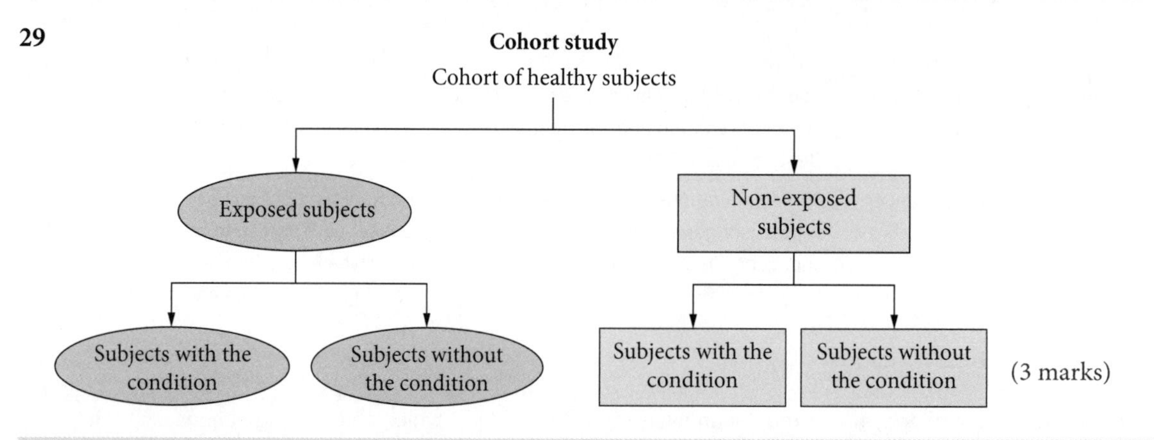

(3 marks)

You could also draw a flow chart for a case-control study or randomised controlled trial.

30

	Benefit	Limitation
Individual	Individual satisfaction from contributing to knowledge about health for future generations (½ mark)	Individuals are unlikely to benefit directly from participation (½ mark)
Society	Society benefits from the addition of knowledge about exposures and risk factors and their contribution to disease (½ mark)	Large studies can be expensive with many years of follow-up required (½ mark)

31 ©NESA 2019 MARKING GUIDELINES SII Q23

Name of disease	Program to prevent disease
Melanoma (½ mark)	Slip, slop, slap, seek, slide (½ mark)
Lung cancer (½ mark)	Quit (smoking) (½ mark)

Public education programs can raise awareness of the risk of exposure to various harmful environmental agents. For example, UV radiation can cause melanoma and tobacco smoke increases the risk of lung cancer. As a result of the programs, people can alter their behaviour to reduce their exposure to harmful situations. For example, not everyone can avoid the sun in their daily lives, but the program encourages them to wear a hat, shirt and sunscreen so that exposure to UV radiation is reduced. This reduces the risk of melanoma. (3 marks)

32 a A: cochlea, B: tympanic membrane, C: pinna (1 mark)

b Cochlear implant (1 mark)

c Cochlear implants are small electronic devices that can partially restore a sense of hearing in people with sensorineural hearing loss (1 mark). Cochlear implants bypass any damage that exists in the outer and middle ear and deliver the sound vibrations directly to the auditory nerve (1 mark). Signals travel via the auditory nerve and the brain interprets the electrical signals as sound (1 mark).

33 ©NESA 2020 MARKING GUIDELINES SII Q24 (ADAPTED)

a

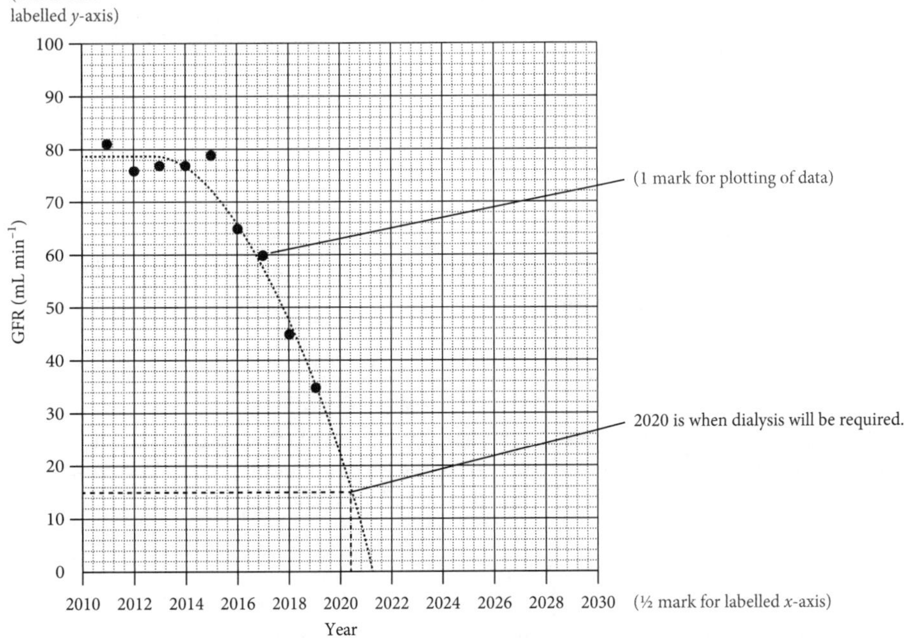

b Loss of kidney function may result in a failure to remove urea from blood (1 mark). In dialysis, blood from the patient passes through selectively permeable dialysis tubing (1 mark). Because the urea diffuses from the high concentrations in the blood to the low concentrations in the dialysate, the urea is removed from the blood. (1 mark)

SOLUTIONS

34 a The kidneys (1 mark)

b Dialysis is a medical technique that replaces the normal filtering function of the kidneys (1 mark). Haemodialysis is one type of dialysis. The filtering of blood occurs outside the body. A fistula is used to connect a tube to blood vessels in the arm. Blood is then pumped through an external machine that contains a special filter called a dialyser (artificial kidney), which removes waste and excess fluid from the blood. Clean blood is then returned to the body. (2 marks).

35

a Vision disorder	b Structure of the eye affected	c Technology
Myopia (1 mark)	Eyeball or cornea (1 mark)	Glasses, contact lenses or corrective laser surgery are used to direct the focal point of light onto the retina (1 mark)
Cataracts (1 mark)	Lens (1 mark)	Surgery is used to replace the clouded lens with an artificial lens (1 mark)
Glaucoma (1 mark)	Optic nerve (1 mark)	Treatments to lower the pressure in the eye, including eye drops, medication and laser surgery (1 mark)

Three sample answers for parts a–c are provided in the table. Only one disorder is needed to answer the question.

CHAPTER 5

The scientific method

Multiple-choice solutions

1 B

Validity is about ensuring the necessary controls over variables are in place so that the procedure tests only the effect of the independent variable on the dependent variable.

A is incorrect because it relates to accuracy. **C** is incorrect because it relates to precision. **D** is incorrect because it relates to repeatability.

2 D

The control plate was not exposed to any water sample. The appearance of microbes in the control agar indicates that it was contaminated prior to the experiment.

A is incorrect as the control plate was not touched with an inoculation loop. **B** is incorrect because it is not the water source that is contaminated. **C** is incorrect as there is not sufficient detail to determine whether the microbes are the same in each water sample.

3 B

The data is discrete and categorical, so a column graph is the most appropriate.

A is incorrect because a pie chart is typically used to represent the percentage or proportion of a population represented. **C** is incorrect because a line graph is used to represent continuous data. **D** is incorrect because a histogram is used for continuous data to examine the distribution of values.

4 D

It is possible to have two independent variables. This example compares both age and sex for type 2 diabetes.

A is incorrect because the percentage population is the dependent variable. **B** is incorrect because age group is also an independent variable. **C** is incorrect because sex is also an independent variable.

5 D

A trend is the general tendency of a set of data to move in a certain direction.

A is incorrect because it is a single data point from which a trend cannot be determined. **B** is incorrect because the risk actually increases, not decreases. **C** is incorrect because the trend is present for both men and women; however, men actually experience a greater increase than women.

6 D

The stimulus indicates that the plant lives in a dry environment and that the plants transpire more water when the stomata are larger. Therefore, to conserve water in the hotter periods of the day (noon) the plant will close its stomata.

A and **C** are incorrect because they have the stomata largest at noon, which means they would be losing too much water at this time of the day. **B** is incorrect because it shows the stomata closed at the cooler periods of the day (midnight), which is when they would be open to maintain transpiration.

7 C

Given the repetition and the high values for the other four tests it is obvious that test 4 is an outlier and should be removed from the data set.

A is incorrect because 159 is the average of the four tests after the outlier is removed. **B** is incorrect because the removal of an outlier is justifiable. **D** is incorrect because 159 is the average of the four tests after the outlier is removed and only five tests were performed for each sample set.

8 C

The untreated sample is the experimental control. Any observed results will be due to the chlorine.

A is incorrect because control variables are those that are kept constant, e.g. volume tested in each sample. **B** is incorrect because accuracy is improved by ensuring the best available equipment is used. **D** is incorrect because increasing the number of tests for each concentration, not increasing the doses themselves, will improve reliability.

9 D

2.5 mg L^{-1} chlorine had 14 *Giardia* cysts L^{-1}. This is above the recommended 0–10 cysts L^{-1}.

A is incorrect because it represents a single data point and cannot be relied on. **B** is incorrect because the 2.5 mg L^{-1} chlorine is not safe for consumption. **D** is incorrect because even at 5 mg L^{-1} chlorine there are some *Giardia* cysts detected. This could cause an infection in some people.

10 A

As soil water content increases, the plant is under less water stress and the stomates will open more.

B is incorrect because this graph indicates that the stomates will close with increased soil water content.
C is incorrect because it indicates that the opening of stomates does not change with soil water content.
D is incorrect because it indicates that the stomates close initially as soil water content increases, then remain closed independent of the water content, before opening up again at even higher water content.

11 A

The disease is autosomal recessive. Therefore, because the mother is not a carrier of the recessive allele, her eggs will only ever carry the 'normal' allele.

B is incorrect because no child can have the disease. **C** is incorrect because no child can have the disease.
D is incorrect because the disease is not sex-linked. Neither male nor female offspring can have the disease.

12 D

It is the only peer-reviewed article that has been cited in other scientific journals.

A is incorrect because YouTube videos, while useful, cannot be categorised as a valid source of research, no matter how many views they get. **B** is incorrect because slideshows published on the Internet are not peer-reviewed and are therefore not a valid source. **C** is incorrect because newspaper articles are not a valid research source, even if claiming to quote scientists.

Short-answer solutions

13 a Based on the method provided, the independent variable: antibiotic type **or** antibiotic concentration (1 mark)

Dependent variable: zone of inhibition (mm) (1 mark)

b Risk: ingestion of bacteria (1 mark)

Safety measure: wear the correct PPE (1 mark)

c For each condition, the experiment is only performed once (1 mark). Therefore, the reliability of the experiment would need to be improved; for example, by repeating the experiment for each condition three times and averaging the result (1 mark).

Validity is achieved by ensuring all variables except the independent variable are kept constant. In this experiment the same volume of agar is added to each plate, and they are each kept at their relevant temperatures for 48 hours. However, important variables are not controlled. For example, all plates should have been kept at the same temperature and the same concentration should have been used for all the antibiotics (1 mark). As a result, it is not a valid experiment (1 mark).

d Antibiotics are drugs that only act against bacterial infection (1 mark). As a result, no matter what conditions are changed in the experiment, using antiviral drugs will have no impact on bacterial growth (1 mark).

14 ©NESA 2020 MARKING GUIDELINES SII Q27

a

For 2 marks, your answer should identify **two** features of the method that contribute to the validity of the study. They could include:

- identifying the age of participants
- identifying the sex of participants
- calculating each participant's exposure to arsenic
- including a large sample size
- identifying each individual's socioeconomic status.

b Survival is highest in those exposed to less than 90 $\mu g\,L^{-1}$ arsenic in both males and females. This group serves as a control showing that most young people in the study survived over the 11-year period. The level of arsenic to which they were exposed is higher than recommended by WHO but survival was high nevertheless (1 mark).

In both males and females, increasing doses of arsenic led to decreased survival, which suggests that arsenic is causing the decline in survival (1 mark). The increasing response to increasing doses was most clearly seen in males. In females, all doses over 90 $\mu g\,L^{-1}$ led to a similar survival decrease, which suggests there may be other factors that interact with the dose of arsenic to produce this result. Other factors could include nutritional state or genes (1 mark).

Survival declined progressively over the 11 years, which supports the idea that as arsenic exposure increases over the years, survival declines. However, although the numbers in the study were large, survival only dropped by 0.1% or less (1 mark).

15 a Independent variable(s): time and conditions (1 mark)

Dependent variable: volume of water loss from transpiration (mL) (1 mark)

b Type of plant used (1 mark)

Number of leaves (1 mark)

c Repeat five times for each condition. (1 mark)

Remove outliers and average the results. (1 mark)

d

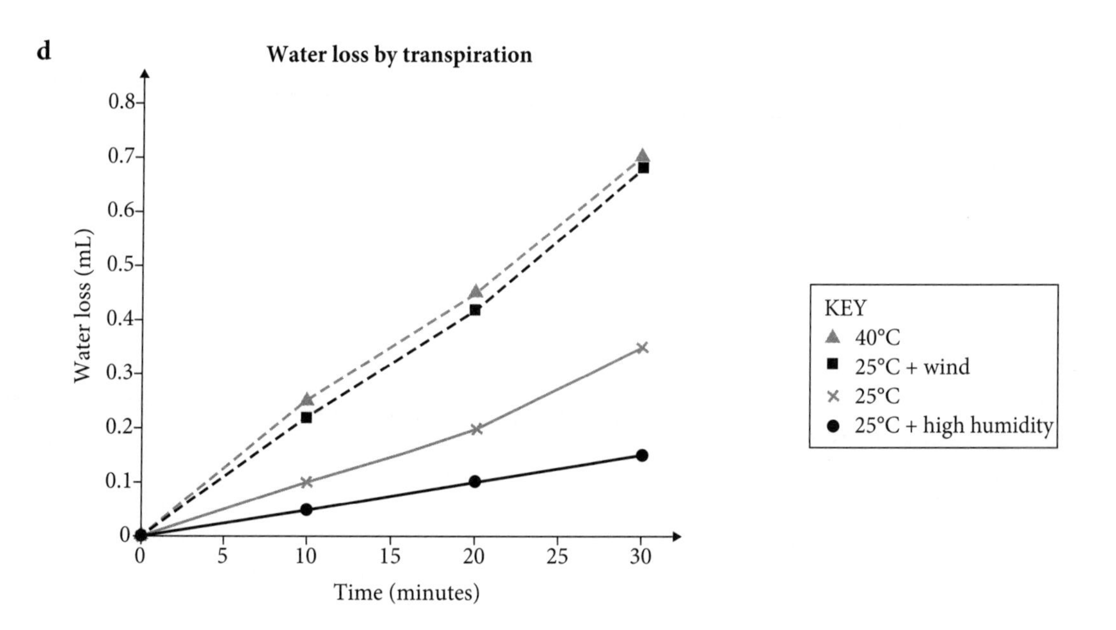

For 4 marks, the graph must include:

- both axes correctly labelled with units
- four lines correctly plotted
- a key.

e At 40°C, water loss by transpiration increases with time (1 mark)

For 1 mark, identify any appropriate trend.

f

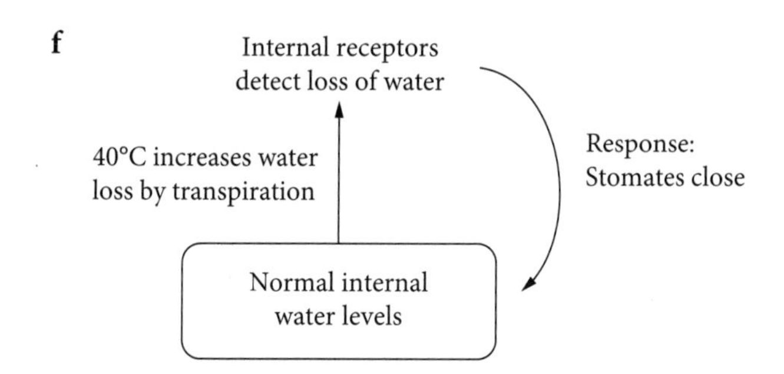

For 3 marks, you must correctly draw a diagram (1 mark) that shows detection of internal water loss (1 mark) and a response to bring it back to the normal state (1 mark).

16 a Since the Industrial Revolution in the 1800s the amount of CO_2 in the atmosphere has continually increased (1 mark). Over the past 400 000 years, atmospheric CO_2 and temperature have fluctuated together (1 mark).

b There is a relationship between atmospheric CO_2 levels and temperature. Historically, temperature increases precede the increase in atmospheric CO_2 levels. For example, between 140 000 and 130 000 years ago, temperatures peaked at 3°C above 'normal' (1 mark). The increase in atmospheric CO_2 levels lagged slightly behind the temperature increase. There is an overall trend that there have been ongoing fluctuations in atmospheric CO_2 and temperature levels going back 400 000 years. This means that these fluctuations are also natural events, as it was not until the Industrial Revolution in the 1800s that human-induced CO_2 emissions began to increase significantly (1 mark). There is a clear link between human activity and atmospheric CO_2 as the levels are at 400 ppm (1 mark), the highest in recorded history. It is therefore likely, given the relationship, that there will be a significant increase in global temperatures (1 mark).